Kant and the Systematicity of the Sciences

This book provides the first comprehensive discussion regarding the role that Kant ascribes to systematicity in the sciences. It considers not only what Kant has to say on systematicity in general, but also how the systematicity requirement for science is specified in different fields of knowledge.

The chapters are divided into three thematic sections. Part I is devoted to historical context. The chapters explore precursors of Kant's account of the systematicity of the sciences. Part II addresses the application of systematicity to the special sciences – cosmology, physics, chemistry, logic, mathematics, the life sciences, and history. Finally, Part III explores the systematicity of philosophy.

Kant and the Systematicity of the Sciences will be of interest to scholars and advanced students working on Kant and the history and philosophy of science.

Gabriele Gava is an Associate Professor of Theoretical Philosophy at the University of Turin. He works on Kant, 18th-century German philosophy, pragmatism, and epistemology. He is the author of *Peirce's Account of Purposefulness: A Kantian Perspective* (2014) and *Kant's Critique of Pure Reason and the Method of Metaphysics* (2023).

Thomas Sturm is ICREA Research Professor at the Universitat Autònoma de Barcelona. He works on Kant, theories of rationality in philosophy and the sciences, and as editor for the new Academy edition of Kant's *Gesammelte Schriften*. Selected publications: *Kant und die Wissenschaften vom Menschen* (2009), articles in *Kant-Studien*, *Kantian Review*, and *Synthese*.

Achim Vesper is an Assistant Professor at Goethe University Frankfurt. He has published extensively on Kant's ethics and aesthetics. Together with Gabriele Gava, he wrote *Kants Philosophie* (2024). Another book, *Kant über Schönheit und systematische Einheit der Natur*, will be published in 2025.

Routledge Studies in Eighteenth-Century Philosophy

Metaphysics as a Science in Classical German Philosophy
Edited by Robb Dunphy and Toby Lovat

Kant on Freedom and Human Nature
Edited by Luigi Filieri and Sofie Møller

Condillac and His Reception
On the Origin and Nature of Human Abilities
Edited by Delphine Antoine-Mahut and Anik Waldow

Consciousness, Time, and Scepticism in Hume's Thought
Lorne Falkenstein

Baumgarten's Legacy in Kant's Ethics
Toshiro Osawa

Hume and Contemporary Epistemology
Edited by Scott Stapleford and Verena Wagner

Kantian Citizenship
Grounds, Standards and Global Implications
Edited by Mark Timmons and Sorin Baiasu

Kant and the Systematicity of the Sciences
Edited by Gabriele Gava, Thomas Sturm, and Achim Vesper

For more information about this series, please visit: https://www.routledge.com/Routledge-Studies-in-Eighteenth-Century-Philosophy/book-series/SE0391

Kant and the Systematicity of the Sciences

Edited by Gabriele Gava, Thomas Sturm, and Achim Vesper

NEW YORK AND LONDON

First published 2025
by Routledge
605 Third Avenue, New York, NY 10158

and by Routledge
4 Park Square, Milton Park, Abingdon, Oxon, OX14 4RN

Routledge is an imprint of the Taylor & Francis Group, an informa business

© 2025 selection and editorial matter, Gabriele Gava, Thomas Sturm, and Achim Vesper; individual chapters, the contributors

The right of Gabriele Gava, Thomas Sturm, and Achim Vesper to be identified as the authors of the editorial material, and of the authors for their individual chapters, has been asserted in accordance with sections 77 and 78 of the Copyright, Designs and Patents Act 1988.

All rights reserved. No part of this book may be reprinted or reproduced or utilised in any form or by any electronic, mechanical, or other means, now known or hereafter invented, including photocopying and recording, or in any information storage or retrieval system, without permission in writing from the publishers.

Trademark notice: Product or corporate names may be trademarks or registered trademarks, and are used only for identification and explanation without intent to infringe.

Library of Congress Cataloging-in-Publication Data
Names: Gava, Gabriele, 1981- editor. | Sturm, Thomas, 1967- editor. | Vesper, Achim, editor.
Title: Kant and the systematicity of the sciences /
edited by Gabriele Gava, Thomas Sturm, and Achim Vesper.
Description: New York : Routledge, 2025. | Series: Routledge studies in eighteenth-century philosophy | Includes bibliographical references and index.
Identifiers: LCCN 2024051760 (print) | LCCN 2024051761 (ebook) | ISBN 9780367756888 (hbk) | ISBN 9780367763299 (pbk) | ISBN 9781003166450 (ebk)
Subjects: LCSH: Kant, Immanuel, 1724-1804. | System theory.
Classification: LCC B2799.S32 K36 2025 (print) |
LCC B2799.S32 (ebook) | DDC 193--dc23/eng/20250127
LC record available at https://lccn.loc.gov/2024051760
LC ebook record available at https://lccn.loc.gov/2024051761

ISBN: 978-0-367-75688-8 (hbk)
ISBN: 978-0-367-76329-9 (pbk)
ISBN: 978-1-003-16645-0 (ebk)

DOI: 10.4324/9781003166450

Typeset in Sabon
by KnowledgeWorks Global Ltd.

Contents

List of Contributors	*vii*
Citations of Kant's Writings	*xi*

Introduction: The Significance of Kant's Account
of Scientific Systematicity ... 1
GABRIELE GAVA, THOMAS STURM, AND ACHIM VESPER

PART I
Systematicity: Historical Backgrounds ... 19

1 Kant and Crusius on the Hierarchy of Human Ends ... 21
   GABRIELE GAVA

2 Lambert's System of the Sciences ... 39
   HENNY BLOMME

PART II
The Systematicity of Special Sciences ... 67

3 Kant's Early Cosmology, Systematicity, and Changes in
   the Standpoint of the Observer ... 69
   FABIAN BURT AND THOMAS STURM

4 Kant and the Idea of a System of Logic ... 95
   CLINTON TOLLEY

5 Mathematics: Systematic Unity and Construction in the
   Theory of Conic Sections   114
   KATHERINE DUNLOP

6 Kant's Conception of the *Metaphysical Foundations of
   Natural Science:* Subject-Matter, Method, and Aim   148
   THOMAS STURM

7 Systematicity, the Life Sciences, and the Possibility of
   Laws Concerning Life   173
   HEIN VAN DEN BERG

8 Kant's Aethereal Hammer: When Everything Looks Like a Nail   192
   MICHAEL BENNETT MCNULTY

9 Systematicity and the Definition of a Science: Physics in
   Kant's *Opus postumum*   215
   STEPHEN HOWARD

10 Systematicity in Kant's Philosophy of History   234
   ANDREE HAHMANN

11 Systematicity with a Worldly Orientation? On Kant's
   Theory and Practice of Gazing with an "Eye of Philosophy"   254
   HUAPING LU-ADLER

PART III
The Systematicity of Philosophy   275

12 The Systematicity of Natural Science: Logical and Real   277
   ERIC WATKINS

13 What Is a System of Moral Philosophy for? Systematicity
   in Kant's Ethics   296
   STEFANO BACIN

14 Kant's System of Systems   315
   PAUL GUYER

   Index   336

Contributors

Stefano Bacin is a Professor of History of Philosophy at the University of Milan. He is the author of *Il senso dell'etica. Kant e la costruzione di una teoria morale* (2006), *Fichte in Schulpforta (1774–1780)* (2007), and *Kant e l'autonomia della volontà* (2021). He is also co-editor of a three-volume *Kant-Lexikon* (with Marcus Willaschek, Georg Mohr, and Jürgen Stolzenberg, 2015), of *Kant's Lectures on Moral Philosophy 1784–85: The Mrongovius II Notebook* (with Jens Timmermann, forthcoming), of *The Emergence of Autonomy in Kant's Moral Philosophy* (with Oliver Sensen, 2019), and *Fichte's System of Ethics: A Critical Guide* (with Owen Ware, 2021).

Henny Blomme studied philosophy and psychology in Leuven, Lausanne, and Köln and obtained his PhD from the Universities of Paris-Sorbonne and Wuppertal. He was a postdoctoral research fellow at the Max Planck Institute for the History of Science in Berlin and at the University of Edinburgh, and a guest researcher at the University of Bochum. From 2015 until 2022, he was a researcher funded by the Flemish Research Organization FWO and affiliated with the Institute of Philosophy at KU Leuven. In 2015, he was a guest professor at the University of Rio Grande do Norte (Natal). Since October 2022, he is a lecturer in philosophy at the Université libre de Bruxelles (ULB).

Fabian Burt is a doctoral candidate at Goethe-University Frankfurt and scientific assistant at the Berlin-Brandenburg Academy of Sciences and Humanities (DFG-project 'Neuedition, Revision und Abschluss der Werke Immanuel Kants'). Together with Thomas Sturm he is editor of Kant's *Universal Natural History and Theory of the Heavens* in Vol. I (2023) of the revised Academy Edition of Kant's Writings.

Katherine Dunlop is an Associate Professor of Philosophy at the University of Texas, Austin. She has written a number of papers on Kant's views of mathematics and natural science and on their background in eighteenth-century German philosophy.

Gabriele Gava is an Associate Professor of Theoretical Philosophy at the University of Turin. He has published articles in leading philosophical journals on Kant, Peirce, pragmatism, and epistemology. His first book, *Peirce's Account of Purposefulness: A Kantian Perspective*, was published in 2014 by Routledge. His second book, *Kant's* Critique of Pure Reason *and the Method of Metaphysics*, was published in 2023 by Cambridge University Press. He is an Associate Editor of the journal *Studi Kantiani*.

Paul Guyer is the Jonathan Nelson Professor emeritus of Humanities and Philosophy at Brown University and the Florence R.C. Murray Professor emeritus of Humanities at the University of Pennsylvania. He is the author, editor, and/or translator of more than 30 books. Recent works include *Reason and Experience in Mendelssohn and Kant* (2020), *Idealism in Modern Philosophy* (2023), with Rolf-Peter Horstman), *Kant's Impact on Moral Philosophy* (2024) and *Kant on the Moral Foundations of Right* (2025). He was General Co-editor of the Cambridge Edition of Kant, with Allen Wood. He is a Fellow of the American Academy of Arts and Sciences and a former president of both the American Philosophical Association Eastern Division and the American Society for Aesthetics.

Andree Hahmann is an Associate Professor at Tsinghua University in Beijing. His research covers a wide range of topics in ancient and modern philosophy. He is the author of *Aristoteles' »Nikomachische Ethik«: Ein systematischer Kommentar* (2022), *Aristoteles gegen Epikur* (2017), *Aristoteles' »Über die Seele«: Ein systematischer Kommentar* (2016), and *Kritische Metaphysik der Substanz. Kant im Widerspruch zu Leibniz* (2009).

Stephen Howard is a Senior Research Fellow at KU Leuven, Belgium. He is the author of *Kant's Late Philosophy of Nature: The Opus postumum* (2023) and various articles on early modern and modern natural philosophy and metaphysics, in journals including the *Southern Journal of Philosophy*, the *European Journal of Philosophy*, and *Kantian Review*. From the autumn of 2025 he will be leading a DFG project on "Kant's Regulative Cosmology" at the Goethe-Universität Frankfurt.

Huaping Lu-Adler is a Professor of Philosophy at Georgetown University. She is the author of *Kant and the Science of Logic* (2018) and *Kant, Race, and Racism* (2023). She has also published articles on various aspects of the history of Western philosophy, from epistemology and philosophy of language to metaphysics and philosophy of science.

Michael Bennett McNulty is an Associate Professor of Philosophy at the University of Minnesota, Twin Cities. His research focuses on Immanuel Kant's philosophy of nature and his views on the special sciences. He has published articles on these topics in journals such as *HOPOS, Kant Yearbook, Kant-Studien, Kantian Review, Studies in History and Philosophy of Science*, and *Synthese* and edited *Kant's* Metaphysical Foundations of Natural Science: *A Critical Guide*.

Thomas Sturm is Research Professor in Philosophy and History of Science, Catalan Institution for Research and Advanced Studies (ICREA) and Universitat Autònoma de Barcelona (UAB), and a member of the Academia Europaea. His research focuses on Kant's philosophy, theories of rationality, and philosophy and history of science. He is the author of *Kant und die Wissenschaften vom Menschen* (2009) and *How Reason Almost Lost Its Mind: The Strange Career of Cold War Rationality* (2013, with P. Erickson, J. Klein, L. Daston, R. Lemov, and M. Gordin), and he has edited selected Kantian texts on natural science for the new Academy edition of Kant's *Gesammelte Schriften* and numerous articles in journals such as *Erkenntnis, European Review, Kant-Studien, Kantian Review, Studies in History and Philosophy of Science, Synthese, Philosophical Psychology,* or *Journal of the History of the Behavioural Sciences*.

Clinton Tolley is a Professor of Philosophy at the University of California, San Diego. He is the author of numerous essays on Kant and later modern German philosophy, including German Idealism, phenomenology, and early analytic philosophy. He is also the co-editor and co-translator (with Sandra Lapointe) of *The New Anti-Kant* (2014).

Hein van den Berg obtained his PhD at the VU Amsterdam in history and philosophy of science in 2011, with a prize-winning dissertation on Kant's conception of proper science and Kant's philosophy of biology. After obtaining a postdoctoral grant from the Royal Netherlands Academy of Arts and Sciences (KNAW) for conducting research on the history of biology at the Technical University Dortmund, he became assistant professor at the Institute for Logic, Language, and Computation of the University of Amsterdam in 2016. He does research on the history and philosophy of logic, biology, and psychiatry. As a member of the *Concepts in Motion* group since 2011, he has been involved in a large number of computational and data-driven history of ideas projects.

Achim Vesper is an Assistant Professor at Goethe University Frankfurt. He has published extensively on Kant's ethics and aesthetics. He has also worked on other Enlightenment philosophers such as Wolff, Meier,

Reimarus, Lambert, and Feder. Together with Gabriele Gava, he wrote *Kants Philosophie* (2024). Another book, *Kant über Schönheit und systematische Einheit der Natur*, will be published in 2025.

Eric Watkins is a Distinguished Professor of Philosophy at the University of California, San Diego. He is the author of *Kant and the Metaphysics of Causality* (2005) and *Kant on Laws* (2019). In addition, he has edited *Kant and the Sciences* (2001) and *Immanuel Kant: Natural Science* (2012) and co-edited (with Ina Goy) *Kant's Theory of Biology* (2014). He has translated *Kant's* Critique of Pure Reason: *Background Source Materials* (2009). He has published numerous articles on Kant and the history of modern philosophy more generally in leading international journals. He is currently the President of the Board of Trustees of the North American Kant Society and a member of the Kant-Kommission at the Berlin-Brandenburg Academy of Sciences and Humanities.

Citations of Kant's Writings

Throughout the book, we have adopted a simple system of citation of Kant's writings. Citations from the *Critique of Pure Reason* refer to its first and second editions by only using A and B, respectively (for example, A191/B236). Citations from the *Akademie-Ausgabe* (*Kants gesammelte Schriften*, Berlin: Reimer, De Gruyter, 1900–) only indicate volume (in Roman numerals) and page (in Arabic numerals) number (for example, V:146–48). Some authors, while adhering to this mode of citation, also use additional abbreviations in their chapters. These are clarified in endnotes within the relevant chapters.

Introduction
The Significance of Kant's Account of Scientific Systematicity

Gabriele Gava, Thomas Sturm, and Achim Vesper

What defines science? In what way is scientific knowledge different from other kinds of knowledge? How can science make progress? Such questions, which lie at the center of much of general philosophy of science, certainly do not have agreed-upon answers.[1] Is it the way in which claims are justified in science, or the method it uses, that characterizes scientific knowledge? Is it the way in which its theories are ordered and related to one another, and how they help us to explore, explain, or understand reality? Or, still, are it specific goals that define what science is, such as truth, instrumental adequacy, or practical usefulness?

Philosophers of science have mostly focused on one or the other of these topics – epistemology, theory, or goals – in order to determine the nature of science and scientific progress. Somewhat surprisingly, Kant has a distinctive and unified answer to the nature of science: what is fundamental in science and what distinguishes science from other forms of knowledge is its *systematicity*. This is indeed the most general way in which he speaks about the methods in which scientists gather and justify knowledge, in which they develop and connect their theories, and the goals they pursue in this. It is the framework within which he thinks about the methods and goals of science, as well as the order and connection of its theories and other knowledge-claims. His concept of scientific systematicity is, then, a complex and rich concept that becomes explicated and applied in different ways to different sciences. It is also one that most scholars have not adequately analyzed, let alone seriously discussed.

But before one can see this, one needs to start with certain basic assumptions that most familiar with Kant's philosophy of science know well enough. Each and every body of knowledge that aspires to deserve the title of a science must be systematic, that is, not just consist in what Kant calls a mere "aggregate" of knowledge. For Kant, that means at least that every science requires an "idea of reason" and a "schema" to execute that idea, thus producing a body of knowledge that possesses "unity" and "completeness" (A832f./B860f.). Kant's point here is that in developing a particular doctrine as part of a science, we have to rely on a systematic order into which all particular cognitions can and will be integrated.

Do these Kantian views constitute an unnecessary demand for a "procrustean bed" (Schopenhauer 1988, 509) or a "baroque" architecture (cf. Strawson 1966, 23f.)? Or are they worthy of a fresh look, analysis, and discussion, and perhaps defensible and valuable even today? Our working hypothesis goes in these latter directions, partly inspired by the advent of renewed thinking in current philosophy about the systematicity of science (Rescher 1979; Hoyningen-Huene 2008, 2013; for studies on Kant, cf. Rescher 2000; Sturm 2009).[2] In addition, we believe that the new approach presented by this volume should also help to improve our understanding of how Kant's views are truly related to later accounts of scientific knowledge.

Surely, Kant's mentioned general views are rather abstract and thus require explanation and discussion. They can be understood quite differently. For instance, they might express the view that the relationships between the different cognitions of a science are not arbitrary and are instead determined by objective features that are independent of personal perspectives (see Gava 2014; Gava 2023, Ch. 1). Alternatively, they might be related to a traditional rationalistic, axiomatic ideal of science (De Jong and Betti 2010; Van den Berg 2014). Finally, they might be interpreted as expressing Kant's claim that all human cognition stands under the "government of reason" (A832/B860). In other words, the systematicity of science is a normative demand that follows if we view science as a rational enterprise.[3] This demand places different requirements on science. For example, these concern scientific progress, such that each science needs a coherent framework that underlies its investigations. In addition, different sciences must be clearly distinguished from one another so that they do not become conflated. To achieve this, not only must each science itself be developed systematically; we must also aim to integrate all sciences into a whole system, a task which Kant calls an "architectonics." In this vein, one can distinguish between an "internal" and an "external" systematicity of the sciences (see Sturm 2009, Ch. 3, 2020).

Such different interpretations of the systematicity of science do not necessarily contradict one another. They might be complementary and perhaps each necessary for a better understanding of what is special about science. Still, they point in different directions. In Kant's work, one finds expressions and illustrations of all of these interpretations. This volume provides ample evidence for that.

While aspects of Kant's notion of systematicity have been investigated by some scholars, no investigation exists that would connect his general conceptual views with their concrete applications and their historical contexts as well, with the aim of achieving a complete and unified picture – what in

Kant's terms would itself be a systematic analysis of scientific systematicity. Let us give a brief overview of the main existing directions of previous research.

First, some have examined how Kant's notion of systematicity is related to views held by his predecessors and contemporaries. In this respect, some have claimed that Kant's views are original, especially when the Wolffian school of thought is considered. It has been argued that while Christian Wolff's model of the systematicity of science is deductive, Kant's model is teleological instead (Hinske 1991). This reading has however been criticized and his views on the systematicity of science has been interpreted as in continuity with Wolff and other historical figures, like Johann Heinrich Lambert and Jean-Jacques Rousseau (Manchester 2003), or as expression of a traditional axiomatic model of science (De Jong 1995; De Jong and Betti 2010; Van den Berg 2014, 2021). Yet, as this volume shows, there are still other figures and traditions that might have influenced Kant's thinking on the systematicity of science or aspects of the aforementioned authors that have been neglected.

Second, scholars have examined Kant's notion of the systematicity of science in general (e.g. McRae 1957; Hinske 1991; Kitcher 1994; Rescher 2000; Fulda and Stolzenberg 2001; La Rocca 2003; Sturm 2009, Ch. 3; Gava 2014, 2023, Ch. 1; Van den Berg 2014; Sturm 2015, 2020. But how does the notion of systematicity get specified for special sciences, such as logic, mathematics, cosmology, geology, physics, history, anthropology, or still others that Kant himself was interested in – not the least his own transcendental philosophy or critical metaphysics? Only some case studies exist. To make room for more, we built on research that has rejected a number of common but mistaken views. For a start, it is not true that Kant's concept of science refers only to what are sometimes called the "hard", "mature," or "exact" sciences – such as mathematics or physics – and that his concept of systematicity is made only for these. As several contributions in this volume show, Kant had also an interest in disciplines that are *potential* sciences, even if in his times they seemed to make no progress. Increasing their systematicity was, for him, a central tool for making progress possible (see Sturm 2009). His concept of science has therefore a partly *normative* meaning (cf. Breitenbach 2022), and systematicity is supposed to help to promote and approximate the ideal. Furthermore, not even with respect to what Kant calls the "rational sciences" or "sciences of reason" (*Vernunftwissenschaften*) – logic, mathematics, the "pure" part of physics, and a critically refined metaphysics – is it correct to say that there is only one concept of systematicity which grounds the unity of each of them. For example, in logic, systematicity is – at least in part – constituted by the table of the logical forms of judgment (cf. e.g. Lu-Adler 2018). As is well known, Kant also uses this table as a necessary fundamental element of his critical philosophy as a whole, most clearly with its systems of categories and principles of the pure understanding in the first *Critique of Pure Reason*. These, in turn, are said to be extended into the *Metaphysical Foundations of Natural Science* (MFNS) of 1786 (see IV:473–4; Sturm, this volume). However, it is not easy or even possible to derive exact systems of, say, metaphysics of nature and morals, or any other discipline which Kant views as an actual or potential science from these tables. Moreover, beyond these areas Kant applies the notion of systematicity to other rational disciplines, such as mathematics, without referring back to the logical forms of judgment anyhow. More than logic is needed for explaining the systematicity of all sciences – except of logic. Therefore, the *Vernunftwissenschaften* do not display only one and the same kind of systematicity for Kant. A fortiori, he must think it unlikely or impossible to determine from the pure armchair the systematicity of empirical disciplines such as cosmology, geography, natural history, anthropology, or history. We have to look at them case by case.[4] What is more, Kant sometimes hints that achieving the systematicity of each science requires that one already has gathered much knowledge of its subject-matter and that developing a special science's system is no easy work (X:242; A848f./B876f.). As this volume shows, special sciences have their own kinds of systematicity, rooted in conceptual and theoretical frameworks specific to different domains of reality, the methods of scientific inquiry and reasoning adequate and available to these domains, and sometimes also constrained by diverse epistemic and practical goals. Scientific systematicity has some common structural features, but it also has many different faces.

Third, among Kant scholars, the application of the notion of systematicity to philosophy and metaphysics has already attracted some considerable attention and will be addressed in this volume as well. Thus, it has been noted how Kant often suggests that the systematicity of metaphysics must reflect the systematicity of human reason (see, e.g., Neiman 1994; Ypi 2022). Important for our collection is also that Kant's claims concerning the systematicity of philosophy have consequences for the systematicity of the sciences: in his view, metaphysics grounds the "possibility" of some special sciences, and "the use of all" of them (cf. A850-1/B878-9). When discussing the relation between metaphysics, reason, and the sciences, Kant uses metaphors – for instance, architectural and technical ones, such as "edifice" and "plan," or biological ones, such as "animal body" and "germ" (A832-5/B860-3) – that point in opposite directions (see Ferrarin 2015). While the metaphors have been explained against their historical backgrounds (e.g. Zöller 2001; Manchester 2003; Mensch 2013), a critical assessment of their adequacy is still lacking. In addition, there are also less-discussed metaphors in this context, such as "the eye of philosophy" that deserve closer attention (see Lu-Adler, this volume).

Fourth, scholars have also studied Kant's claim that the natural sciences must assume the systematicity of nature in order to carry out their investigations. One important question in this respect is how strong Kant's claim in this context is (see Gava 2022 for an overview). Some have taken Kant to make a bold contention, which holds that it is in fact certain that nature is systematic, even though we do not know in what way and to what extent this nature is systematic (see Wartenberg 1979, 1992; O'Shea 1997). Others read Kant more modestly as only claiming that we must treat natural laws as if they were systematically ordered (see Kitcher 1994; La Rocca 1999; Rauscher 2010; for a third way between these options, see, e.g., Horstmann 1997a; Thöle 2000). Another issue is what form of assent or taking-to-be-true is at stake when we assume an idea of reason regulatively, as for example the idea of the systematicity of nature. Some have taken this assumption as "belief "(*Glaube*), in Kant's technical sense of the term (Chignell 2007). Others have linked it to "opinion" (*Meinung*) (Willaschek 2018, Ch. 4). Scholars have also presented different views concerning how Kant's position on the systematicity of nature evolved from the first to the third *Critique* (Horstmann 1989; Guyer 1991; Horstmann 1997b; La Rocca 1999; Ginsborg 2015). For our concerns in this volume, a relevant question is how the systematicity of science and nature are related to one another in various contexts of Kant's philosophy (see the contributions by Tolley, McNulty, Watkins, and Guyer in this volume).

As one can see from this overview of the state of the art, Kant's notion of the systematic nature of science has been scattered among different areas of scholarship. It raises a number of related questions that have never been addressed in connection. The present collection of essays is the first comprehensive discussion regarding Kant's account of the systematicity of the sciences across many disciplines and against its historical background. Its chapters focus on three main topics, corresponding to its three main parts. To indicate to the reader the complexity and richness of the account, we will mark in **bold face** its core aspects discussed in the chapters of this volume.

0.1 The Essays of the Collection

First, two chapters focus on historical precursors of Kant's account of the systematicity of the sciences. In the first, **Gabriele Gava** turns to Kant's Architectonic of Pure Reason of the first *Critique*. While there has been a surge of studies on this Architectonic in the last few years, we still lack a comprehensive account of the historical sources that have influenced Kant's emphasis on its moral orientation. Gava argues that Christian August Crusius is one of them. In particular, Crusius's account of ends and their relationships in the *Anweisung vernünftig zu leben* (1744) is a key source for understanding Kant's important claims concerning the **necessary hierarchy of human ends** in the Architectonic that are a crucial part of his account of scientific systematicity. As Gava shows, reading Kant from this perspective allows us to get a better grasp on two issues: whether Kant thinks that obtaining cognition (*Erkenntnis*) or knowledge (*Wissen*) is an essential end of reason, and what kinds of subordination between ends are at stake in Kant's hierarchy.

Henny Blomme's contribution turns to another figure often seen as influential for the Kantian account, Johann Heinrich Lambert, and looks at his systematic classification of the sciences especially in his own voluminous *Architectonic* (1771). Lambert and Kant coincide that systematicity is the central hallmark of science and, among other things, that another important task here is **to classify the sciences in a complete system**. As Blomme shows, Lambert's *Architectonic* is intended to provide a metaphysical foundation for several fundamental sciences. In the second chapter of this work, Lambert offers a table showing the correlations between a list of eight simple fundamental concepts and sixteen fundamental sciences. The goal of Blomme's contribution is to gain a better understanding of this systematic classification of fundamental sciences by (1) analyzing the table of correlations. In the literature on Lambert's philosophical writings, the table that links fundamental concepts with fundamental sciences is mostly missing or conceived as an end point of the investigation. Consequently, we lack an analysis of its concrete form. (2) Then, Blomme considers the disciplinary status of the table. Contrary to prevailing interpretations, he argues that Lambert conceives of it as the center of an alternative ontology intended to ground the systematic connection between all the fundamental sciences. Blomme explains that this alternative ontology constitutes Lambert's response to the search for a metaphysical system, which implies that he thought that such a system should not be established independently of demonstrating its utility for bringing about foundational unity to scientific knowledge.

The second topic that the collection addresses, and the one to which the majority of the chapters are dedicated, is Kant's own account of the systematicity of the special sciences, providing case studies of sciences as different as logic and cosmology, mathematics and physical anthropology, physics, the life sciences, and history. As we will see, these will often involve further comparisons with Kant's forerunners and contemporaries as well.

In the first article, **Fabian Burt and Thomas Sturm** discuss **epistemological aspects of Kant's concept of systematicity** from their origin in his most elaborate scientific theory, presented in the *Universal Natural History and Theory of the Heavens* (*UNH*, 1755), to his critical writings. In *UNH*, they find a twofold meaning of the term "system": a doctrinal meaning (related to the German term *Lehre*) and an ontic meaning. As they explain, these are connected through modes of observation and reasoning crucial for progress in cosmology and cosmogony. Central for this is the technique of changing imagined standpoints of observation. According to Burt and Sturm, Kant drastically expands this Copernican technique, thereby **achieving a more unified and comprehensive theory**, another crucial aspect of Kant's notion of systematicity. In addition, Burt and Sturm show how changing imagined standpoints remains important for Kant's thinking in the critical period. They argue that the technique is a necessary methodological, heuristic device; no more, and no less. However, Kant now uses it in two quite different ways: on the one hand, to reconcile seemingly incompatible theories; on the other hand, to decide between actually incompatible theories. Thereby, the notion of standpoint changes leads to new problems. Still, it shows that his concept of scientific systematicity is not only about the definition and the classification of the sciences, nor only about the ends of each science and the demand for maximal comprehensiveness of theories, but also concerns **the dynamical progress of research.**

In his chapter, **Clinton Tolley** reminds us that Kant appeals to the results of logic to justify the systematicity of his own doctrines at key points in the critical philosophy – perhaps most infamously in his metaphysical deduction of the categories at the outset of the first *Critique*'s Transcendental Logic. Tolley focuses on **two important respects in which Kant takes logic itself to relate to the idea of systematicity**: on the one hand, insofar as **logic (*qua* science) itself *constitutes* a system** of cognitions; on the other hand, insofar as **logic is *about* a system**, due to the fact that the object of its cognitions itself forms a system – namely, a system of activities which define the nature of the faculty of the understanding. Having highlighted two distinct systems pertaining to logic, he then argues that Kant takes the system inherent in the activities of the understanding (logic's subject-matter) to have a priority in relation to logic itself, *qua* system of cognitions which constitutes the science of this faculty. Tolley concludes by considering a worry of circularity which arises on the resulting picture: the kind of thing that logic itself is – a system of cognitions – belongs to logic itself as part of its subject-matter (in its Doctrine of Method). It might seem to force the allegedly prior system of the activities of the understanding to anticipate too closely (or even already contain) the subsequent system of the science of logic.

In the next chapter of this part, **Katherine Dunlop** explains Kant's view that mathematics not only proves the possibility of synthetic *a priori* cognition, but exemplifies the surpassing **unity sought by reason** and does so in a way **that guides the study of nature.** This unity is manifested not directly in intuition, but rather by systematic relationships among concepts. Insofar as mathematical theories specifically illustrate these relationships, they serve to schematize reason's principles. Dunlop points to Alison Laywine, who brought out the importance of the theory of conic sections, as such an illustration. Laywine, however, argues that Kant's view of mathematical reasoning as constructive cannot easily accommodate the algebraic advances by which this theory became more general and powerful. Dunlop shows that Kant can be understood to allude to Newton's treatment of conic sections. In deliberate opposition to Descartes's algebraic treatment, Newton follows a constructive approach, which has still greater unifying power. In closing, Dunlop suggests that we can understand Newton's constructive technique as a hypothetical origin for the curves, and thus as illustrating the manner in which reason's idea can mediate the application of its principle(s) to the understanding's cognition.

The next contribution, by **Thomas Sturm**, discusses aspects of Kant's theory of scientific systematicity in one crucial area of his mature philosophy: namely, the **systematic relation between his transcendental philosophy, as presented in the first** *Critique,* **and the** *Metaphysical Foundations of Natural Science* (*MFNS*, 1786). How are his conceptions of these two fields related to one another? Some scholars have argued that the relation is a very close one: that in Kant's view, the real point of his transcendental philosophy is to provide a systematic – meaning here, a unified and complete – framework or foundation for natural science, or even that transcendental philosophy requires to be completed by the *MFNS*. Others insist that Kant's transcendental philosophy is in no such way related to the *MFNS*. In his paper, Sturm starts by describing the entanglement between metaphysics and the sciences in Kant's thought more generally. Then, he turns to Kant's views on the relation between metaphysical and transcendental principles of cognition, considering especially the "Analogies of Experience" and the basic laws of mechanics. Neither of these discussions decides the debate over the relation between the first *Critique* and the *MFNS*, as Sturm shows. By analyzing **the systematically defining features of the aims, methods, and subject-matter of the concept of natural science** given in the Preface to *MFNS*, he develops a complete, unified, and moderate interpretation: the *MFNS* presents important concretizations of Kant's doctrines of synthetic *a priori* conditions of empirical cognition, but it is not necessary to complete transcendental philosophy.

Hein van den Berg discusses **in what sense physics, chemistry, and the life sciences each constitute a systematic unity** according to Kant. He starts by discussing Christian Wolff's views on the hierarchy of the sciences and argues that, according to Kant, in one specific sense physics, chemistry, and several life sciences constitute a unity: physics and chemistry provide statements that can be used to provide proofs in the life sciences. However, the unity of physics, chemistry, and the life sciences is limited in scope, since Kant claims that the purposeful unity of organisms is mechanically inexplicable. Van den Berg finally discusses whether there are laws within the life sciences according to Kant. He argues that the fact that Kant acknowledged that physics and chemistry ground the life sciences does not imply that there are laws of life. The reason is that the life sciences of Kant's time were concerned with explaining the purposeful unity of organisms, which is mechanically inexplicable according to Kant, and the regularities discussed by life scientists in Kant's time lack *a priori* grounding.

Michael Bennett McNulty shows **how Kant utilizes an omnipresent aether to explain a wide variety of physical phenomena throughout his works on natural philosophy**, including optical, thermodynamical, chemical, and magnetic phenomena. This provides another detailed and rich **illustration of his theory of the systematicity of nature and scientific inquiry**. Kant even goes as far as claiming that the existence of an omnipresent physical aether can be deduced *a priori* (without appeal to experience, observation, or experiment), in the notorious "aether proof" of his *Opus postumum*. In retrospect, these commitments are widely seen as a blunder, especially after the demise of the luminiferous aether at the turn of the 20th century. McNulty situates Kant's theory of the aether in the context of the physics of his day. He shows that the 17th and 18th centuries witnessed a staggering proliferation of aethereal explanations in natural philosophy: those appealing to subtle substances as causal grounds for classes of physical phenomena. Kant was a part of this tradition in physics, and his aether theory was no embarrassment, but rather revealed a keen understanding both of coeval aethereal theories and their shortcomings.

In the next chapter, **Stephen Howard** seeks to clarify **the role of definitions in Kant's theory of the systematicity of the sciences**, a topic that has been discussed in recent literature but without consensus. Two opposed interpretations have been put forward: Thomas Sturm considers definitions to be key to Kant's account of the systematicity of science, while Katharina Kraus considers them to play a minor and dispensable role. Howard first presents passages in which Kant denies that definitions can be legitimately used in philosophy. These passages seem to provide further evidence in support of Kraus's position. However, according to Howard, a different picture emerges if we turn to Kant's late reflections on the systematicity of empirical physics in the *Opus postumum*. Here, Kant repeatedly attempts to define "physics." After providing a brief overview of these definitions, Howard argues that Kant's application of his theory of definition in his late drafts validates the general direction of Sturm's interpretation, while, at the same time, encouraging us to amend some of the details. Kant's definitional attempts in the *Opus postumum*, on Howard's reading, usefully illuminate some nuances of the canonical statements about systematicity and definition in the Architectonic and Discipline chapters of the first *Critique*.

Andree Hahmann's contribution deals with the question of **whether history can become a systematic science according to Kant**. First, Hahmann examines how the historical sciences of the 18th century understood a systematic approach and relates Kant's account of history to methodological considerations concerning the historical sciences of his time. It turns out that a crucial question for historians was how exactly a scientific system can be generated from an aggregate of historical facts. As Hahmann argues, **the assumption of an end point of historical development is the key to answering this question**. Second, Hahmann discusses the role that teleology plays in the construction of a system, especially for Kant. He examines Kant's pre-critical lectures on anthropology in order to gain insight into how the different goals of history as expressed in his various texts can be related to one another. Third, he examines how history understood in this way is integrated into the overarching system of Kant's critical philosophy.

Huaping Lu-Adler shows that Kant uses the image of a Cyclops to depict someone who looks at the world without the **eye of philosophy**. Analyzing Kant's various remarks about this metaphor, Lu-Adler teases apart **three layers of his theory of systematicity – the systematic unity of a given science, a systematic interconnection among the sciences, and systematicity with a worldly orientation**, whereby all sciences are referred to the final (moral) ends of humanity. With this theory in mind, Lu-Adler then uses two cases to illustrate how Kant, in practice, looks at things with the eye of philosophy. One is his scientific theory of race, which shows how he sees "race" only by taking the viewpoint of a philosophical investigator of nature. The other is his long silence about colonial slavery as an institution. This case shows how a philosopher's desire to fix his gaze resolutely on the final ends of humanity (*Menschheit*) as such, which on Kant's account is irreducible to an aggregate of concrete human beings, can lead him to look away from present human sufferings. This raises the question of whether we should reconsider Kant's insistence on looking at the world with the eye of philosophy, as Lu-Adler elaborates.

Finally, the third part of this volume takes into account **the systematicity of philosophy**, as a science within Kant's classification of the sciences, but also one that has **a particular role for understanding the special sciences in a unifying perspective**. In his chapter, **Eric Watkins** starts from Kant's philosophical distinction between different uses of reason, specifically the "logical" and the "real" uses of reason. He then argues that understanding systematicity as a purely logical requirement does not fully capture the conception of systematicity that Kant requires for natural science. Instead, he takes systematicity to include a grasp of the systematic unity of the objects of nature itself in addition to the kind of logical unity that is provided by syllogisms. In short, for Kant, **systematicity in natural** science must be both logical and real. For the logical use of reason, which concerns our mere forms of thought and the logical conditioning relations on which syllogisms are based, does not capture what reason's most fundamental interest is in natural science. In addition to its logical use, reason has a real use, one that concerns the real conditioning relations among objects in nature and that is ultimately oriented toward what is fundamental therein, both metaphysically and in the order of explanation. What's more, it is precisely this aspect of Kant's conception of systematicity that helps to explain the strictness of his narrow conception of natural science "properly so called", for the standards that must be met to attain cognition of the systematic unity of nature are so demanding that it turns out to be impossible for us ever to achieve them and we are left with the task of approximating them as best we can, as Watkins argues.

As **Stefano Bacin** highlights in the penultimate chapter, Kant repeatedly stresses that **moral philosophy, too, should find completion in the form of a system**. The chapter focuses on **three main aspects** that characterize his view of such a system. First, Bacin argues that, in Kant's view, the system of ethics does not pursue a merely coherentist project of systematizing moral beliefs. Systematicity in moral philosophy is, for Kant, about unfolding the obligations that are **justified by the fundamental principle**, following a Pufendorfian paradigm widespread in the eighteenth century. Second, Bacin shows that Kant's systematic treatment of moral philosophy is neither a logically consistent arrangement of demands, nor a concluded collection of ethical duties, but a system of ends that yields **an open-ended system of duty types**. Third, Bacin suggests that in Kant's view a systematic treatment of ethics is supposed to provide orientation for moral thinking through a broader perspective from which perplexing cases should be considered by emphasizing **the holistic character of ethical demands and clarifying their connections.**

In the last chapter, **Paul Guyer** turns to **Kant's view of the systematicity of Critical philosophy as a whole**. There, Guyer had argued that Kant intended the systems of the laws of nature and freedom to join together into **a single system of nature and freedom**. Here, Guyer argues that even prior to their unification, each of these systems should be regarded both objectively and subjectively, that is, as comprising a system of *representations*, on the one hand, such as concepts, laws, and maxims, and a system of *objects*, on the other, that is, comprising objects in nature and nature as a whole and also the empire of ends, the moral world into which the natural world is to be transformed. According to Guyer, only thus can we understand the full force of Kant's conception of systematicity.

* * *

12   *Kant and the Systematicity of the Sciences*

In sum, the contributions collected here show the many faces and facets of the idea of scientific systematicity in Kant's works. Let us now summarize the main aspects of the Kantian account in a more systematic overview, indicating also which contributors to this volume deal with which aspects:

1 Systematicity is the central hallmark of science (Blomme) or the defining characteristic that distinguishes science, *qua* systematized knowledge, from both common knowledge and from mere "aggregates" of knowledge as well.
2 For the systematization of knowledge, concepts and principles of the understanding as well as ideas and schemata of reason are required. These function for the development of unified and complete frameworks for knowledge in special sciences (Sturm, Hahmann).
3 The systematization of knowledge is a goal that we must aspire to achieve in a stepwise manner, aiming at ever more unified and more comprehensive theories or doctrines. Thereby, the systematization of knowledge through the sciences is connected to a historical and dynamic understanding of the progress of research (Burt and Sturm, McNulty).
4 The systematization of different sciences can follow different principles (Tolley, Dunlop, Burt and Sturm, Van den Berg, Hahmann), though there may be familiarities.
5 The sciences must (ideally) be classified within a complete system of all sciences (Blomme, Sturm).
6 This classification of the sciences requires working on the definition of each particular science in terms of its object, method or "mode of cognition," and its ends or goals (Sturm, Howard).
7 The ends of the sciences must be reflected and organized systematically in a hierarchical manner (Gava), with the ultimate goal of serving the essential ends of humankind (Hahmann, Lu-Adler).
8 Philosophy is a discipline that itself aspires to be systematic in all of its parts (Bacin, Guyer) – and thereby a science.
9 Philosophy also plays special roles in helping other disciplines on their way toward ever higher degrees and qualities of systematization through a battery of higher-order considerations: by providing, where possible, *a priori* frameworks for the special sciences (Sturm); by moving from mere logical to "real" uses of reason for the reconstruction of the connections of knowledge-claims (Watkins), or by understanding both the connections between representations and the connections between the objects of such representations (Guyer); by reflecting the goals or ends of research and their hierarchies, up to practical and even moral ends

(Gava, Lu-Adler, Bacin); by reflecting and improving methods and tools of research (Burt and Sturm); and by helping the sciences to (re-)define themselves better (Howard, Sturm).

The picture that emerges here of Kant as a philosopher of science is obviously richer than interpretations with an exclusive focus on his views concerning the *a priori* presuppositions of science have it. There may be still other aspects of scientific systematicity in Kant, and surely the ones we have expressed in these nine claims might be formulated differently. We have aimed to show both central similarities of his applications of the basic concept and its many case-sensitive and flexible adaptations. They reveal that Kant's concept of science is rich, complex, and undogmatic – but not without general constraints either. These explain his views about the difference between science and common knowledge, and about conditions for the possibility of scientific progress, without excessively restricting his concept of science to, say, the natural or "exact" or "mature" sciences. On the contrary, the general constraints as well as their case-sensitive applications present ideas and ways for disciplines that have yet to make progress in their epistemic or rational qualities. That is the true and lasting significance of the Kantian concept of the systematicity of science. This should open new paths for scholarship on Kant's philosophy of science and, ideally, for reconsiderations and reassessments of his relation to later science and philosophy.[5]

This volume developed out of a conference we organized in Frankfurt in July 2019. A lot has happened since then and its contents do not exactly reflect the structure of that conference. Some people that presented their work in 2019 did not submit a chapter for the volume, and we invited others to join the team along the way. We are grateful to the participants of the conference and the contributors of this volume (where many of course belong to both groups). In addition, we wish to thank the institutions that financially supported the 2019 conference, namely the *Vereinigung von Freunden und Förderern der Goethe-Universität*, the *Stiftung zur Förderung der internationalen wissenschaftlichen Beziehungen*, and the Institute of Philosophy of Goethe University Frankfurt. Thomas Sturm's chapter "The conception of Kant's *Metaphysical Foundations of Natural Science*: Subject-matter, method, and aim" was originally published in *Kant's "Metaphysical Foundations of Natural Science": A Critical Guide* (ed. by Michael Bennett McNulty, Cambridge: Cambridge University Press, 2022, pp. 13–35) and is reprinted here with kind permission from Cambridge University Press. Last but not least, we thank Andrew Weckenmann and Rosaleah Stammler from Routledge for supporting the project and for their great patience.

Notes

1 We say "general" philosophy of science since much philosophy of science has in recent years focused increasingly on philosophical topics concerning the special sciences. Questions about the general nature of science have moved into the background, so much that at some points, general philosophy of science was declared "dead" (Fine 1988; on this development, see Kitcher 2023, Ch. 2). However, Hoyningen-Huene's work on systematicity, developed since the mid-2000s, tried to revive the discussion about the general nature of science.

2 In earlier correspondences, Hoyningen-Huene and Sturm discussed the relation between Kant's account and Hoyningen-Huene's (Hoyningen-Huene 2013, 155–9, 250n). This led him to revise his earlier interpretation of Kant as defending the traditional rationalistic, axiomatic ideal of science (Hoyningen-Huene 2008, 178 – a reading still defended, e.g. by de Jong and Betti 2010 and van den Berg 2014). Back then, discussions on Kant's philosophy of science centered on the possibility and the role of the synthetic *a priori* in natural science, on the relations between the first *Critique* and the *Metaphysical Foundations of Natural Science* (e.g. Plaass 1965; Brittain 1978; Friedman 1992, 2001; Pollok 2001; Watkins 1998), the role of regulative ideas and maxims of reason for inquiry (e.g. Morrison 1989; Friedman 1991; Kitcher 1994), and issues in the philosophy of mathematics (e.g. Posy 1992) and the life sciences (e.g. McLaughlin 1990; Guyer 2001). That Kant's philosophical thought on the sciences is more complex, more flexible, and that it encompasses many disciplines beyond natural or "exact" science is by now fully acknowledged (some diversity was already present in Watkins 2001; see also Sturm 2015; Sturm and De Bianchi 2015; Breitenbach and Massimi 2017; Watkins and Stan 2023).

3 This characterization is also only a first approximation. To add more flesh to it requires, for instance, drawing on Kant's account of different kinds of the "use of reason" (*Vernunftgebrauch*) in the sciences. Different uses of reason include the hypothetical use (for scientific explanations) or the intuitive use (for demonstrations in mathematics; and there are many others (see Sturm and Meer 2024). Sometimes these uses help to characterize the method of a special science, but some also cut across disciplinary distinctions. In any case, the plurality of uses of reason means that there is no simple Kantian answer to the question of what constitutes the rationality of science. Thus, insofar as scientific systematicity depends on scientific reason, systematicity must be plural too. This does not mean that anything goes. It means that the analysis must be complex.

4 For a close analysis of one unusual example and type of scientific systematicity, Kant's own attempt to establish a new science called "pragmatic anthropology", see Sturm (2009, esp. Chs. 7 and 8; for the case of history, Ch. 6). Again, Kraus (2018) presents an attempt to revive a Kantian empirical psychology as systematic science. We therefore left these examples aside in the present volume.

5 In different ways, the concept of systematicity comes even into play even in Kant's *empirical* account on science, i.e. his remarks and reflections on psychological, historical, and social conditions of scientific research. Here, the idea of systematicity becomes connected to, for instance, the cognitive processes of researchers for expanding their observations, forming new hypotheses and models, or structuring one's first ideas when taking new directions in research (see Sturm and Wallner 2025).

Bibliography

Breitenbach, A. 2022. "Kant's Normative Conception of Natural Science." In *Kant's "Metaphysical Foundations of Natural Science": A Critical Guide*, edited by M.B. McNulty, 36–53. Cambridge: Cambridge University Press.
Breitenbach, A., and M. Massimi, eds. 2017. *Kant and the Laws of Nature*. Cambridge: Cambridge University Press.
Brittain, G.J. 1978. *Kant's Theory of Science*. Princeton: Princeton University Press.
Chignell, A. 2007. "Belief in Kant." *Philosophical Review* 116: 323–60.
De Jong, W. R. 1995. "How Is Metaphysics as a Science Possible? Kant on the Distinction between Philosophical and Mathematical Method." *The Review of Metaphysics* 49: 235–74.
De Jong, W. R., and A. Betti. 2010. "The Classical Model of Science: A Millennia-Old Model of Scientific Rationality." *Synthese* 174: 185–203.
Ferrarin, Alfredo. 2015. *The Powers of Pure Reason*. Chicago: University of Chicago Press.
Fine, A. 1988. "Interpreting Science." *PSA: Proceedings of the Biennial Meetings of the Philosophy of Science Association 1988*, vol. 2, 3–11.
Friedman, M. 1991. "Regulative and Constitutive." *Southern Journal of Philosophy* 30: 73–102.
Friedman, M. 1992. *Kant and the Exact Sciences*. Cambridge, MA: Harvard University Press.
Friedman, M. 2001. *Dynamics of Reason*. Stanford: CSLI Publications.
Fulda, H.F., and J. Stolzenberg, eds. 2001. *Architektonik und System in der Philosophie Kants*. Hamburg: Meiner.
Gava, G. 2014. "Kant's Definition of Science in the Architectonic of Pure Reason and the Essential Ends of Reason." *Kant-Studien* 105(3): 372–93.
Gava, G. 2022. "Kant on the Status of Ideas and Principles of Reason." *Open Philosophy* 5(1): 296–307.
Gava, G. 2023. *Kant's Critique of Pure Reason and the Method of Metaphysics*. Cambridge: Cambridge University Press.
Ginsborg, H. 2015. *The Normativity of Nature: Essays on Kant's Critique of Judgement*. Oxford: Oxford University Press.
Guyer, P. 1991. "Reason and Reflective Judgment: Kant on the Significance of Systematicity." *Noûs* 24: 17–43.
Guyer, P. 2001. "Organisms and the Unity of Science." In *Kant and the Sciences*, edited by E. Watkins, 259–81. New York: Oxford University Press.
Hinske, N. 1991. "Die Wissenschaften und ihre Zwecke. Kants Neuformulierung der Systemidee." In *Akten des 7. Internationalen Kant-Kongress, Mainz 1990*, edited by G. Funke, Vol. I, 157–77. Bonn: Bouvier.
Horstmann, R.-P. 1989. "Why Must There Be a Transcendental Deduction in Kant's Critique of Judgment?" In *Kant's Transcendental Deductions*, edited by Eckart Förster, 157–76. Stanford: Stanford University Press.
Horstmann, R.-P. 1997a. "Die Idee der systematischen Einheit. der *Anhang zur transzendentalen Dialektik* in Kants *Kritik der reinen Vernunft*." In *Bausteine kritischer Philosophie*, 109–30. Bodenheim: Philo.
Horstmann, R.-P. 1997b. "Zweckmäßigkeit als transzendentales Prinzip – ein Problem und keine Lösung." In *Bausteine kritischer Philosophie*, 165–80. Bodenheim: Philo.
Hoyningen-Huene, P. 2008. "Systematicity: The Nature of Science." *Philosophia* 36: 167–80.

Hoyningen-Huene, P. 2013. *Systematicity*. Oxford: Oxford University Press.
Kitcher, P. 1994. "The Unity of Science and the Unity of Nature." In *Kant and Contemporary Epistemology*, edited by P. Parrini, 253–72. Dordrecht: Reidel.
Kitcher, P. 2023. *What's the Use of Philosophy?* New York: Oxford University Press.
Kraus, K. 2018. "The Soul as the 'Guiding Idea' of Psychology: Kant on Scientific Psychology, Systematicity, and the Idea of the Soul." *Studies in History and Philosophy of Science* 71: 77–88.
La Rocca, C. 1999. *Esistenza e giudizio: linguaggio e ontologia in Kant*. Pisa: ETS.
La Rocca, C. 2003. *Soggetto e mondo. Studi su Kant*. Venezia: Marsilio.
Lu-Adler, H. 2018. *Kant and the Science of Logic: A Historical and Philosophical Reconstruction*. Oxford: Oxford University Press.
Manchester, P. 2003. "Kant's Conception of Architectonic in Its Historical Context." *Journal of the History of Philosophy* 41: 187–207.
McLaughlin, P. 1990. *Kant's Critique of Teleology in Biological Explanation*. Lewiston/Queenston/Lampeter: Edwin Mellen Press.
McRae, R. 1957. "Kant's Conception of the Unity of the Sciences." *Philosophy and Phenomenological Research* 18: 1–17.
Mensch, J. 2013. *Kant's Organicism: Epigenesis and the Development of Critical Philosophy*. Chicago: University of Chicago Press.
Morrison, M. 1989. "Methodological Rules in Kant's Philosophy of Science." *Kant-Studien* 80: 155–72.
Neiman, S. 1994. *The Unity of Reason*. New York: Oxford University Press.
O'Shea, James R. 1997. "The Needs of Understanding: Kant on Empirical Laws and Regulative Ideals." *International Journal of Philosophical Studies* 5: 216–54.
Plaass, P. 1965. *Kants Theorie der Naturwissenschaft*. Göttingen: Vandenhoek & Ruprecht.
Pollok, K. 2001. *Kants 'Metaphysische Anfangsgründe der Naturwissenschaft'. Ein kritischer Kommentar*. Hamburg: Felix Meiner Verlag.
Posy, C., ed. 1992. *Kant's Philosophy of Mathematics: Modern Essays*. Dordrecht: Kluwer Academic Publishers.
Rauscher, Frederick. 2010. "The Appendix to the Dialectic and the Canon of Pure Reason: The Positive Role of Reason." In *The Cambridge Companion to Kant's Critique of Pure Reason*, edited by Paul Guyer, 290–309. Cambridge: Cambridge University Press.
Rescher, N. 1979. *Cognitive Systematization: A Systems-Theoretic Approach to a Coherentist Theory of Knowledge*. Oxford: Backwell.
Rescher, N. 2000. *Kant and the Reach of Reason: Studies in Kant's Theory of Rational Systematization*. Cambridge: Cambridge University Press.
Schopenhauer, A. 1988. *Die Welt als Wille und Vorstellung I*. In *Arthur Schopenhauer, Sämtliche Werke*, edited by Arthur Hübscher, Vol. 2. Wiesbaden: Brockhaus.
Strawson, P. F. 1966. *The Bounds of Sense*. London: Methuen.
Sturm, T. 2009. *Kant und die Wissenschaften vom Menschen*. Paderborn: Mentis.
Sturm, T. 2015. "Wissenschaft." In *Kant-Lexikon*, edited by S. Bacin, G. Mohr, J. Stolzenberg, and M. Willaschek, Vol. 3, 2711–16. Berlin: De Gruyter.
Sturm, T. 2020. "Kant on the Ends of the Sciences." *Kant-Studien* 111: 1–28.
Sturm, T., and S. De Bianchi. 2015. "Naturwissenschaft." In *Kant-Lexikon*, edited by S. Bacin, G. Mohr, J. Stolzenberg, and M. Willaschek, Vol. 2, 1669–76. Berlin: DeGruyter.

Sturm, T., and R. Meer. 2024. "Kant on the Many Uses of Reason in the Sciences: A Neglected Topic." *Studies in History and Philosophy of Science* 106: 54–59.
Sturm, T., and L. Wallner. 2025. "Kant's empirical account of science: Psychological, social, and historical conditions of research." *Annals of Science*.
Thöle, B. 2000. "Die Einheit der Erfahrung. Zur Funktion der regulativen Prinzipien bei Kant." In *Erfahrung und Urteilskraft*, edited by R. Enskat, 113–34. Würzburg: Königshausen und Neumann.
Van den Berg, H. 2014. *Kant on Proper Science: Biology in the Critical Philosophy and the Opus Postumum*. Dordrecht: Springer.
Van den Berg, H. 2021. "Kant's Ideal of Systematicity in Historical Context." *Kantian Review* 26: 261–86.
Wartenberg, T. E. 1979. "Order through Reason." *Kant-Studien* 70: 409–24.
Wartenberg, T. E. 1992. "Reason and the Practice of Science." In *The Cambridge Companion to Kant*, edited by P. Guyer, 228–48. Cambridge: Cambridge University Press.
Watkins, E. 1998. "The Argumentative Structure of Kant's *Metaphysical Foundations of Natural Science*." *Journal of the History of Philosophy* 36: 567–93.
Watkins, E., ed. 2001. *Kant and the Sciences*. New York: Oxford University Press.
Watkins, E., and M. Stan. 2023. "Kant's Philosophy of Science." *Stanford Encyclopedia of Philosophy*. https://plato.stanford.edu/entries/kant-science/#Bib
Willaschek, M. 2018. *Kant on the Sources of Metaphysics: The Dialectic of Pure Reason*. Cambridge: Cambridge University Press.
Ypi, L. 2022. *The Architectonic of Reason*. Oxford: Oxford University Press.
Zöller, G. 2001. "Die Seele des Systems': Systembegriff nd Begriffssystem in Kants Transzendentalphilosophie." In *Architektonik und System in der Philosophie Kants*, edited by H.-F. Fulda and J. Stolzenberg, 53–72. Hamburg: Meiner.

Part I
Systematicity
Historical Backgrounds

1 Kant and Crusius on the Hierarchy of Human Ends

Gabriele Gava

After having been almost neglected for a very long time, Kant's Architectonic of Pure Reason is attracting more and more attention (see Goy 2007; Sturm 2009; Mensch 2013; Ferrarin 2015; De Boer 2020; Ypi 2021; Gava 2023 – to cite only some of the monographs where the Architectonic plays a crucial role). This has resulted in an improved understanding of the role of the Architectonic in defining Kant's project in metaphysics and, more generally, in identifying what kind of systematicity is required in science. Scholars have also made progress in determining the relationship between Kant's view and accounts of the systematicity of the sciences defended by his predecessors and contemporaries. In this respect, there have been discussions of Kant's account of the systematicity of science in comparison with the Wolffian model.[1] Within the Wolffian tradition, the importance of Meier's influence on Kant has been emphasized, too (see Sturm 2009, Ch. 3; van den Berg 2021). Moreover, it has been shown that Johann Heinrich Lambert exerted a relevant influence on Kant's conception of Architectonic (see Manchester 2003; Sturm 2009, Ch. 3; Blomme 2015; van den Berg 2021).[2]

Even though our understanding of Kant's Architectonic has significantly improved, both philosophically and historically, there is still an aspect of it for which we do not have a satisfying account of its historical sources. Kant's concept of Architectonic possesses an important *practical* dimension. Kant argues that it is only with reference to reason's "essential ends," and to the "final end" that sits on top of these, that the system of human cognition can attain "architectonic unity." The final end is *moral* according to Kant. It is "nothing other than the entire vocation of human beings, and the philosophy of it is called moral philosophy" (A840/B868). This practical aspect in Kant's Architectonic has been already emphasized (see Ypi 2011, 2021; Gava 2014, 2023). However, studies on its historical sources are still sparse. A valuable exception in this respect is Paula Manchester. She has argued that the relevant source of Kant's emphasis on moral orientation in his Architectonic is Jean-Jacques Rousseau

DOI: 10.4324/9781003166450-3

(Manchester 2003). According to Manchester, "Lambert's conception of architectonic is from the perspective of an artisan striving to be an architect with the tools of mathematical logic. Kant's conception of architectonic is developed from the perspective of a cosmopolitan notion of philosophy" (Manchester 2003, 205). In her view, "Rousseau is the only modern philosopher who aspired to being an architectonical philosopher" in this sense (Manchester 2003, 205).

Without challenging the view that Rousseau was an important figure for the development of Kant's moral philosophy, I suggest that he is not the relevant source when it comes to characterizing Kant's emphasis on the need of moral orientation in the Architectonic. To see this, one only needs to consider how important the notion of a *hierarchy of human ends* is in order to spell out that orientation. I argue that Kant's use in the Architectonic of concepts like "final" and "highest" end, in the singular, and "subalternate" ends, in the plural, together with the emphasis on moral orientation they convey, was influenced by Christian August Crusius. The impact of Crusius on various aspects of Kant's thought has already been documented (see Hogan 2009; Chance 2019; Gava 2019; Oberst 2019; Mileti Nardo 2021). It has been shown that this influence concerns a key concept in Kant's moral philosophy, namely, his concept of the highest good (Schmucker 1961, 83). But since it is the highest good that crowns Kant's hierarchy of human ends in the Architectonic, and since Crusius similarly places the unification of virtue and happiness at the top of his hierarchy of ends, it is most likely that Crusius influences Kant on this aspect of the Architectonic, too.

I believe that seeing this connection between Crusius's and Kant's accounts of the system of human ends has two main advantages for our understanding of Kant. While it is relatively easy to determine what is the final or highest end for Kant, it remains an open question, first, what subalternate essential ends are and, second, how we should describe their relationships to one another and to the final or highest end.

I will start in Section 1.1 by presenting how Kant describes the hierarchy of human ends in the Architectonic chapter, identifying some questions that remain open. Section 1.2 will be dedicated to Crusius's account of ends and their relationships in the *Anweisung vernünftig zu leben* (Crusius 1744, hereafter *Anweisung*). As we will see, Crusius makes some interesting distinctions between types of ends and between types of relationships among ends. In turn, these distinctions might prove useful for resolving some interpretative problems in Kant. Section 1.3 will be dedicated to determining how cognition or knowledge can contribute to achieving the chief end of God and humanity according to Crusius. Finally, Section 1.4 will use the tools we have collected through our analysis of Crusius to make progress in our understanding of Kant's hierarchy of human ends.

In particular, I take the relationship between cognition and the highest good into account and consider two issues: whether obtaining cognition or knowledge can be considered an essential end for Kant and what kind of subordination is at stake in the relationship between cognition and the highest good.

1.1 Kant on the Hierarchy of Human Ends in the Architectonic

Famously, in the Architectonic of Pure Reason, Kant claims that systematicity is essential to science. More precisely, a science cannot be considered a science if it does not have "architectonic unity."[3] Elsewhere, I have argued that architectonic unity actually requires more than mere systematicity. It requires that a science realizes *one* particular system of cognitions, namely the system that corresponds to the "idea" of that science. I take this idea to be the *correct* description of the body of cognitions belonging to a science and the parts–whole relationships within it (see Gava 2023, Ch. 1). Up to this point, there is nothing that suggests that talk of "ends" is essential to defining what architectonic unity is. If, in order to have architectonic unity, a science needs to realize the system that corresponds to its idea, this could simply mean that we need to determine relationships of dependence and subordination among its cognitions that correspond to that idea.

Yet, Kant insists that the relation to "ends," and to "essential ends" in particular, is key to architectonic unity. For example, he distinguishes between architectonic and "technical" unity by saying that the latter arises "from the similarity of the manifold or the contingent use of cognition *in concreto* for all sorts of arbitrary external ends" (A833/B861), whereas the former arises "for the sake of its affinity and its derivation from a single supreme and inner end, which first makes possible the whole" (A833/B861). The emphasis on essential ends becomes more central when Kant discusses the architectonic unity of philosophy, in particular. He distinguishes between two concepts of philosophy, namely the "school concept" (*Schulbegriff*) and the worldly concept (*Weltbegriff*) of philosophy. While the former is only interested in the systematic ordering of our cognitions (A838/B866),[4] from the perspective of the latter "philosophy is the science of the relation of all cognition to the essential ends of human reason (*teleologia rationis humanae*)" (A839/B867).

Kant does not spell out what these essential ends (*wesentliche Zwecke*) are. He only submits that there is a hierarchy of these ends and that the final (*Endzweck*) or highest end (*höchster Zweck*) among these is moral.

> Essential ends are on this account not yet the highest, of which (in the complete systematic unity of reason) there can be only a single one. Hence they are either the final end, or subalternate ends, which

necessarily belong to the former as means. The former is nothing other than the entire vocation of human beings, and the philosophy of it is called moral philosophy.

(A840/B868)[5]

In the closing paragraphs of the Architectonic, Kant specifies that philosophy "in the genuine sense," namely, according to its worldly concept, "relates everything to wisdom, but through the path of science" (A850/B878). This path rests on the essential contribution of metaphysics, which, Kant continues, is the culmination of human culture, even though its main achievements have more to do with disciplining reason to avoid errors, than with obtaining cognition. Here, the "chief end" (*Hauptzweck*) of reason, for the achievement of which metaphysics is instrumental, is characterized as "general happiness" (A851/B879).

That Kant relates the worldly concept of philosophy to wisdom and that he characterizes the final or chief end as general happiness are indications that this end is the highest good, namely the proportion of virtue and happiness according to which the happiness we get is a consequence of our moral merits. Kant explicitly equates philosophy as a "doctrine of wisdom" (*Weisheitslehre*) to a "doctrine of the highest good" (V:108). Moreover, the fact that, in characterizing the chief end of reason, he speaks of "general happiness" with no reference to virtue should not mislead us, for virtue can be seen as the condition according to which happiness can be "general," in the sense that the pursuit of happiness of each of us does not get in the way of the happiness of others.

Therefore, it is plausible to submit that the final end of reason is the highest good. However, there are many questions that remain open when we consider the relation of this end to subalternate essential ends. First of all, it is not clear what these subalternate essential ends are. For example, the fact that Kant describes science as a necessary means to achieving wisdom suggests that obtaining knowledge is a (subalternate) essential end of reason. This is also suggested by Kant's account of reason's three fundamental questions in the Canon. Famously, these questions are: "1. What can I know? 2. What should I do? 3. What may I hope?" (A805/B833). Kant stresses that these questions unite all interest of reason (both practical and theoretical) (A804/B832). Defining what an interest of reason is and how it relates to reason's essential ends is challenging. One possible way to do that is to claim that essential ends of reason describe the states of affairs that would result from a perfect and complete application of reason. In this respect, the achievement of complete rational knowledge can be characterized as the end of theoretical reason, while virtue or the highest good could be the ends that arise from the practical use of reason. By contrast, "interests of reason" could designate conditions to achieve reason's ends. In this sense, reason has an interest in God and immortality because they are conditions for achieving the highest good (for this suggestion, see Gava 2023, Ch. 1; Gava and Willaschek 2023; On Kant's teleological account of reason, see Kleingeld 1998; Ferrarin 2015, Ch. 1). According to this picture, achieving rational knowledge is an essential end of reason because it is what would result from a perfect and complete application of theoretical reason.

Putting the difficulty of specifying which the subalternate essential ends of reason are aside, Kant does not clarify how we should understand the relationship of subordination between the final end and subalternate essential ends, either. Are the latter subordinate to the former because they are means to achieving it? Or are they subalternate simply because they are less important than the final end? If we focus again on the end of obtaining complete rational knowledge, it is indeed plausible to suggest that it is less important than the highest good, as the final end. But Kant also stresses that essential ends are instrumental for achieving the final end of reason (A840/B868). Therefore, we need to determine how importance and instrumentality are connected in defining relationships of subordination in Kant's hierarchy of ends.

1.2 Crusius on Types of Ends and Their Relationships

Before inquiring whether Crusius's account of the ends of God and humanity can help us in making sense of Kant's views on essential ends of reason and their relationships, it is useful to note that Crusius introduces some valuable distinctions between types of ends and types of relationships among ends. Of course, he was not the only one to do so and other figures certainly exerted an influence on Kant's differentiations among types of ends. For example, in the lectures collected in the *Mrongovius Metaphysics* (dated 1782–1783), Kant defines a final end as that to which all ends are subordinated (XXIX:847). As is well known, Kant used Baumgarten's *Metaphysics* as a textbook for his lectures on metaphysics. In fact, it is possible to trace this definition back to §343 of Baumgarten's text (Baumgarten 1757). I will provide some reasons to link Kant's discussion of essential ends of reason to Crusius in the next section.[6] But even before we are able to establish this link, Crusius's discussion of ends and their relationships is interesting for a very simple reason. Namely, it is much more refined and specific than competing accounts in his contemporaries. Accordingly, we can use it to consider whether the distinctions he makes can help us to illuminate some of Kant's claims.

First of all, Crusius distinguishes between a broad and a narrow sense of the concept of a "final end" (*Endzweck*). According to the former, a final end is that which a spirit wants and is a synonym of "intention" (*Absicht*). According to its narrow sense, a final end must be wanted with consciousness (*Bewußtsein*) and distinct cognition (*deutliche Erkenntniß*)

(Crusius 1744, §13; see also Crusius 1753, §456). Even with this narrower characterization of the final end in mind, Crusius clearly uses the word quite broadly, to include any object or state of affairs that we want to realize while being conscious of what we are doing and having a distinct representation of it.[7] Therefore, final end is simply another name for "end" in general. Crusius further distinguishes between the "subjective final end" (*subjectivischer Endzweck*), understood as the effective force in us that grounds our striving for something, the "objective final end" (*objectivischer Endzweck*) as the thing that we represent as the object of our willing, and the "formal final end" (*formaler Endzweck*) as the thought of a certain relationship of the objective final end with the willing subject (Crusius 1744, §13; see also Crusius 1753, §456). Crusius uses Alexander the great's actions in the Persian wars as an example to clarify these distinctions. Accordingly, he submits that when Alexander initiated the war, the Persian empire was his objective end, the submission of that empire under his power his formal end, and his lust for power his subjective end (Crusius 1744, §13).

While these distinctions are interesting, it is Crusius's characterization of the relationships between ends that is more relevant for our purposes. In this context, a "middle final end" (*mittlerer Endzweck*) is an end that is pursued for the sake of another end, where the latter is called the "more distant final end" (*fernerer Endzweck*). Since a "more distant final end" might in turn be a means to another end, Crusius calls an end that is pursued only for its own sake an "absolute" (*absoluter*) or "ultimate end" (*letzter Zweck*) (Crusius 1744, §15). Interestingly, Crusius insists that a middle final end should not be confused with a mere means. For the middle final end is something that we want and has some value for us independently of its instrumental relationship with the more distant final end. By contrast, a mere means is something that we only pursue for its instrumental relationship with an end. Accordingly, we would be happy to avoid it if we could attain the more distant final end without it. For example, a trip to reach a certain destination can be a mere means for a person that does not like traveling, while it would be a middle final end for a person who does (Crusius 1744, §15; see also Crusius 1753, §457). This suggests that ends can be in a relationship of subordination such that we would still pursue a subordinate end if the superordinate end would fail: if we like traveling, we would still look for occasions to travel even though the need to reach a particular destination at a particular time would vanish.

Yet, Crusius provides a different picture when he introduces the distinction between "chief ends" (*Hauptzwecke*) and "auxiliary ends" (*Nebenzwecke*). While also in this case the relationship of subordination is instrumental, such that the auxiliary end is a means for attaining the chief end, what defines an auxiliary end is that it dissolves if the chief end is abandoned (Crusius 1744, §16). But this seems to contradict what Crusius says on the middle final end. Since, as we saw, the middle final end is something that we value independently of the end to which it is instrumentally subordinated, it seems that we could have reasons to continue to pursue the former even when the latter ceases to be. This problem is confirmed if we look at what "coordinate final ends" (*coordinierte Endzwecke*) are for Crusius. These are ends that we would still continue to pursue even if any other end in this relationship of coordination would fail (Crusius 1744, §16). According to this definition, a middle final end that we value independently of any relationship with other ends and the chief end for which it is a means would be coordinate final ends, since we would continue to pursue either of them even if the other end would fall apart. But this is counterintuitive, since the fact that one is the "middle" and the other the "chief" end is clearly intended to capture a relationship of subordination.[8]

Besides instrumentality, Crusius identifies another criterion for determining relationships of subordination among ends, namely, strength. Accordingly, the "highest end" (*höchster Zweck*) is the end that we crave for most strongly among certain given ends. If it is the strongest in comparison to all ends, it can be called the "absolutely highest end" (*schlechterdings höchster Zweck*) (Crusius 1744, §18; but see Crusius 1753, §456, where the absolutely highest end is equated with the ultimate end).[9]

Finally, Crusius hints at another type of subordination in his distinction between a positive and a negative subordination. His general characterization of "subordinate final ends" (*subordinierte Endzwecke*) suggests that this is simply another name to capture a relationship of instrumentality. Accordingly, he stresses that a subordinate end is pursued insofar as one makes another end its guideline (*Richtschnur*) (Crusius 1744, §17). But when he further distinguishes between positive and negative subordination, it becomes clear that the relationship of subordination that is at stake is not necessarily one of instrumentality. Instrumentality is key to defining what positive subordination is. Accordingly, an end is in a relationship of positive subordination with another end when it is a means to achieving the latter. By contrast, negative subordination seems to capture a relationship of importance between ends, such that we subordinate the pursuit of an end to its not hindering the pursuit of another end that is more important (Crusius 1744, §17). This means that we would avoid pursuing the former if this pursuit would hinder our achievement of the latter (Crusius 1753, §456).

Let's take stock and see what is relevant for our discussion. As far as the relationships between ends is concerned, Crusius identifies three different kinds of subordination. This can identify a relationship of either instrumentality, such that pursuing an end is a means to achieving another end, strength, such that we want an end more strongly than another, or importance, such that we only pursue an end if doing so does not hinder the pursuit of another, which is more important. Therefore, one way to make Crusius's distinctions relevant for our discussion of Kant is to ask which among these types of subordination obtains in Kant's account of the hierarchy of human ends.

1.3 Crusius on Cognition and the Chief End of God and Humanity

But let's see if Crusius has something relevant to say not only on ends in general, but on the specific ends that are fundamental for defining what human beings are. Importantly, it is when we specifically consider what for Crusius the chief end of God and humanity is that we can draw a connection between Kant and Crusius. We have seen that Kant identifies the final end of reason with the highest good, as the proportion of virtue and happiness. Now, it has already been shown that Crusius's view on the relationship between virtue and happiness has exerted a major influence on Kant's conception of the highest good (see Schmucker 1961, 83; for a criticism of this view, see Schwaiger 1999). In the *Anweisung*, Crusius characterizes God's chief end (*Hauptzweck*)[10] in this life as virtue. In creating a world with rational beings endowed with freedom, God's end is that these beings act virtuously. But given God's goodness, God also wants that virtuous human beings are happy. Accordingly, God creates a world in which, first, rational beings can exercise their free virtue and, second, happiness is granted to them in proportion to their moral worth (see Crusius 1744, §213; see also §208. In §295, Crusius uses the term "highest good" to describe the proportion between virtue and happiness). Clearly, Kant's "final end" closely resembles Crusius's "chief end" in that they both characterize these ends through the idea of a proportion between virtue and happiness. Of course, in Kant's account the final end is an end of reason, whereas Crusius ascribes the chief end to God. However, we should not forget that Kant identifies the idea of the highest *original* good with God, since only God can ground the proportion between virtue and happiness represented in the idea of the highest *derived* good (see A810-11/B838-9). Moreover, Crusius believes that the proportion between virtue and happiness is the chief end of human beings, too (see Crusius 1744, §234), but only insofar it is first of all the chief end of God.

There are thus good reasons to believe that Crusius influenced Kant's account of the final or chief end of reason. My aim here is not to either support or challenge this view. I will rather simply assume that it is correct. What I want to do is to determine whether Crusius provides some instruments to illuminate Kant's views on essential ends that are not the final and on the relationships obtaining among all these ends. In Section 1.1, I suggested that obtaining knowledge, and more specifically complete rational knowledge, can be considered an essential end of reason for Kant. Moreover, we have seen that there is some ambiguity regarding whether the relationship of subordination between the final end of reason and its subalternate essential ends is one of either instrumentality or importance. So, let's see if Crusius can help us making some progress with respect to these issues. In particular, I will consider his remarks on the relevance of cognition for the chief end of reason.

In §216 of the *Anweisung*, Crusius explicitly rules out that either the cognition of truth or happiness considered in isolation can be the chief end of human life. As far as the former is concerned, Crusius argues that, given the very nature of our understanding, cognition cannot be an end in itself. Our understanding pursues cognition for the sake of our will, so that we can better achieve given final ends (Crusius 1744, §216). Clearly, Crusius defends here an instrumentalist account of cognition, according to which it is in the nature of cognition to serve some practical end. Moreover, he explicitly links cognition to usefulness and stresses that knowledge (*Wissen*) should never be sought for its own sake (Crusius 1744, §233).[11]

From this instrumentalist perspective, cognition of truth gains importance when it is a means to attaining virtue. Accordingly, Crusius places the cognition of truth among "nobler" goods (*edlere Güter*), which are goods that are not the highest, as virtue is (see Crusius 1744, §295), but can nonetheless positively contribute to virtue when additional conditions are satisfied (Crusius 1744, §296).[12] In this framework, Crusius adds that some *specific* cognitions are fundamental for their instrumental role in relation to achieving virtue. Accordingly, there are three essential objects (*wesentliche Objecte*) toward which every human being should direct their understanding in order to obtain cognition of them. These are God, virtue, and oneself (Crusius 1744, §234). With cognition of oneself, Crusius means cognition of our states (*Gemüthszustände*), final ends, and opinions (*Meinungen*). Here, self-knowledge is required because virtue must be exercised with consciousness. This suggests that one should be conscious of one's motivation for acting in a certain way. Cognition of God is required because the obligation to act virtuously is dependent on God. It might sound paradoxical to claim that cognition of virtue is necessary to achieving virtue. But clearly, Crusius uses virtue to refer to different things here. When it is described as an end, virtue is a property that human beings or their actions can have as they act in the world. But when Crusius says that we should "cognize" virtue, he means that we should have a grasp of moral principles and how to enact them. Accordingly, he clarifies that cognition of virtue has two components. First, one must know what good or bad, allowed or prohibited is. But second, one must also know how to apply these principles to every occurring case (Crusius 1744, §234).

Crusius's claim that cognition of virtue is necessary for achieving our chief end should not be confused with the contention that one cannot become virtuous unless one studies moral philosophy. In fact, Crusius explicitly rejects this view. He argues that the uneducated understanding must have a grasp of moral laws, too, otherwise we would not regard it as subject to these laws (Crusius 1744, §135). Accordingly, a person with an ordinary understanding (*mittelmäßiger Verstand*) can discern what is good and what is bad, even in difficult and complex cases (Crusius 1744, §136).[13] Therefore, the cognition of virtue that is necessary to act virtuously is within the reach of the untrained understanding. This applies to the other two cognitions that are deemed necessary for virtue, namely cognition of oneself and cognition of God. It is plausible that "cognition of oneself" does not require any particular philosophical insight. Insofar as Crusius maintains that the ordinary human understanding is sufficient for grasping moral laws, we should expect that the kind of self-knowledge that is necessary for properly enacting those laws should be in its power, too. That cognition of God is also something obtainable by the untrained understanding might sound more dubious. Yet, Crusius maintains that we have an innate cognition of God and the moral laws that depends on him (Crusius 1744, §137). The thought seems to be the following: since the untrained understanding can cognize moral laws, but these laws in turn depend on God, we should similarly be able to cognize God.

Crusius's account of the relevance of cognition for our chief end is interesting for at least three reasons. First of all, it is interesting in its own right because it is multi-layered. While it is true that cognition in general is not valuable in itself, some specific cognitions are indeed very important for their instrumental relationship with God's chief end. Second, assuming that Crusius's account of the value of cognition might have influenced Kant, this could give us some reasons to challenge our suggestion above that knowledge is an essential end of reason for Kant. Third, Crusius's account of how cognition of God, virtue, and oneself is essential in view of God's chief end is useful because it specifies a way in which achieving cognition or knowledge can be instrumentally subordinated to our final or chief end.

1.4 Cognition and the Final End of Reason in Kant

Let's see whether Crusius's discussion of ends and their relationships can indeed help us in making progress in our analysis of Kant's hierarchy of human ends. Following my analysis of Crusius in the last section, I will take the relationship between cognition or knowledge, on the one hand, and virtue and the highest good, on the other, as paradigmatic for the question regarding how ends can be subordinated to one another. I will consider two issues in particular. First, I will consider whether cognition or knowledge is an essential end according to Kant. Second, I will spell out in which senses cognition or knowledge is subordinated to the pursuit of the highest good in Kant.

1.4.1 Cognition as an Essential End of Reason

I have just suggested that Crusius's account of the relationship of cognition in general with our chief end might give us some reasons to challenge the view that knowledge is an essential end of reason for Kant, for Crusius rules out that cognition has a value independently of its relationship with ends that we pursue through our action. Indeed, Kant claims that pursuing philosophical cognition for its own sake only provides a "doctrine of skill" (*Lehre der Geschicklichkeit*), where cognition can be subordinated to any arbitrary end one might fancy (see A839n/B867n; 9:24). These contentions bear important similarities with Crusius's instrumentalist account of cognition. Moreover, Crusius uses the word "skill" (*Geschicklichkeit*) to define what prudence (*Klugheit*) is, namely, the capacity to rationally pursue our contingent practical ends by selecting and applying good means (Crusius 1744, §161). Kant's use of the word "skill" in this context might have a Crusian origin.

And yet this seems insufficient for abandoning the view that attaining knowledge is an end in its own right for Kant. Think of his characterization of reason in the narrow sense as the faculty of inference.[14] The idea of a complete explanation is essential for characterizing how this faculty works, where a complete explanation is an explanation that is able to identify the complete series of conditions for any given conditioned cognition. It is true that this idea lies at the basis of reason's dialectical inferences. However, Kant also identifies a legitimate regulative use for this idea, in which we employ it as an ideal point toward which our cognition should converge, even though it remains unachievable (see A616/B644). No matter how we characterize this legitimate use, it is clear that it is connected to an account of reason in which cognition is essential to it. In an important sense, the idea of complete explanation is characteristic of reason exactly because it indicates what its end is. Moreover, even if we contend that reason must give up this end as an end that is actually achievable, its regulative use is justified because it contributes to furthering cognition, which confirm that cognition (even though possibly not complete explanation) is an essential end of reason.

1.4.2 Three Ways in Which Cognition Is Subordinated to the Highest Good

If Crusius's instrumentalist account of cognition is not enough to challenge the idea that obtaining knowledge is an essential end of reason for Kant, his views might nonetheless help us in making progress in defining in which sense cognition or knowledge is subordinated to the highest good in Kant's hierarchy of human ends. I suggest that, following our analysis of Crusius's views, we can distinguish three ways in which this subordination should be characterized.

In a first sense, some non-philosophical cognitions are subordinated to our final end simply because they are *conditions* to pursuing it. As we saw, Crusius maintained that cognition of God, virtue, and oneself are conditions for attaining virtue. Here, the subordination of cognition to the pursuit of virtue simply is a consequence of a cognitive requirement on moral actions. The thought is that my action cannot be judged as morally apt if I fail to have some cognition. For our purposes, two types of cognitions are key. For Crusius, I must be able, first, to recognize relevant moral principles and, second, to properly assess the motivation of my action and determine whether it is depended on my recognition of the relevant moral principles. Again, Crusius submits that I do not depend on philosophical insight for obtaining any of these cognitions.

Kant seems to agree with Crusius that there is a cognitive requirement on moral action. In a way similar to Crusius, he stresses that "common human reason" (IV:404) has a grasp of the principle of morality, which for Kant is the categorical imperative, even in absence of a philosophical analysis of the latter. Famously, Kant argues that actions, in order to be moral, must be not only *in conformity with* duty but also performed *from* duty (V:81). This requires that we are able to recognize the moral law and that the law is the determining ground of our action.

Kant appears to be more skeptical than Crusius on our capacity to become conscious of our motivation for action. Our actions, in order to be moral, should originate from the right motivation, but this does not mean that we cannot easily go wrong in assessing whether our motivation is in fact morally apt (IV:407). Since Kant denies that we have an immediate and transparent grasp of our motivation, it is not the case that self-knowledge is a condition for moral action. What is important is that our motivation is in fact moral, even though we cannot be completely certain that it is.

Yet, in the *Metaphysics of Moral*, Kant famously stresses that there is a duty of self-knowledge (VI:441). This duty demands that we determine whether the sources of our actions are "pure or impure" (VI:441), which seems another way to say that we should determine whether they are performed from duty or not. This demand does not necessarily conflict with Kant's earlier position.[15] In this sense, self-knowledge is not a requirement for an action to be moral. Rather, self-knowledge is required for morally perfecting ourselves and attain a virtuous disposition. Since we are subject to self-deception and sometimes believe that we act morally even when we do not, improving self-knowledge regarding our motivation is key to improving our capacity to act morally over the course of time. If this is right, we can stress the following: cognition of the moral law is a condition for pursuing virtue because it is first of all a condition to act morally. By contrast, self-knowledge regarding our motivation is a condition for pursuing virtue because it furthers the development of a moral disposition, where this is in turn essential for virtue because it improves our capacity to act morally over the course of time. In both cases, it is not the case that philosophical training is necessary to obtain the relevant cognitions.

A second sense in which cognitions are subordinated to the pursuit of the final end specifically concerns *philosophical* cognitions. This relationship of subordination is captured by those passages in which Kant stresses that we should not pursue wisdom while neglecting science (A850/B878; IX:26). While not indispensable for moral action and virtue, philosophical cognition still improves our capacity to act morally and to develop a virtuous disposition. Accordingly, in the *Groundwork*, Kant stresses that "wisdom – which otherwise consists more in conduct than in knowledge – still needs science, not in order to learn from it but in order to provide access and durability for its precepts" (IV:405). In a way similar to the duty of self-knowledge, we can act morally even without knowing any science of morals. And yet, having a philosophical cognition of moral principles can have an impact on our capacity to avoid self-deception, which in turn furthers our capacity to pursue virtue as a component of our final end, the highest good. It is not clear whether Crusius leaves room for the utility of a philosophical study of moral principles in view of our pursuit of virtue.[16] Here, Kant's emphasis on the positive impact of this study is much clearer.[17]

Until now we have identified two senses in which cognition of *moral* principles, either ordinary or philosophical, is subordinated to the pursuit of virtue because it is either a condition for it or a way of furthering it. In both these cases, the subordination can be broadly characterized as instrumental. The cognitions in questions are means to obtaining our end, where they can be either necessary means or not. In the Architectonic, Kant clearly speaks of a subordination of *theoretical* cognition to the pursuit of our final end, too. For example, Kant describes the mathematician, the naturalist, and the logician as "artists of reason" and claims that the philosopher uses the cognitions developed by them as "tools to advance the essential ends of human reason" (A839/B867). Kant's talk of "tools" suggests that even in the case of theoretical cognition the relationship of subordination to the final end is one of instrumentality. But in which sense can mathematical cognition or cognition in natural science be a means to the pursuit of virtue or the highest good? I suggest that in this case the relationship of subordination is not one of instrumentality, but rather one of importance, and that Crusius's notion of negative subordination can help us to illuminate the latter.

Recall that negative subordination captures a relationship of importance between ends, such that we subordinate the pursuit of an end to its not hindering the pursuit of another end that is more important (Crusius 1744, §17). A consequence of this subordination is that we would avoid pursuing the former if it would hinder our achievement of the latter (Crusius 1753, §456). Clearly, seeing the subordination of theoretical cognitions to the pursuit of the highest good from this perspective is much more plausible. Taking again mathematical cognition and cognition in natural science as an example, subordinating them to the highest good would simply mean making sure that pursuing the former would not in any way hinder the realization of the latter. And of course, it seems that we do not have any reason to think that pursuing cognitions in those sciences could have a negative effect on our moral disposition. Moreover, at the end of the Architectonic, Kant describes the positive effect of the boundaries set by the *Critique of Pure Reason* on metaphysics exactly along these lines. He writes:

> That as mere speculation it [metaphysics] serves more to prevent errors than to amplify cognition does no damage to its value, but rather gives it all the more dignity and authority through its office as censor, which secures the general order and unity, indeed the well-being of the scientific community, and prevents its cheerful and fruitful efforts from straying from the chief end, that of the general happiness.
> (A851/B879)

We know that given the results of the *Critique*, we should avoid seeking cognition of objects of pure reason like God, freedom, and the soul. Here, Kant emphasizes that setting these boundaries on our cognition has a positive effect on our pursuit of reason's chief end, too, since attempting to achieve those cognitions could "stray" us from that end. Therefore, it seems that we have a further reason to avoid striving for those cognitions in addition to the fact that they are not in our reach. Simply, striving for those cognitions can get in the way of our pursuit of virtue and the highest good. Of course, we can ask *why* seeking to cognize God, freedom, and the soul can hinder that pursuit. As an answer to this question, we might suggest the following: given that, according to Kant, the existence of God and the immortality of the soul are conditions for attaining the highest good, pursuing and failing to obtain theoretical cognition of God and immortality could bring us to doubt that God exists and that our soul is immortal, which in turn could cause despair and inhibit our efforts to contribute to the realization of a moral world. It is true that Kant does not explicitly make this suggestion in order to explain why seeking cognition of God and immortality might hinder our pursuit of virtue. However, notwithstanding how we interpret Kant's claim here, it seems that it could be interpreted along the lines of Crusius's negative subordination.

1.5 Conclusion

My aim in this chapter was to determine whether Crusius's account of final ends and their relationship could help us to shed light on Kant's description of the hierarchy of human ends in the Architectonic of pure reason. In particular, I have addressed two issues: which essential subalternate ends Kant identifies and what sorts of relationship of subordination can be identified among essential ends. As far as the former issue is concerned, I have considered whether cognition or knowledge is an essential end according to Kant. While Crusius's instrumentalist account of cognition gives us some reasons to challenge the view that cognition or knowledge is an essential end for Kant, it seems that we have more reason to defend it than to give it up. Still, Kant's claim that the pursuit of cognition for its own sake only provides a "doctrine of skill" can be related to Crusius's account on cognition as instrumental.

In relation to the second issue, Crusius has helped us to differentiate among different kinds of subordination among ends in Kant's Architectonic as far as cognition is concerned. Accordingly, I have distinguished among three senses in which cognition is subordinated to the pursuit of the highest good in Kant. First, some non-philosophical cognitions are conditions for either acting morally or for improving our capacity to act morally over the course of time. The cognitions in question are respectively cognition of the moral law and self-knowledge regarding our motivation. Second, philosophical cognitions regarding moral principles are instrumental to avoid self-deception and to build a moral character. In all these cases, the relationship of subordination to the final end is instrumental. Third, and finally, theoretical cognition is subordinated to the highest good because it is less important than the latter. In this sense, we must make sure that seeking that type of cognition does not get in the way of our pursuit of the highest good. Indeed, it seems that one reason to put a stop to philosophical inquiries regarding God, the soul, and freedom rests exactly on the need to avoid hindering this pursuit.

Notes

1 The results of these discussions are, however, sometimes conflicting. While Hinske (1998) insists on the differences between the Kantian and the Wolffian model, Gava (2018) and van den Berg (2021) see much more continuity between the two.
2 As is well known, Lambert published his *Architektonic* in 1771 (Lambert 1771) and anticipated the account he develops there to Kant in their correspondence.
3 On architectonic unity, see also La Rocca (2003, Ch. 6), Manchester (2003), Manchester (2008), Sturm (2009, Ch. 3), Ferrarin (2015, Ch. 1), Mensch (2013, Ch. 7), Fugate (2019), and Ypi (2021).
4 In Gava (2014, 2023, Ch. 1), I argue that philosophy according to this concept cannot achieve architectonic unity. For different interpretations regarding the status of philosophy according to its school concept, see Tonelli (1994, 272), La Rocca (2003, 221), Ypi (2011, 144), and Ferrarin (2015, 81).
5 Kant's reference to the "vocation of the human being" (*Bestimmung des Menschen*) has to be read on the background of a discussion in Germany that started with Johann Joachim Spalding's *Betrachtung über die Bestimmung des Menschen* (1748). On this debate, see Macor (2013). On Kant's specific use of the concept, see Brandt (2007).
6 Baumgarten's definition of a final end in his *Metaphysics* does not seem to be Kant's relevant source when we take into account Kant's specific concept of a final end *of reason*. Baumgarten provides his definition of the final end in the context of his discussion of types of causes, where the concept of end is key to characterizing *final* causes. But, as we have seen, Kant essentially links his concept of a final end of reason to the idea of the moral vocation of humankind, whereas Baumgarten's definition of the final end to which Kant refers in his lectures is not morally characterized.
7 Clearly, Crusius uses the word "final end" in a much broader way in comparison to Kant.
8 One way to remedy this tension between what Crusius says on middle final ends and auxiliary ends, respectively, is to claim that we would indeed stop pursuing the former if the ends for which they are instrumental are taken away. Still, in order to maintain their difference from mere means, it is necessary that in pursuing them we value them and really want them, even though they depend on an instrumental relationship with other ends.
9 In fact, Crusius seems to conflate the instrumentality and strength criterion. When he defines what an absolutely highest end is, he goes back to using instrumentality. This is the case because he claims that in comparison to the absolutely highest end, all other ends are subordinated through either positive or negative subordination, where, as we will see, at least positive subordination is defined in instrumental terms (see Crusius 1744, §18 and §17).
10 Recall that this is the same word that Kant uses at the very end of the Architectonic (A851/B879).
11 Having said that, Crusius also argues that human beings have a basic drive (*Grundtrieb*) for truth (Crusius 1744, §117).
12 More precisely, Crusius claims that nobler goods contain the "matter" of virtue (Crusius 1744, §296). This is described as a doing or omitting (*thun oder lassen*) that agrees with divine or human perfection, but can be distinguished from obedience to God (*Gehorsamkeit gegen Gott*) as the "form" of virtue (Crusius 1744, §177).

13 Schneewind (1998, 450) argues that these claims directly attack the Wolffians, who, according to Crusius, are not so keen to recognize a capacity for moral evaluation to the ordinary understanding. Shell and Velkley (2017) and Callanan (2019) emphasize that Kant also thinks that the ordinary understanding is capable of making apt moral judgments. They maintain that it was Rousseau who influenced Kant's view in this regard. Given Crusius's insistence on the capacities of the ordinary understanding, he might have influenced Kant, too.
14 As is well known, reason in the narrow sense is also an essential component of reason in the broad sense as the faculty of a priori cognition.
15 On the problems posed by this claim, see Ware (2009).
16 He claims that the task of moral philosophy is that of providing distinction and order, certainty and completeness to moral truths, as well as insight into their grounds and connection (Crusius 1744, Preface), but he is not explicit on the effect of this study on our pursuit of virtue.
17 On the role of moral philosophy as a remedy to self-deception, see Callanan (2019).

Bibliography

Baumgarten, Alexander. 1757. *Metaphysica*. 4th ed. Halle: Hemmerde.
Blomme, Henny. 2015. "La Notion de 'Système' Chez Wolff, Lambert et Kant." *Estudos Kantianos* 3(01): 105–26.
Brandt, Reinhard. 2007. *Die Bestimmung Des Menschen Bei Kant*. Hamburg: Meiner.
Callanan, John J. 2019. "Kant on Misology and the Natural Dialectic." *Philosophers' Imprint* 19(47): 1–22.
Chance, Brian A. 2019. "Kantian Non-Evidentialism and Its German Antecedents: Crusius, Meier and Basedow." *Kantian Review* 24(3): 359–84.
Crusius, C.A. 1744. *Anweisung vernünftig zu leben*. Leipzig: Gleditsch.
Crusius, C.A. 1753. *Entwurf der nothwendigen Vernunft-Wahrheiten*. 2nd ed. Leipzig: Gleditsch.
De Boer, Karin. 2020. *Kant's Reform of Metaphysics: The Critique of Pure Reason Reconsidered*. Cambridge: Cambridge University Press.
Ferrarin, Alfredo. 2015. *The Powers of Pure Reason*. Chicago: University of Chicago Press.
Fugate, Courtney. 2019. "Kant's World Concept of Philosophy and Cosmopolitanism." *Archiv Für Geschichte Der Philosophie* 101(4): 535–83.
Gava, Gabriele. 2014. "Kant's Definition of Science in the Architectonic of Pure Reason and the Essential Ends of Reason." *Kant-Studien* 105(3): 372–93.
Gava, Gabriele. 2018. "Kant, Wolff, and the Method of Philosophy." *Oxford Studies in Early Modern Philosophy* 8: 271–303.
Gava, Gabriele. 2019. "Kant and Crusius on Belief and Practical Justification." *Kantian Review* 24(1): 53–75.
Gava, Gabriele. 2023. *Kant's Critique of Pure Reason and the Method of Metaphysics*. Cambridge: Cambridge University Press.
Gava, Gabriele, and Marcus Willaschek. 2023. "Transcendental Doctrine of Method." In *The Kantian Mind*, edited by S. Baiasu and M. Timmons. London: Routledge.
Goy, Ina. 2007. *Architektonik oder Die Kunst der Systeme: Eine Untersuchung zur Systemphilosophie der Kritik der reinen Vernunft*. Paderborn: mentis.

Hinske, Norbert. 1998. *Zwischen Aufklärung und Vernunftkritik: Studien zum Kantschen Logikcorpus*. Stuttgart: Frommann-Holzboog.
Hogan, Desmond. 2009. "Three Kinds of Rationalism and the Non-Spatiality of Things in Themselves." *Journal of the History of Philosophy* 47(3): 355–82.
Kleingeld, Pauline. 1998. "The Conative Character of Reason in Kant's Philosophy." *Journal of the History of Philosophy* 36(1): 77–97.
Lambert, Johann Heinrich. 1771. *Anlage zur Architectonic oder Theorie des Einfachen und des Ersten in der philosophischen und mathematischen Erkenntniss*. Riga: Hartknoch.
La Rocca, Claudio. 2003. *Soggetto e mondo. Studi su Kant*. Venezia: Marsilio.
Macor, Laura Anna. 2013. *Die Bestimmung des Menschen (1748–1800): eine Begriffsgeschichte*. Stuttgart-Bad Cannstatt: Frommann-Holzboog.
Manchester, Paula. 2003. "Kant's Conception of Architectonic in Its Historical Context." *Journal of the History of Philosophy* 41(2): 187–207.
Manchester, Paula. 2008. "Kant's Conception of Architectonic in Its Philosophical Context." *Kant-Studien* 99(2): 133–51.
Mensch, Jennifer. 2013. *Kant's Organicism: Epigenesis and the Development of Critical Philosophy*. Chicago: University of Chicago Press.
Mileti Nardo, Lorenzo. 2021. *Forme della certezza: Genesi e implicazioni del Fürwahrhalten in Kant*. Pisa: Edizioni ETS.
Oberst, Michael. 2019. "Kant and Crusius on Causal Chains." *Journal of the History of Philosophy* 57(1): 107–28.
Schmucker, Josef. 1961. *Die Ursprünge der Ethik Kants in seinen vorkritischen Schriften und reflektionen*. Meisenheim am Glan: Anton Hein.
Schneewind, Jerome B. 1998. *The Invention of Autonomy: A History of Modern Moral Philosophy*. Cambridge: Cambridge University Press.
Schwaiger, Clemens. 1999. *Kategorische und andere Imperative: zur Entwicklung von Kants praktischer Philosophie bis 1785*. Stuttgart-Bad Cannstatt: Frommann-Holzboog.
Shell, Susan, and Richard Velkley. 2017. "Rousseau and Kant: Rousseau's Kantian Legacy." In *Thinking with Rousseau*, edited by Helena Rosenblatt and Paul Schweigert. 192–210. Cambridge: Cambridge University Press.
Spalding, Johann Joachim. 1748. *Betrachtung über die Bestimmung des Menschen*. Greifswald: Weitbrecht.
Sturm, Thomas. 2009. *Kant und die Wissenschaften vom Menschen*. Paderborn: mentis.
Tonelli, Giorgio. 1994. *Kant's Critique of Pure Reason within the Tradition of Modern Logic*. Hildesheim: G. Olms.
Van den Berg, Hein. 2021. "Kant's Ideal of Systematicity in Historical Context." *Kantian Review* 26(2): 261–86.
Ware, Owen. 2009. "The Duty of Self-Knowledge." *Philosophy and Phenomenological Research* 79(3): 671–98.
Ypi, Lea. 2011. "Practical Agency, Teleology and System in Kant's Architectonic of Pure Reason." In *Politics and Metaphysics in Kant*, edited by Sorin Baiasu, Sami Pihlström, and Howard Williams, 134–52. Cardiff: University of Wales Press.
Ypi, Lea. 2021. *The Architectonic of Reason: Purposiveness and Systematic Unity in Kant's Critique of Pure Reason*. New York: Oxford University Press.

2 Lambert's System of the Sciences

Henny Blomme

2.1 Introduction

Johann Heinrich Lambert (1728–1777) devoted the majority of his scientific activity to studies in mathematics, physics, geographical projection, and astronomy. He was exceptionally disciplined and productive: During his rather short life, he wrote more than a thousand texts. Between 1762 and 1764, he also wrote two major philosophical treatises.[1] Standing in the tradition of similarly titled works by Aristotle and Bacon, Lambert's *New Organon*, written between October 1762 and November 1763, offers an overview of the tools (*Instrumente*) that human understanding must use to explore truth (Lambert 1764a, iv) (see also Lambert 1965–2020, vols. 1–2).[2] Written in 1764 but first published in 1771, the *Plan for an Architectonics* is an original endeavor to unify the sciences by assigning them a place in a foundational philosophical system.

The aim of this contribution is to provide a better understanding of Lambert's systematic classification of fundamental sciences. In the second chapter of the *Architectonics*, he presents a table that depicts the correlations between a list of eight simple fundamental concepts and sixteen fundamental sciences (Lambert 1771, 46 – see a copy on page! 49 here). This table is often overlooked or considered the end point of investigation in the literature on Lambert's philosophical writings.[3] Consequently, while many scholars offer valuable insights into the argumentative steps leading to the creation of this table,[4] there is a lack of analysis regarding its specific structure.

Moreover, the status of the table itself is a point of contention in the literature. Wolters conceives of it as the center of a new discipline (*Architectonics*) intended to provide a foundation for the natural sciences. He argues that this foundation is entirely different and independent from Lambert's conception of a metaphysical system, which Lambert deemed impossible (Wolters 1980, 23). In contrast to Wolters, Wellmann (2018) conceives of the table of correlations as the center of a propaedeutic for the

DOI: 10.4324/9781003166450-4

future establishment of a possible metaphysical system.[5] She thus identifies in Lambert's work a prefiguration of the Kantian distinction between a critical investigation into the possibility of metaphysics and a subsequent metaphysical system that neither Lambert nor Kant ultimately achieved. Contrary to these interpretations, I take Lambert to conceive of the table of correlations as the center of an alternative ontology intended to ground the systematic connection between all fundamental sciences. This alternative ontology, in my view, represents Lambert's response to the quest for a metaphysical system, suggesting that he believed such a system should not be established independently of demonstrating its utility in unifying scientific cognition at a foundational level.[6]

Before providing an overview of the sections in my contribution, I would like to point to an important issue that will mostly remain untouched here. Investigating the concrete form of Lambert's system of fundamental sciences may seem to presuppose an answer to the question of why Lambert believes that scientific cognition must be systematically ordered. In other words, why is it that all sciences must establish a systematic connection among their findings? Lambert surely follows generally shared convictions when he correlates true cognition with scientific cognition, and scientificity with systematicity. However, philosophers differ on the reasons behind such correlations. While Lambert does provide a pragmatic reason for the link between scientificity and systematicity, he appears to remain silent on the metaphysical presuppositions that guarantee the truthfulness of this connection.[7] To understand the perspective adopted here, it is important to note that I consider the question of what leads Lambert to classify the sciences as he does in the *Architectonics* to be separate from the more general question of why, for Lambert, science must be systematic.[8]

In the first two sections, I investigate Lambert's construction of an a priori foundation for the sciences based on fundamental concepts. In Section 2.2, I show how, driven by the quest for metaphysical truth, Lambert discovers a new method to arrive at certain fundamental self-evident concepts. In Section 2.3, I explain why these concepts are important for the sciences and how they play a role in the systematic ordering of the sciences. In Section 2.4, I analyze the details of the system of the sciences that Lambert presents in his *Anlage zur Architectonic*. And in Section 2.5, I reflect on possible influences of Lambert on Kant. Although this question has become a common topic in Lambert scholarship, I will avoid frequent references to Kant's work until the last section. I have two reasons for this. First, constantly comparing Lambert to Kant would obscure the presentation of Lambert's ideas in Sections 2.2–2.4.[9] Second, as I will explain in Section 2.5, I hold a more reserved view regarding the potential influence of Lambert's philosophical work on Kant than what is typically suggested in the literature.

2.2 The Quest for Metaphysical Truth and Lambert's Fundamental Concepts

In June 1761, the Berlin Academy of Sciences launched a prize question that asked the following question:

> Whether metaphysical truths in general, and in particular the first principles of Theologiae naturalis and morals, are capable of the same clear proof as geometrical truths, and, if they are not capable of the said proof, then what is the real nature of their certainty, to what sort of degree can one bring their certainty, and whether this degree is sufficient for complete conviction?[10]

The question aptly reflects concerns that were widely shared by German philosophers at the time: Should we apply the mathematical method to philosophy? If so, will it yield cognitions as certain as mathematical truths? If not, how should we then obtain metaphysical truth? Does metaphysical truth exist at all? The deadline for the receipt of essays was January 1, 1763. Five months later, the Academy voted to award the prize to Moses Mendelssohn's contribution (Mendelssohn 1764). The commission also gave special commendation to the contribution by Immanuel Kant (1764).

In November 1761, Lambert noted in his diary: "Treatise on the *Criterium Veritatis*" (Lambert 1915a, 24). This treatise lays the foundation of his philosophical contributions and contains a first sketch of the doctrines that he would soon develop more extensively in his *New Organon* and his *Architectonics*. Although it is uncertain whether Lambert started to work on the *Criterium Veritatis* only after learning about the Prize Question,[11] it clearly tackles philosophical questions about truth. The opening paragraph of the *Criterium Veritatis* is programmatic:

> Every proposition that one passes off as true, must be proved to us to be true by the fact that the words by which it is expressed have a correct meaning in every respect, that the concepts which we connect with these words are exactly correct, and contain in their scope neither what is erroneous nor contradictory, and that finally the one can be affirmed or denied by the other in the way in which they are presented in the proposition.[12]

What we see here is that Lambert effectuates a transition from truth as a property of propositions to truth as a property of concepts. It is a transition from what he later calls logical truth to metaphysical truth.[13] The problem, however, is that it is not always immediately clear whether a concept entails a contradiction or not. Therefore, one needs a criterium to distinguish between concepts without contradiction and concepts that turn out to be contradictory. However, such a criterium is not available:

> It is not yet possible to distinguish according to a general rule those concepts in which there is still a contradiction from those in which there is no contradiction at all, and which are therefore fully correct.[14]

According to its opening paragraphs, the endeavor to secure the truthfulness of our cognitions still motivates the project of the *New Organon*, Lambert's first major philosophical work, published in two volumes in 1764. Lambert observes that there is a serious discrepancy between truth and what the philosophers have taught. While truth is uniform and immutable, the doctrines of philosophers are forever changing and follow certain trends, just as the style of clothing changes with the current fashion (Lambert 1764a, ii). In the wake of Bacon, Lambert points to four possible sources of error: (1) Human understanding is not strong enough to follow the path of truth without obstacles. (2) Human understanding is not capable of recognizing truth and distinguishing it from falsity. (3) The languages which human understanding necessarily must use to express what it recognizes as truth are not precise enough and often polysemous. (4) Human understanding is blinded by mere appearance, which prevents it from reaching truth (Lambert 1764a, iii).

These four possible sources of error are investigated in the four philosophical sciences that constitute the *New Organon*:

1. Dianoiology mostly concerns logical truth and studies the laws that human understanding must obey when moving from one true proposition to another. In the wake of Wolff, Lambert not only studies the nature of concepts and judgments here but also the kinds of inferences and proofs that are in conformity with human understanding. The dianoiology could be easily dismissed as just another handbook of formal logic. However, Lambert innovates by proposing a method to immediately grasp the validity or invalidity of a given syllogism by symbolically representing the extensions of the used concepts with shorter or longer lines.[15]

2. Alethiology is the science of truth insofar as the latter is opposed to falsity. Lambert justifies such a science by positing that human understanding must somehow know how to determine whether a proposition is true. That is why we must start with some kind of basic truths and proceed with them. In the wake of Locke, Lambert situates these basic truths in a list of fundamental simple concepts. The simple concepts must be necessarily true and form the basis from which we can judge whether a proposition is true. The truth we find in simple concepts

is said to be material and allows us to judge whether a given proposition is true. By reducing complex concepts to simple ones, human reason becomes capable of self-evaluation. This allows it to reflect on the false steps taken while attempting to increase the number of true propositions.
3 Semiotics studies the meaning of thoughts and things and must provide the tools that allow for the translation of that meaning into a symbolic language. The goal of symbols is to provide clarity when our consciousness of words and concepts remains obscure.[16] This is the case if those concepts cannot be clarified by reference to sensation.
4 Phenomenology is the study of mere appearance (*Schein*) as opposed to truth and falsity (*Irrtum*). It must discover mere appearances (that are mistakenly accepted as truths) and teach how to avoid them.

It is essentially in the Dianoiology that Lambert discusses what is scientific cognition, how we arrive at it, and how it relates to experience. Under the influence of Locke, Lambert observes that the primary source of concepts is in sensation (Lambert 1764a I, 6). By being attentive to the input of the senses, we can have representations and become conscious of things and their characteristics.[17] The characteristics or marks (*Merkmale*) are those properties of a thing that allow us to recognize it and to distinguish it from other things. The concept of a thing is distinct if we have clear representations of its marks; that is, if we can represent them both individually and in their connection with other marks (Lambert 1764a I, 7). But these marks can be simple or composite. If a mark M is composite, then M itself has certain marks (e.g. m1, m2, m3) that allow us to recognize M and to distinguish M from other marks. But we can distinguish between proper (or inner) marks and common marks of a thing (Lambert 1764a, 11–12). The common marks are those that belong to its species and genus, while the proper marks are what makes the thing different from other things of the same genus (they are constitutive of the specific difference).

This is, to be sure, nothing new, and in large portions of the *Dianoiology*, Lambert indeed just provides an exposition of quite traditional views on the logic of concepts that we also find in Wolff. However, the distinction between proper marks and common marks ultimately leads Lambert to adopt a new, anti-wolffian conception of fundamental concepts.

In his *Ontology*, Wolff taught that we should use the method of abstraction to arrive at ever more general concepts (Wolff 1730, 14) and that from their definition we can deduce other true propositions. Lambert in contrast holds that general or common concepts (*Gemeinbegriffe*) cannot serve as first concepts in a philosophical system. There are two reasons for this: (1) The most general concepts are formal, whereas the first concepts on which a philosophical system can be founded must be directly linked to sensation. That is why Lambert calls them – in an analogous sense – "material." (2) First concepts must be simple, whereas, according to Lambert, common concepts are always composite. Take for example the concept of a "thing in general," with which Wolff begins his ontology.[18] In a letter to Holland, Lambert writes that this concept is far from being simple and that upon analysis it may turn out to be the most complex among all concepts.[19]

Thus, contrary to Wolff's conviction that ontology should deal with the most abstract notions, Lambert proposes analyzing our experience to isolate those elements that are immediately clear to us, either as representations or as sensations. According to Lambert, this allows the complex concepts that we draw from experience to be dissolved into more simple concepts until we arrive at a fundamental simple concept that contains only one proper mark.[20] Such simple concepts (*einfache Begriffe*) are called foundational concepts (*Grundbegriffe*) because they express universal and unconditioned possibility and thus "constitute the foundation (*Grundlage*) of all cognition" (Lambert 1764a I, 420).[21] However, since we inevitably have to start from what is given to us in experience, we cannot be certain of finding all foundational concepts.

On this occasion, Lambert remarks that if we had all foundational concepts, sciences would be entirely a priori. Since the possibility of simple concepts is included in their representation, they remain true apart from experience. In other words: Although the simple concepts are given via experience, and although we can become conscious of them by paying attention to what we sense, once we are conscious of them, we do not need experience to build a philosophical system on their basis. Indeed, the mere conscious representation of simple concepts involves their possibility (Lambert 1764a I, 422). Lambert provides an interesting explanation of why simple, foundational concepts stand above all doubt as regards their (metaphysical or real) possibility: Any concept of something impossible must involve a composition of at least some A and some non-A in order to show contradiction. Hence, impossible concepts can never be simple (Lambert 1764a I, 421).[22]

In the Alethiology, Lambert then provides a list of such "simple concepts," making explicit reference to Locke's notion of "simple ideas." Locke distinguishes between three kinds of simple ideas, depending on whether they stem from (1) sensation – and then they refer to either primary (<u>extension</u>, <u>solidity</u>, figure) or secondary qualities of bodies (color, sound, smell, taste, <u>motion</u>, rest), (2) reflection – in this case they refer to basic acts of inner faculties (perception, retention, discerning, comparing, compounding or enlarging, abstraction, <u>volition</u>), or (3) sensation and reflection jointly – here we find the simple ideas of pleasure, pain, <u>existence</u>, <u>unity</u>, <u>power</u>, and <u>succession</u>).[23] From these, Lambert retains the ones I have underlined above: extension, solidity, motion, existence, (duration and) succession, unity, power (*Kraft zu bewegen*), volition, and adds consciousness (Lambert 1764a I, 477).[24]

Why does Lambert add consciousness? Although certain passages in the *New Organon* suggest that fundamental concepts are simple because their content is delivered immediately through sensation, this is not entirely correct: While the contents of those concepts are indeed immediately given when sensation occurs, they are immediately given *to consciousness* rather than in the sensation itself. Thus, the fundamental concept of consciousness is at the same time the fundamental condition of all other fundamental concepts and always "appears together with them" (Lambert 1771, 45). Without consciousness, we would not even have "clear sensations" (*klare Empfindung*), representations, or concepts. Lambert in this regard subscribes to a more fundamental notion of consciousness than Locke. Locke takes the qualities that affect our senses to produce ideas in the mind (see Locke 1690, e.g. Book 2, chapter II §1), but for him the simple idea of "consciousness" is the result of reflection rather than a first condition of *all* simple ideas.[25] Notice, however, that being a presupposition of our representations of simple concepts makes consciousness fundamental but not necessarily a simple concept itself. The recognition of consciousness being simple indeed derives from consciousness being its own and sole mark.

2.3 The Usefulness of Simple Concepts for the Sciences

Why do we need fundamental or simple concepts? Lambert's *Plan for an Architectonics*[26] builds further on the Alethiology of the *New Organon* and carries out a new investigation into the basic or fundamental metaphysical doctrines (Lambert 1771 I, iii) that can be established on the material basis offered by simple concepts. Against Locke's mere anatomy of human knowledge, Lambert states:

> It is not enough to select simple concepts, but we must also investigate in which cases general possibilities can arise from their composition.[27]

What exactly does Lambert have in mind here? He refers to §692 of the Dianoiology, where we read that geometry searches for "the simplest case that does not limit the possibility" (Lambert 1764a I, 442). This simplest case is that a line could be drawn between any two points (and extended beyond them) and that around each point one can describe a circle of any size (or at least represent such a circle as drawn) (Lambert 1764a I, 443). These were Euclid's first two postulates, and Lambert refers to them to oppose anyone who denies certain geometrical possibilities. One example here is the possibility of an equilateral triangle: The two postulates allow Euclid to construct such a triangle. Lambert proposes Euclid's method as an alternative to Locke's merely anatomic method and to Wolff's "definitional" method, to secure a metaphysical ground for the sciences through a number of postulates derived from first concepts.[28]

The fact that in geometry postulates state primitive possibilities or "feasibilities"[29] leads Lambert to the insight that postulates are even more necessary in his fundamental doctrine, because the latter cannot rely on figuration (Kant would say: construction) of concepts.[30] For Lambert, postulates and axioms are even more important than simple concepts because it is only through the former that we can guarantee the possibility of things themselves. They must guarantee (1) that it is feasible to form a certain concept and (2) to determine the limitations on the possibility of complex concepts that result from the composition of simple concepts.

Apart from postulates, simple concepts also give rise to principles or fundamental propositions (*Grundsätze*) that express laws of thinkability. Only these principles can truly serve as a basis of all a priori sciences. Lambert states that while postulates indicate "certain possibilities concerning simple concepts," principles indicate "certain modifications concerning simple concepts" (Lambert 1764a I, 519). This may sound a bit vague, but it enables Lambert at least to conclude that the postulates and principles are also simple, and hence also clear and evident.[31]

As an example of the relation between a simple concept and the principles and postulates that can be established on its basis, let us consider the simple concept "extension" (*Ausdehnung*). Lambert explains that we gain the concept "extension" immediately through "feeling" (*Gefühl*)[32] and mediately through vision. The principles (*Grundsätze*) that are given by means of the simple concept of extension are the following:

(Gr1) The parts of the extension exist, or are thought to exist, at the same time.
(Gr2) The parts continue in all three dimensions.
(Gr3) No part is different from another part, except by being before, behind, above, under, etc.
(Gr4) No part of the extension is there where another part is, and so each part is external from each other.

The following postulates are given by the simple concept of extension:

(P1) In the extension, one can take a point wherever one wishes.
(P2) From such a point, one can draw a line of any length in each direction.

The simple concepts provide a basis in this way to develop a priori sciences. Since it essentially uses a procedure of combination, the development of a priori knowledge is said to be independent from experience. Separated from experience and henceforth unrelated to the conditions of existence, the simple concepts form the logical framework of sciences that develop, in a formally necessary way, under the sole determination of so-called eternal truths. To build further on our example: the simple concept of "extension" can be combined with "space" and "time" and in this way yields three a priori sciences: geometry, chronometry, and phoronomy.

It is not always clear, however, how exactly Lambert conceives of such combinations. Take for example "space" and "time." These concepts must be *simple*, says Lambert, if we agree that the three sciences mentioned above are a priori.[33] However, "space" and "time" are not among the list of *fundamental* simple concepts and are thus somehow derivative simples. Both "duration" and "succession" are among the list of fundamental simple concepts, but according to Lambert's own elucidations, succession should be considered still more fundamental than duration. While "succession" is immediately present in consciousness, because "we think one thought after the other" (Lambert 1764a I, 501), it is only because we relate this succession of thoughts to our simple concept of "existence" that we arrive at the simple concept of "duration": "As long as we think, we continue to exist. Accordingly, we attribute a duration to our existence" (Lambert 1764a I, 501). The genealogy of the simple concept "time" is then presented in simple terms as follows: "The beginning, continuation and cessation of individual representations gives us the concept of time and of its individual parts" (Lambert 1764a I, 501).

But what about empirical sciences – sciences whose development depends on empirical data?[34] According to Lambert, by virtue of the original provenance of the simple concepts whose possible combinations they unfold, the principles and postulates of the a priori sciences can also be used to build physical theories. However, Lambert understands that such theories involve concepts that cannot be reduced to mere elements of formal representation. One the one hand, Lambert honors the physically given by connecting its materiality with sensation, but, on the other hand, he claims that the relations in which it necessarily stands or could possibly stand are the object of a fundamental doctrine and that these relations can be deduced a priori from the simple concepts that correspond to the sensible given. Thus, while the status of the simple concepts is somehow both a posteriori and a priori, the fundamental doctrines that can be deduced from them without recourse to experience are a priori. In this sense, physical theories translate a purified cognition of experience[35] into propositions of which the "thinkability" can been made transparent through mere deduction from the a priori fundamental doctrine.[36] More concretely, Lambert claims that the true propositions of physics are ultimately grounded on the fundamental concepts of "extension," "solidity," and "motion."[37]

2.4 The Architectonics and Its System of the Sciences

In his *Architectonics*, Lambert also adds the concept of "magnitude" to the list of fundamental concepts, resulting in a total of ten: (1) Solidity, (2) existence, (3) duration, (4) extension, (5) force, (6) consciousness, (7) volition, (8) mobility, (9) unity, (10) magnitude.

However, because of considerations of which the status is at times a bit unclear, Lambert further modifies this list, stating that not all these fundamental concepts can serve within an architectonics of the fundamental sciences:

1 As we have seen above, Lambert maintains that "consciousness" is a fundamental presupposition of all fundamental concepts and therefore cannot belong to the systematic account of the sciences that the *Architectonics* must establish. Insofar as it is linked with veracity, consciousness manifests as conscience, and this concept must be studied in the *Organon*. As a presupposition of reflection and knowledge of the self, consciousness must be studied in a theory of the thinking subject and is therefore part of psychology.
2 "Volition" is a fundamental concept of morality and must therefore be studied in Lambert's agathology (doctrine of the good). In another respect, however, it is also an object of study for psychology.[38]
3 Since "magnitude" can be related to unity (the measurement of magnitude consists in counting a number of unities), Lambert removes it from the list;[39]
4 The simple concept "identity" must be added, although according to Lambert it is derived from the adverb "identically" (*einerlei*).[40]

As a result of these additional considerations, Lambert's final list contains eight concepts: solidity, existence, duration, extent, force, mobility, unity, and identity.

We must now establish Lambert's account of the fruitful relationships between these eight concepts to arrive at an architectonic grasp of the fundamental sciences. Very few combinations are actually fruitful: Of the 250 combinations that are theoretically possible, only 16 lead to the a priori development of the conceptual foundations of a particular fundamental science (Lambert 1771, 47). This result can be found in Figure 2.1. What is surprising here is that of the 16 fundamental doctrines, only 11 seem to properly indicate scientific disciplines, while in five cases Lambert just mentions (again) a certain concept. Moreover, he also lists one of the fundamental concepts (existence) as a discipline.

Correlata der Grundlehren.

	1	2	3	4	5	6	7	8	9	10	11	12	13	14	15	16
Solidität	+	–	*	*	*	=	–	–			–	–	=	=		–
Existenz	–	–	–	–		*	–	–								
Dauer	–	–	–	–		=	*	*	*			–		–	=	=
Ausdehnung	–	–	=		–			–		*	*	*		–	=	=
Kraft	+	–	–	–	–	=	–				–	–	*	*		–
Beweglichkeit	–	–	–									–		–	*	*
Einheit	–	*	–		–	=			–	–	–	–	–	–	–	–
Identität	*	–	–			=		–								
	Calcul der Qualitäten. §. 55.	Allgemeine Mathesis. §. 56.	Ding, Realität, Ontologie. §. 57.	Substanz. §. 58.	Systematologie. §. 59.	Existenz. §. 60.	Subsistenz. §. 61.	Zeit und Raum. §. 62.	Chronometrie. §. 63.	Geometrie. §. 64.	Static. §. 65.	Dichtigkeit, allgemeine Hydrostatic. §. 66.	Kraft der Substanzen. §. 67.	Bewegende Kraft, Mechanic. §. 67.	Phoronomie. §. 68.	Dynamic. §. 68.

Figure 2.1 The correlations between fundamental concepts and fundamental scientific doctrines.

Notes

* the fundamental concept (*der zum Grunde gelegte Begriff*)

= a concept that is necessarily connected to the fundamental concept

\+ the object of the fundamental concept

– the compared concept

A careful reading of Lambert's explanations reveals, however, that even in cases where he merely mentions a concept, he envisions the possibility of a doctrine or theory centered on that concept, such as a "theory of substance." As Lambert remarks in §75, the system of sciences that he sketches in the *Architectonics* is a wider-scope alternative to traditional ontologies (Lambert 1764, 58). One can wonder, then, why we also encounter "ontology" among the 16 fundamental doctrines. The explanation is that ontology in its traditional sense is only a part of Lambert's broader conception of ontology that is the *Architectonics*.

Below is a descriptive overview of Lambert's 16 fundamental sciences, containing the explanation of which fundamental simple concept is central and which simple concepts the discipline relates with the central one.

2.4.1 Quality-Calculation

The fundamental concept that grounds this discipline is "identity." Because this concept can relate to each of the other fundamental concepts, no fundamental concept is considered to be disconnected from the discipline. However, "solidity" and "force" are the objects of the discipline, and together with the concept of "identity," they constitute the foundation of a discipline that concerns the calculation of qualities.

According to Lambert, the concept of "identity" leads to a first class of general a priori principles, of which I quote the first three in order to provide an idea of what Lambert takes to be their (self-evident) nature. I also quote the ninth principle, because it is less evident, especially when considering the importance of mereological relationships in, for example, Kant's work:

1 Everything is equal, similar, identical to itself.
2 If two things are in equal parts similar, identical to a third thing, then they are in these parts equal, similar, identical.
3 If two things are in equal parts and in the same way different from a third thing, then they are in these parts identical (not different).
[...]
9 The whole is equal to its parts taken together.
(see Lambert 1771 I, 98)

These general (ontological) principles of identity are then specified in more particular scientific disciplines. But because the sciences are concerned with systems of things that are taken together or that belong together, there is a supplementary principle that is applicable to scientific systems. We can call this Lambert's fundamental principle of *identity in scientific systems*:

If in a system that which can be different or changed or determined otherwise is identical, then also that which depends on it (either as a consequence or as a presupposition or as necessarily connected) is identical
(Lambert 1771 I, 99–100)

As examples, Lambert gives the application of this principle in optics and photometry: "Seeing is identical if an identical eye is affected in an identical manner." And in mechanics: "If an identical body colludes with equal speed and direction, then also the motion, the shock, the force etc. are identical." Concerning the principle in optics and photometry, Lambert remarks that it is not to be mistaken as a principle that states that from an equal sensation one can conclude an equal thing that causes the sensation.

This latter principle would be invalid, because, as Lambert shows in the *New Organon* (more specifically in its fourth part: the *Phenomenology*), the domain of senses and sensations is also the domain of semblances. There, he states the principle as follows: "if a same sense receives an identical impression, it produces an identical sensation" (see Lambert 1764a II, 243) and remarks that it only concerns sensations as such – and thus not the consciousness of what these sensations contain.

2.4.2 General Mathematics ("Organon quantorum")

Here the fundamental concept is "unity":

(a) On the one hand, through repetition, unity leads to numbers, and the latter have already been scientifically studied within arithmetic.[41]
(b) On the other hand, "unity" relates to all other fundamental concepts.

2.4.3 Ontology

According to Lambert, the fundamental concept of ontology is "solidity." In other words, "solidity" is fundamental for metaphysical truth. How so? For Lambert, metaphysical truth allows us to make a transition from what is only thinkable to what is actually real. Possibility in the sense of thinkability is only possibility in view of the laws of understanding and does not say anything with respect to the possibility to exist. Now, since, "without solid and without forces we cannot think anything that exists, the solid is, together with the forces, the foundation of metaphysical truth" (Lambert 1771 I, 287).

This is somewhat surprising, considering that Lambert simultaneously subscribes to the traditional definition of ontology as "the theory of a thing in general" (Lambert 1771, 48). While Lambert admits that the word "thing" is "used in a more general sense" (meaning: in a more general sense than the physical things for which "solidity" is evidently a necessary predicate), he stresses that the concept "solidity" is equally fundamental for things in this broad sense since they indicate something real. Lambert thus seems to think that because things in general contain at least a reference to something real, their concept presupposes solidity. To use Kantian terms, we could say that Lambert conceives of "solidity" both as a non-schematized category (as an ontological predicate of things in general, whereby the latter are not necessarily physical) and as a schematized category, applying objectively to extended bodies. Lambert indeed remarks a bit obscurely that we can make a distinction between extended and unextended solidity (Lambert 1771, 49).[42]

2.4.4 The Theory of Substance

Lambert explains that "solidity" is the fundamental concept that, connected with the concept of "forces" generates the concept "substance." More precisely, the forces guarantee both the existence and the persistence of the solid, and this leads to our concept of "substance." The theory of substance deals with substances considered in isolation (apart from other substances).

2.4.5 Systematology (Theory of Systems)

If the fundamental concept of "solidity" is connected with "extension," "force," and "unity," we get the conceptual basis for the establishment of a metaphysical or physical system that deals with reality. The theory of systems is known as "systematology." Its "transcendent" part concerns intellectual systems. The systematology contains a theory of order and of perfection, which is somehow presupposed by the notion of a system. The systematology deals with substances insofar as they are connected. This is in a nutshell what Lambert explains in his *Fragment for a Systematology* (see Lambert 1988).

2.4.6 Doctrine of Existence

The concept "existence" (*Existenz*) refers to an absolute unity that does not allow for degrees. Existence not only presupposes solidity and forces (like substances), but also duration, i.e. something cannot exist without persisting over time. What exists must also remain one and the same. The simple concept of "existence" is thus necessarily connected with the other simple concepts of "solidity," "duration," "force," "unity," and "identity" as co-foundational elements. It is strange that one of the simple concepts appears again here as a foundational theory (*Grundlehre*). Can Lambert continue to assert that "existence" is a simple concept? Again, we could solve the issue by distinguishing between a merely logical or unschematized concept of "existence" (which just contains itself as a mark and thus is both unexplainable and self-evident) and a schematized concept, through which we think about how existence is possible. Only in the latter case do we find that it appears in connection with other simple concepts.[43]

2.4.7 Theory of Subsistence

The theory of subsistence is the first of three fundamental doctrines that has "duration" as its central foundational concept. If connected with "solidity" and "force," the concept of "duration" generates the concept "subsistence." It is not clear, however, what kind of content it is that Lambert associates with this theory.[44]

2.4.8 Theory of Time and Space

Connecting the central foundational concept of "duration" with "solidity," "existence," "extension," and "identity" give rise to special principles that limit the composition or connection of *in se* unconditioned possibilities. These special principles belong, says Lambert, to "Systematology or to the theory of the compossible" (Lambert 1741, 51). While in the eighth column of the overview of foundational doctrines, Lambert merely writes "Time and Space," he mentions neither of them in his explanation in §62. However, the concepts of "time" and "space" are implied by the special principles he gives as examples, such as: "that one and the same solid cannot be at the same time in different places," or "that different solids cannot be at the same time in the same place" (Lambert 1741, 51).

2.4.9 Chronometry

When the central foundational concept of "duration" is coupled with "unity," this leads to the discipline of chronometry as a theory of the measurement of time. Lambert remarks that the measurement of time would be especially difficult if we could not use motion (e.g. the escapement and the hands of a clock). In Lambert's view, however, chronometry must also inquire about the subjective length of time.[45]

2.4.10 Geometry

Three disciplines draw on the foundational concept "extension." When the concept of "extension" is connected with "unity," we get geometry. Lambert does not give an explanation here, probably because it is clear for him that geometry studies and measures space or spatial figures by means of a (freely chosen) fixed unity.

2.4.11 Statics

When we connect the central foundational concept of "extension" with "solidity," "force," and "unity," we are dealing with general statics. It is a "theory on the state of rest and permanence within systems" (Lambert 1741, 52). Again, these systems can be either physical or intellectual (see Section 2.2).

2.4.12 Hydrostatics

In a first step, we can connect the central foundational concept of "extension" with both (extended) "solidity" and "unity." We are then dealing with the study and measurement of the degrees of density of physical bodies. If we additionally connect "force," "duration," and "motricity," we get "hydrostatics." This science "determines to what degree the inequality of forces in a system does not allow for each order" (Lambert 1741, 52). Hydrostatics then studies how a system changes itself so that it reaches a state of rest or permanence.

2.4.13 Force of Substances

Here again, Lambert somewhat obscures the notion of a fundamental simple concept when he states that the concept of "force" presupposes the concept of "solidity." But he thinks that it is nonetheless possible to conceive of a theory that focuses exclusively on the notion of force. If we consider "force" as a ground for real or positive possibility, we are dealing with the force of substances.

2.4.14 Motive Force – Mechanics

If we consider "force" as a cause of motion, we are dealing with mechanics. This is very common and does not seem to require further explanation.

2.4.15 Phoronomy

We now turn to "motricity" as a central foundational concept. Connected with "extension," "duration," and "unity," we get phoronomy, as a science that is concerned with local motion as such, insofar we can study it with respect to time, space, and speed.

2.4.16 Dynamic

If we add the concepts of "solidity" and "force" to the conceptual connection of phoronomy, we get the theory of dynamics. It studies the forces at work in motion and is concerned with the changes that a system undergoes when forces are exerted on it.

2.5 Lambert and Kant

When finishing the editing of the *Architectonics* in 1764, Lambert had high expectations, but he struggled to find a publisher. His request to Kant, in a letter from November 1765, to recommend the work for publication to the latter's publisher, Johann Jacob Kanter, marked the beginning of his correspondence with the already famous philosopher from Königsberg. The correspondence touches upon several interesting questions, but its central theme is clearly the possibility of reforming metaphysics: How can we transform metaphysics so that it acquires the status of a science? Many of Lambert's proposals refer to what he had himself already undertaken in his (unpublished) *Architectonics*. Without doubt, the correspondence marked the beginning of a new period of enthusiasm about his philosophical work. Recalling the very meagre reception of his *New Organon*, Lambert thought it wise to write his own review of his *Architectonics*[46] as soon as the latter had been published by Johann Friedrich Hartknoch, who had started his printing career with Kanter before settling in Riga. Kant never replied to Lambert's last letter to him, from 1770, so the correspondence stopped before the *Architectonics* was published.

With the publication of the *Critique of Pure Reason* in 1781, Kant develops his own proposal to reform metaphysics, which is very different from Lambert's *Architectonics*. However, because of their correspondence and because of some strong terminological similarities between Lambert's and Kant's work, Lambert's work has been studied for insight into how it could have influenced Kant's thought.[47] As I pointed out in the introduction, I am skeptical when it comes to drawing such lines of influence. Here are three reasons for this skepticism:

1 Since we do not have strong evidence that Kant attentively read Lambert's work, the most obvious source of influence is the correspondence itself. In my view, however, as much as the letters show a programmatic agreement with respect to the reform of metaphysics, they also testify to the extent to which Kant and Lambert were talking past each other. It is striking that Kant never picks up on Lambert's references to concrete arguments within his own work. It is also striking that in his last letter to Kant Lambert does not seem to have understood the radicality of Kant's theory of space and time as subjective forms of sensibility in the *Dissertation*, and his main criticism is that duration, not time, is the equivalent of space (time being the equivalent of place). Referring to his own transition from logical to metaphysical truth, Lambert insists, however, on the question of whether "cognition of the form leads to cognition of the matter of our knowledge" (X:64).[48] For Lambert, the answer had long been clear: No – it doesn't! This was not yet so clear for Kant who, in the *Dissertation*, still allowed room for purely intellectual cognitions. So, the way in which Lambert formulated the problem he had already solved himself may have influenced Kant. But again, Kant's answer to this question in the *Critique* is radically different from Lambert's and requires the – in view of Lambert's critical remarks on Kant's treatment of space and time unchanged – doctrine about the ideality of space and time from the *Dissertation*: We can discover a priori in which sense the forms of discursive understanding have objective reality if we confront them with the (transcendentally material) a priori manifold that is provided by the general form of intuition.

2 The looseness of the terminological similarities: When first reading Lambert, one can easily be struck by the terminological similarities with Kant's work. This is especially the case for Kant's *Metaphysical Foundations of Natural Science*. All chapters of that work refer to disciplines that Lambert had also expounded on ("phoronomy," "dynamics," "mechanics," and "phenomenology"). But the terms "dynamics" and "mechanics" were of course not owned by Lambert. With respect to the less frequent concept of a "phoronomy," it was already used before (and perhaps introduced) by Jakob Hermann (1716). Since Lambert invented the concept of a "phenomenology," Kant must indeed have taken it from him. However, in Lambert it is not one of the sixteen fundamental sciences and thus not on a par with phoronomy, dynamics, and mechanics, but a part of the *New Organon* and thus a purely philosophical theory about the instruments we need to distinguish reality from mere *apparentia*. It would be a mistake to conclude that the terminological similarities indicate instances where Lambert and Kant treat matters of equal content.[49]

3 a The relative absence of references to Lambert in Kant's work:

This in itself is not a very strong argument: Kant may not always have been explicit about his sources. We find, however, more references to Baumgarten, Wolff, Leibniz, Descartes, Newton, etc.

b The nature of the existing references to Lambert:

My thesis here is that whenever Kant does refer to Lambert, it does not show a deep understanding of Lambert's philosophy. Take for example the famous reference to Lambert's views on solidity in Kant's *Metaphysical Foundations of Natural Science* (IV:497). Kant here not only states that Lambert took solidity to be a quality of all things existing, but also that the presence of a real thing in space already analytically implies solidity, and thus resistance to other things being at the same place. Kant then sarcastically adds that the principle of non-contradiction cannot be held responsible for repelling matter that moves forward to penetrate the space occupied by other matter.

As we have seen, Lambert takes both solidity and force to be the fundamental simple concepts that mark a transition to what really exists (and, thus, to metaphysical truth). These concepts do not follow analytically from the concept of something existing. On the contrary, they are taken from experience, and we can combine them in order to a priori derive the concept of substance of which they will then be the marks. As we remarked above, what complicates things, however, is that, following Lambert, existence is also a simple concept. It is not connected only with solidity but also with duration, force, unity, and identity as co-foundational elements. In principle, there is nothing that prevents Lambert from thinking about causes of solidity, such as Kant's metaphysical force of repulsion. But such causes are not immediately derivable from experience and their concepts cannot be simple. Therefore, in Lambert the concept of "repulsion" does not belong to the unificatory philosophical basis that grounds the fundamental sciences.

More specifically, when looking at Lambert's and Kant's takes on systematicity of the sciences, we should not mistake a resemblance in their views for identity in their views. The fundamental elements of Lambert's system of sciences are a list of simple concepts. We just need to compare these with Kant's elementary concepts (space, time, the twelve categories, and the three transcendental ideas) to find some striking differences. (1) While Lambert's simple concepts lead to the establishment of a system of the sciences, Kant's elementary concepts lead to the establishment of a theory of the possibility of synthetic a priori cognitions in mathematics and philosophy and to a new view on objectivity. So, although Kant's systematic division of the sciences is found in a chapter that he also refers to as "architectonics," this third chapter is not a part of Kant's theory of elements but of the doctrine of method and therefore does not pertain to the core conceptual material of his critical metaphysics. The important conclusion is that, while Kant's critical philosophy and critical metaphysics can be established independently of their possible applications in and to the sciences, Lambert's *Architectonics* is exclusively developed with a focus on the philosophy of the sciences. (2) Like Locke, Lambert situates the source of simple concepts in experience itself and argues that if simple concepts were not taken from experience, we could not conceive of them as fundamental. Kant, however, argues that the elementary concepts should not be derived from experience if they are to be apodictic and fundamental.

2.6 Conclusion

At first sight, many parts of the *New Organon* and the *Architectonics* may sound traditional and rather unexciting. However, Lambert does not present us with a mere copy of previous attempts to ground scientific cognition. As we have seen, Lambert claims both that (1) in the wake of Locke, we must derive fundamental concepts from experience, and (2) in the wake of the rationalist tradition, these concepts nonetheless lead to certain a priori principles and postulates that constitute the fundamentals of all kinds of sciences. What is new in Lambert is thus not the search for a list of first concepts and principles, but the way they are found and how they are said to ground scientific cognition. While the origin of those concepts is explicitly empirical, as they are in Locke, Lambert still wants to erect a kind of *mathesis universalis* on their basis a priori. With respect to his *New Organon*, Lambert himself indicates that his position in the Dianoiology is close to Wolff, whereas his position in the Alethiology is close to Locke.[50]

One of the outcomes of scholarship on eighteenth-century German philosophy in the last three decades is that Kant's framing of his own philosophy as pioneering a reconciliation of rationalism and empiricism is exaggerated. First, both the rationalist and empiricist projects that we encounter in eighteenth-century philosophy are far more nuanced than the somewhat caricatured models that Kant put on the scene. Second, independently from Kant, thinkers like Crusius, Tetens, and Lambert already aimed at a coherent integration of rationalist and empiricist elements.

Lambert's *Architectonics* provides a unifying systematic foundation for several fundamental sciences. This system of fundamental sciences, with their postulates and principles, arises out of combinations of eight fundamental simple concepts and is said to be a priori. But since Lambert is not sure whether his list of fundamental simple concepts is complete, his system of the sciences is an open system.[51] While the idea of an a priori philosophical system of sciences is easily associated with rationalist tendencies, Lambert's conviction that such a system must remain open, even where it concerns its most fundamental elements, betrays his empiricist stance. Thus, when compared with Kant's architectonics, which may be open at the bottom but is absolutely closed at the top, Lambert's architectonics offers a less rationalist model of an eighteenth-century systematic philosophy of the sciences.

In a letter to Lambert, Holland writes that by proceeding in Lambert's way and developing a systematic ontology of the sciences, the basic elements of which are originally derived from experience, "we get an experimental metaphysics." Such metaphysics, Holland observes, is "a science nobody has contemplated yet."[52] I contend that Lambert's experimental metaphysics provides us with a more fruitful model for contemporary philosophy of science than eighteenth-century philosophies that pretended to offer, a priori and once and for all, a complete systematic division of the sciences.[53]

Notes

1 Aside from these two books, during his life Lambert published three shorter philosophical texts in Latin (see Lambert 1764b, 1766, 1768). Posthumously, Johann Bernouilli edited two volumes with shorter texts and fragments by Lambert on logic and philosophy. The first volume contains texts that can be seen as preparatory for the *Organon*: First, *Six Attempts at a Drawing-art in the Theory of Reason*, in which Lambert tries to introduce arithmetic symbols in philosophy, and, second, some *Fragments on the Theory of Reason* that

were sketched by Lambert but edited by his close friend Christoph Heinrich Müller (1740–1807), professor for philosophy and history at the Joachimsthalschen Gymnasium in Berlin. The first part of the second volume contains other fragments. The second part contains the reviews of Lambert's writings on logic and philosophy. The third part is a reedition of Lambert (1768). Parts four to six contain texts by Müller and texts on but not by Lambert. Most important is the appendix, which not only contains Lambert's own review of the *Architectonics* but also the notorious *Fragment of a Systematology* (on the latter, see Siegwart's introduction to Lambert 1988). These two volumes were reprinted as volumes 6 and 7 of Lambert (1965–1966).
2 The preface to Lambert's *New Organon* is unpaginated. I use Roman numerals, counting from the first page of text.
3 Among recent publications, I take Basso (2021) to offer the best introduction to Lambert's thought. It does not discuss the table of correlations, however. Fichant (2018), who offers a good introductory survey of the preface and first chapter of the *Architectonics*, quotes Lambert's table of correlations at the end of his contribution but does not discuss it.
4 See especially Leduc (2018) and Pelletier (2018).
5 See Wellmann (2018, 32) for a critical assessment of Wolters' position.
6 While Wellmann (2018) offers a careful analysis of Lambert's thought, I take her more speculative thesis, namely that Lambert wanted to erect a future metaphysical system that would be different from the *Organon* and the *Architectonics*, to be unconvincing.
7 See Sturm (2009, 141) for an account of Lambert's pragmatic explanation of the correlation between scientific cognition and systematicity. According to Sturm, Lambert saw that isolated observations and judgments based on them would never suffice to measure, for example, the size of the earth or other planets. What is immediately given in experience can only be a basis for common or historical knowledge: "that something is, that it is so and not otherwise, and also what it is" (Lambert 1764a I, 387). Sciences, however, are much more than mere collections of isolated cognitions.
8 I grant that such independence is not immediately evident. In the case of Lambert, it could be argued that his conception of sciences as necessarily involving multiple kinds of systematic connections somehow already points to how we must bring together and classify the different sciences. My discussion in what follows will show, however, that this is not the case. In other words: it is not the systematic ordering within a science that defines how it will be connected to other sciences.
9 On this point, I agree with Hammer (2022, 140), who observes: "Der Rezeption der Philosophie Lamberts kam es nicht zugute, dass das Interesse an Lambert meistens durch die Perspektivierung auf Kant präformiert war."
10 I take this translation from Kant (1992, lxii).
11 We do know, however, that he prepared another text as a response to that question. It is entitled *On the Method to More Correctly Prove Metaphysics, Theology, and Morals*, was written in 1762, and contains many references to his *Criterium Veritatis*. The German title is *Über die Methode, die Metaphysik, Theologie und Moral richtiger zu beweisen*. See Lambert (1918). Siegwart (1988, XXII) mistakenly writes the title as "Über die Methode der Metaphysik, Theologie und Moral richtiger zu beweisen," thereby slightly changing its meaning. Lambert did not submit this text, however. Watkins (2018, 176, 180) contends that both *On the Method...* and *Criterium Veritatis* were written in response to the Prize Question of the Berlin Academy. This

is certainly possible, but as regards the *Criterium Veritatis*, I did not find any conclusive evidence for this assumption.

12 Lambert (1915b, 9): "Jeder Satz, den man als wahr ausgiebt, muss sich uns dadurch als wahr an preisen, dass die Wörter, wodurch er ausgedrückt wird, eine nach aller Schärfe richtige Bedeutung haben, dass die Begriffe, so wir mit diesen Wörtern verbinden, genau richtig seyen, und in ihrem Umfange weder irriges noch widersprechendes enthalten, und dass sich endlich der eine von dem andern, auf die Art, wie sie in dem Satze vorgestellt werden, bejahen oder verneinen lassen."

13 On Lambert's conception of truth, see Sturm (2018).

14 Lambert (1915b, 12) writes: "Man kann die Begriffe, in welchen noch ein Widerspruch li[e]gt, von denen, in welchen gar keiner mehr ist, und welche demnach voll kommen richtig sind, noch nicht nach einer allgemeinen Regel unterscheiden."

15 For example, the syllogism (P1) All M are C; (P2) All B are M → (C) All B are C is represented as follows:

```
C ------------ c
  M ------ m
    B - b
```

(see Lambert 1764a I, 128). Today, we would rather make the inclusion of B in C intuitively obvious using Venn-diagrams.

16 Lambert (1764a II, 16): "Die Theorie der Sache auf die Theorie der Zeichen reduciren, will sagen, das dunkle Bewußtseyn der Begriffe mit der anschauenden Erkenntnis, mit der Empfindung und klaren Vorstellung der Zeichen verwechseln." Thus, the background of Lambert's project of a semiotics is constituted by Leibniz' *Meditations on Knowledge, Truth and Ideas* and Descartes' doctrine of the grades of clarity between obscurity and distinctness.

17 Two remarks: (1) For Lambert, representations are always conscious representations; (2) the concept of a "thing" (*Sache*) must be understood in a very large sense here: the movement of the planets, e.g., is a "thing."

18 Wolff (1729, 1): "Ontologia seu Philosophia prima est scientia entis in genere, seu quatenus ens est."

19 Lambert to Holland, April 21, 1765 (Lambert 1965–2020, vol. 9, 34). For Lambert, the concept of a thing entails notions such as unity, veracity, quantity, existence, relation, and thinkability (Lambert 1965–2020, vol. 9, 3).

20 In *Criterium Veritatis*, Lambert still taught that when we reach marks that cannot be further split up in more fundamental marks through which they are determined, we have reached "the most general marks." These are at the same time the basic concepts. On this point, see also Wolters (1980, 67) and Wellmann (2018, 22).

21 Although Lambert is of course first and foremost interested in the systematization and unification of scientific cognitions, in this passage he does not specify that the simple concepts are only foundational for *scientific* cognitions. Because they conceptualize the core material basis of sensible evidence, simple concepts may be said to be foundational for all kinds of cognitions, including everyday knowledge.

22 As Pelletier (2018, 63) remarks, this is similar to what Aristotle had already argued in *De Anima*, III, 6, 430 a 26–27.

23 See Locke (1690, Book 2).

24 In § 68 of the *Alethiology*, Lambert gives a slightly different list, in which he decouples duration and succession and makes a sharper distinction between motion and force: (1) Consciousness, (2) existence, (3) unity, (4) duration,

(5) succession, (6) volition, (7) solidity, (8) extension, (9) motion, (10) force (Lambert 1764a I, 498). In §73, Lambert remarks that "certainty" (*Gewissheit*) is also a fundamental simple concept (Lambert 1764a I, 499).

25 To be fair, though, the difference could be smaller than I make it appear if one grants that the term "consciousness" has a slightly different meaning in the works of Locke and Lambert.

26 Both Sturm (2018, 114) and Watkins (2018, 175) translate "*Anlage zur Architectonic*" as "Appendix on Architectonics." This translation does not reflect the meaning of "*Anlage*" that Lambert had in mind. In my view, Lambert uses the term *Anlage* as resonating with terms such as "construction," "conception," "plan," "system," or "structure." In private conversation, Courtney Fugate suggested "framework" as yet another alternative. A very neutral translation would be "Conception of an Architectonics" – which is also the translation adopted by Wellmann (2018) and Basso (2021) – but it does not square well with the preposition *zur* in the German version. The *zur* (*zu einer*) *Architectonic* should accordingly be read as "in view of (establishing) an Architectonics," and that is why I opt for "Plan for an Architectonics." The fact that "*Anlage*" cannot mean "Appendix" is evident from the end of Lambert's preface, in which he states the following: "[Architectonic] ist in so fern ein Abstractum von der *Baukunst*, und hat in Absicht auf das *Gebäude der menschlichen Erkenntniß* eine ganz ähnliche Bedeutung, zumal, wenn es auf die *ersten Fundamente*, auf die *erste Anlage*, auf die *Materialien* und ihre *Zubereitung* und *Anordnung* überhaupt, und so bezogen wird, daß man sich vorsetzt daraus ein *zweckmäßiges Ganzes* zu machen" (Lambert 1771 I, xxviii–xxix) (see also Lambert 1965, vols. 3–4).

27 Lambert (1764a I, 472): "Es ist nicht genug, einfache Begriffe ausgelesen zu haben, sondern wir muessen auch sehen, woher wir in Ansehung ihrer Zusammensetzung allgemeine Moeglichkeiten aufbringen koennen."

28 To be sure, Wolff had also claimed to use Euclid's method, but Lambert criticizes him for doing so in a defective, incomplete way. In §12 of the *Architectonics*, Lambert writes: "One cannot say that Wolff has completely used the Euclidean method. In his metaphysics, the postulates and tasks are almost completely absent, and the question concerning what is to be defined is not completely decided" (Lambert 1771 I, 9: "Man kann nicht sagen, daß Wolf die Euclidische Methode ganz gebraucht habe. In seiner Metaphysic bleiben die Postulata und Aufgaben fast ganz weg, und die Frage, was man definiren solle, wird darinn nicht völlig entschieden."). Wolff did not resort to postulates and instead started from definitions – even for concepts or things that everyone understands without further ado – to deduce general principles from them. For more details on Lambert's critique of Wolff's use of Euclid's method, see Basso (2008), Dunlop (2009), Blomme (2015), and Wellmann (2017).

29 In the *Architectonics*, Lambert uses the word "Thulichkeiten" (1771 I, 10).

30 On postulates in Lambert, see Laywine (1998, 2010) and Hammer (2022).

31 Lambert (1764a, 519): "Da nun die Grundsätze gewisse Modificationen, die Postulata aber gewisse Möglichkeiten bey den einfachen Begriffen anzeigen, so ist offenbar, daß diese Modificationen und Möglichkeiten an sich auch einfach sind, und zugleich auch mit dem einfachen Begriffe klar und zugegeben werden."

32 "*Gefühl*" must really be translated as "feeling" and not as "touch," because "*Ausdehnung*" refers to immaterial (spatial) extension. For material extension, Lambert has the simple concept "solidity." Excluded from *Ausdehnung* is temporal extension, for which Lambert has the simple concept "duration."

33 Lambert (1764a, 423): "Wenn wir den Begriff der Ausdehnung so wohl dem Raum als der Zeit nach, oder unmittelbar die Begriffe des Raums und der Zeit als ganz einfache Begriffe ansehen: so haben wir drey Wissenschaften, die im strengsten Verstande a priori sind: Nämlich die Geometrie, die Chronometrie und die Phoronomie. Und hinwiederum wenn man zugiebt, daß diese drey Wissenschaften im strengsten Verstande a priori sind, so sind die Begriffe von Raum und Zeit einfache Begriffe."
34 On Lambert's account of sciences that need to partly rely on empirical input, see also Arndt (1971) and Duchesneau (2018).
35 Purified from all errors that are induced by what is mere *apparentia*.
36 Note that in Lambert's work, there is an ambiguity affecting the term "fundamental doctrine." While there is one fundamental doctrine that lists all fundamental simple concepts and that is Lambert's replacement of the traditional ontology (somehow both called *Organon* and *Architectonics*), there are several fundamental doctrines that are grounded in that one fundamental doctrine. As we will see, one of those many fundamental doctrines is yet another conception of "ontology," connected to the simple concept of solidity.
37 Lambert (1764a II, 259): "§72. Dem Gefühl haben wir die drey Grundbegriffe der Ausdehnung, Solidität und Beweglichkeit zu danken, auf welchen die wahre physische Sprache beruht."
38 Concerning (1) and (2), see Lambert (1771, 45): "Ehe ich zur Vergleichung dieser einfachen Grundbegriffe fortschreite, werde ich anmerken, daß ich den sechsten und siebenten, oder das *Bewußtseyn* und das *Wollen* bey dieser Vergleichung weglasse. Denn ersteres kommt bey allen vor, letzteres aber hat ein eigenes Object, nämlich das *gute*, und gehöret daher besonders in die Agathologie, oder die Lehre vom *Guten*, so wie das *Bewußtseyn*, so fern es auf das *Wahre* geht, in dem Organon zum Gegenstande dienet. Beyde aber werden in einer andern Absicht in der Psychologie oder Theorie des denkenden Wesens betrachtet."
39 I give here myself a possible explanation for reducing "magnitude" to "unity." Lambert (1771, 45) laconically suffices with the statement: "*Einheit* (wohin wir auch die *Größe* rechnen)."
40 Lambert (1771, 43): "Aus der vierten Classe aber wird der Begriff *einerley*, Identität, eine besondere Betrachtung verdienen, und kann mit zu den Grundbegriffen gerechnet warden, ungeachtet er von ganz anderer Art, als die Begriffe der ersten Classe ist." To understand this, one should know that Lambert distinguishes between six classes of simple concepts, of which only the first class concerns the *fundamental* simple concepts. See Lambert (1771, 41).
41 According to Lambert, the principles of arithmetic are the following:

1. Every number is the same as itself.
2. Every number is necessarily different from each bigger or smaller number.
3. Every number is related to its unity, from the repetition of which it is generated.
4. Two numbers of which each is equal to a third number are equal.
5. Two numbers of which each is equal to an equal part of a third number are equal.
6. Unity is the basis of gradation.

Concerning principles 1–5, their relationship with corresponding principles of identity (see preceding footnote) is obvious. The postulates of arithmetic are:

1. Every number can be taken as many times as one wishes.
2. Every number can be considered as a bigger unity.

3 To each number, one can add unities and numbers.
4 No matter how big a number, one can take a still bigger number (Lambert 1771 I, 60)

42 In column 3, Lambert indicates that *Ausdehnung* is only partially connected with the concept of a thing (only when we encounter "extended solidity").

43 Lambert's theory of existence is not yet a theory of modality. He remarks that we need a special theory to treat actuality (*Wirklichkeit*), inasmuch as the latter is opposed to the possible and the necessary (Lambert 1771, 50).

44 The *Architectonics* contains a section on "Being something and Being nothing" (§§254–267), but in explaining "being something," Lambert does not use the concept "subsistence."

45 Lambert reminds us that some philosophers – he is likely thinking of Leibniz – take time to be a mere phenomenon. Lambert is not one of them and remarks, without explaining this further, that phenomenalists about time confuse the apparent duration of time with its veritable duration.

46 Lambert (1965–2020, vol. 7, 413–14)

47 An influence of Lambert on Kant is argued for in the following studies: Zimmermann (1879), Lepsius (1881), König (1884), Riehl (1876), Vaihinger (1922), Peters (1968), Beck (1969), Laywine (2001, 2010), Kuliniak (2004), Dunlop (2009), Perin (2016), Hammer(2018), and Meer (2022).

48 Pelletier (2018) draws a link between, on the one hand, Lambert's question and his account of concepts of relation (*Verhältnisbegriffe*), and, on the other hand, Kant's concepts of reflection, as expounded in the chapter on the Amphiboly. There Kant remarks that the most fundamental pair of concepts of reflection is "matter" and "form" (A266/B322).

49 I strongly agree with Hammer (2022, 140) when he writes: "Die Klärung des Verhältnisses von Lambert und Kant durch den Hinweis auf terminologische Übereinstimmungen, die zahlreich sind, birgt [...] die Gefahr, von einer identischen Benennung auf die Identität des dahinterliegenden Begriffs zu schließen. Während der Vergleich der Resultate eher zu negativen Einschätzungen des Einflusses führt, tendiert eine Untersuchung der Terminologie zur voreiligen Feststellung von Übereinstimmungen."

50 Lambert (1764a I, v–vi): "In der Dianoiologie, wo es fürnehmlich um die Methode zu thun ist, komme ich Wolfen näher. Hingegen in dem ersten Hauptstücke der Alethiologie, wo von den einfachen oder Grundbegriffen unserer Erkenntnis die Rede ist, verfalle ich auf die, so Locke als solche angegeben."

51 Wolters (1980, 58) considers Lambert's determination of an axiomatic science as empirically grounded and at the same time a priori to be contradictory. Hammer (2022, 173–85) takes Lambert's a priori to be a strict a priori and compares the way in which fundamental simple concepts are found with Kant's original acquisition. I think that both are wrong and that Lambert is already embracing a relativized a priori: "[Es] läßt sich leicht ausmachen, daß etwas mehr oder minder a priori sey, je nachdem wir es aus entfernteren Erfahrungen herleiten können, und daß hingegen etwas vollends nicht a priori, und folglich unmittelbar a posteriori sey, wenn wir es, um es zu Wissen, unmittelbar erfahren müssen" (Lambert 1764a I, 414–5). For Lambert, a cognition (like the combinatory cognition of postulates and axioms) is a priori if it can be established on the basis of the mere concept of something, but we need experience to give us those concepts.

52 Holland to Lambert, September 22, 1765 (see Lambert 1965–2020, vol. 9, 92).
53 I would like to thank Thomas Sturm, Karin de Boer, and Steffen Ducheyne for generously commenting on an early draft, which helped me to improve my contribution.

Bibliography

Arndt, Hans Werner. 1971. *Methodo scientifica pertractatum. Mos geometricus und Kalkülbegriff in der philosophischen Theorienbildung des 17. und 18. Jahrhunderts*. Berlin/New York: De Gruyter.

Basso, Paola. 2008. "Rien de mathématique dans la methodus mathematica wolfienne. la méthode «mathématique» de Wolff et les objections de lambert." *Lumières* 12(2): 109–21.

Basso, Paola. 2021. "Lambert on Experience and Deduction." In *The Experiental Turn in Eighteenth-Century German Philosophy*, edited by Karin de Boer and Tinca Prunea-Bretonnet, 181–202. New York and London: Routledge.

Beck, Lewis White. 1969. "Lambert und Hume in Kants Entwicklung von 1769–1772." *Kant-Studien* 60: 123–30.

Blomme, Henny. 2015. "La notion de 'système' chez Wolff, Lambert et Kant." *Estudos Kantianos* 3(1): 105–26.

Duchesneau, François. 2018. "Le recours aux modèles dans la philosophie expérimentale selon le *Neues Organon* de Lambert." *Cahiers Philosophiques de Strasbourg* 44: 35–54.

Dunlop, Katherine. 2009. "Why Euclid's Geometry Brooked No Doubt: J. H. Lambert on Certainty and the Existence of Models." *Synthese* 167(1): 33–65.

Fichant, Michel. 2018. "Johann Heinrich Lambert, l'idée de l'architectonique comme philosophie première (*Grundlehre*)." *Cahiers Philosophiques de Strasbourg* 44: 11–34.

Hammer, Martin. 2018. "Lambert als Quelle Kants: Einzelne Urteile und die metaphysische Deduktion der Allheit." In *Natur und Freiheit. Akten des XII. Internationalen Kant-Kongresses*, edited by Violetta L. Waibel et al., 3187–95. Berlin and Boston: De Gruyter.

Hammer, Martin. 2022. "Lamberts Postulate als Quelle der Synthesis Kants." In *Kant's Transcendental Deduction and the Theory of Apperception. New Interpretations*, edited by Giuseppe Motta et al., 133–92. Berlin and Boston: De Gruyter.

Hermann, Jakob. 1716. *Phoronomia sive de viribus et motibus corporum solidorum et fluidorum*. Amsterdam: Wetstenios.

Kant, Immanuel. 1764. *Untersuchung über die Deutlichkeit der Grundsätze der natürlichen Theologie und der Moral. Zur Beantwortung der Frage welche die königl. Academie der Wissenschaften zu Berlin auf das Jahr 1763 aufgegeben hat*. Berlin: Haude & Spener.

Kant, Immanuel. 1992. *Theoretical Philosophy 1755–1770*. Translated and edited by David Wallford with Ralf Meerbote. Cambridge: Cambridge University Press.

König, Edmund. 1884. "Über den Begriff der Objektivität bei Wolff und Lambert mit Beziehung auf Kant." *Zeitschrift für Philosophie und philosophische Kritik* 84: 292–313.

Kuliniak, Radosław. 2004. "Lamberts Einfluss auf Kants Transzendentalphilosophie." In *Transzendentalphilosophie heute. Breslauer Kant-Symposion*, edited by Andreas Lorenz, 153–62. Würzburg: Könighausen & Neumann.

Lambert, Johann Heinrich. 1764a. *Neues Organon oder Gedanken über die Erforschung und Bezeichnung des Wahren und dessen Unterscheidung vom Irrthum und Schein*. Leipzig: bei Johann Wendler.
Lambert, Johann Heinrich. 1764b. "De universaliori calculi idea una cum annexo specimine." *Nova Acta Eruditorum*, 441–73.
Lambert, Johann Heinrich. 1766. "In Algebra Philosophicam Et Richeri Breves Annotations." *Nova Acta Eruditorum*, 335–44.
Lambert, Johann Heinrich. 1768. "De topicis schediasma." *Nova Acta Eruditorum*, 12–33.
Lambert, Johann Heinrich. 1771. *Anlage zur Architectonic, oder Theorie des Einfachen und des Ersten in der philosophischen und mathematischen Erkenntnis*. Riga: bei Johann Friedrich Hartknoch.
Lambert, Johann Heinrich. 1915a. *Monatsbuch*, edited by Karl Bopp. München: Verlag der Königlich Bayerischen Akademie der Wissenschaften.
Lambert, Johann Heinrich. 1915b. "Abhandlung vom *Criterium Veritatis*." In *Kant-Studien Ergänzungshefte No. 36*, edited by Karl Bopp. Berlin: Reuther & Reichard.
Lambert, Johann Heinrich. 1918. "Über die Methode, die Metaphysik, Theologie und Moral richtiger zu beweisen." In *Kant-Studien Ergänzungshefte No. 42*, edited by Karl Bopp. Berlin: Reuther & Reichard.
Lambert, Johann Heinrich. 1965–2020. *Philosophische Schriften*. 10 volumes + supplement, edited by Hans Werner Arndt (continued by Lothar Kreimendahl). Hildesheim: Georg Olms Verlagsbuchhandlung.
Lambert, Johann Heinrich. 1988 (1st 1787). "Fragment einer Systematologie." In *Johann Heinrich Lambert. Texte zur Systematologie und zur Theorie der wissenschaftlichen Erkenntnis*, edited by Geo Siegwart and Horst D. Brandt. Hamburg: Felix Meiner Verlag.
Laywine, Alison. 1998. "Problems and Postulates: Kant on Reason and Understanding." *Journal of the History of Philosophy* 36(2): 279–309.
Laywine, Alison. 2001. "Kant's Reply to Lambert on the Ancestry of Metaphysical Concepts." *Kantian Review* 5: 1–48.
Laywine, Alison. 2010. "Kant and Lambert on Geometrical Postulates in the Reform of Metaphysics." In *Discourse on a New Method. Reinvigorating the Marriage of History and Philosophy of Science*, edited by Mary Domsky and Michael Dickson, 113–34. Chicago/La Salle: Open Court Publishing.
Leduc, Christian. 2018. "Harmonie et dissonance. Lambert et le système des vérités." *Cahiers Philosophiques de Strasbourg* 44: 77–102.
Lepsius, Johannes. 1881. Quellenmässige Darstellung der philosophischen und kosmologischen Leistungen Johann Heinrich Lamberts im Verhältnis zu seinen Vorgängern und zu Kant. Inaugural-Dissertation, München: Ackermann.
Locke, John. 1690. *An Essay Concerning Human Understanding*. London: Thomas Basset.
Meer, Rudolf. 2022. "Von den Grentzen der Sinnlichkeit und der Vernunft zur Idee der Critick der reinen Vernunft. Lamberts Einfluss auf Kants denken zwischen 1770 und 1772." In *Kant's Transcendental Deduction and the Theory of Apperception. New Interpretations*, edited by Giuseppe Motta et al., 275–306. Berlin/Boston: De Gruyter.
Mendelssohn, Moses. 1764. *Abhandlung über die Evidenz in metaphysischen Wissenschaften, welche den von der königlichen Academie der Wissenschaften in Berlin auf das Jahr 1763 ausgesetzten Preis erhalten hat*. Berlin: Haude und Spener.

Pelletier, Arnaud. 2018. "La profondeur et le fond: des concepts simples chez Lambert." *Cahiers Philosophiques de Strasbourg* 44: 55–76.
Perin, Adriano. 2016. "Lambert's Influence on Kant's Theoretical Philosophy." *Con-Textos Kantianos* 3: 44–54.
Peters, Wilhelm S. 1968. "Kants Verhältnis zu Lambert." *Kant-Studien* 59: 448–53.
Riehl, Alois. 1876. *Der philosophische Kriticismus und seine Bedeutung für die positive Wissenschaft. Erster Band: Geschichte und Methode des philosophischen Kriticismus.* Leipzig: Wilhelm Engelmann.
Siegwart, Geo. 1988. "Einleitung." In Lambert (1988), vii–lxxxix.
Sturm, Thomas. 2009. *Kant und die Wissenschaften vom Menschen.* Paderborn: Mentis Verlag.
Sturm, Thomas. 2018. "Lambert and Kant on Truth." In *Kant and His German Contemporaries*, Vol. 1, edited by Corey W. Dyck and Falk Wunderlich, 113–33. Cambridge: Cambridge University Press.
Vaihinger, Hans. 1922. *Kommentar zu Kants Kritik der reinen Vernunft*, 2. Auflage, Stuttgart/Berlin/Leipzig: Union Deutsche Verlagsgesellschaft.
van den Berg, Hein. 2021. "Kant's Ideal of Systematicity in Historical Context." *Kantian Review* 26(2): 261–86.
Watkins, Eric. 2018. "Lambert and Kant on Cognition (Erkenntnis) and Science (Wissenschaft)." In *Kant and His German Contemporaries*, Vol. 1, edited by Corey W. Dyck and Falk Wunderlich, 175–91. Cambridge: Cambridge University Press.
Wellmann, Gesa. 2017. "Towards a New Conception of Metaphysics: Lambert's Criticism on Wolff's Mathematical Method." *Revista de Estudios Kantianos* 2: 135–48.
Wellmann, Gesa. 2018. The idea of a metaphysical system in Lambert, Kant, Reinhold, and Fichte. PhD-Thesis. Institute of Philosophy KU Leuven.
Wolff, Christian. 1730. *Philosophia prima sive Ontologia.* Frankfurt/Leipzig: Renger.
Wolff, Christian. 1732. *Philosophia rationalis sive logica, methodo scientifica pertractata et ad usum scientiarum atque vitae aptata. Praemittitur discursus praeliminaris de philosophia in genere.* Frankfurt/Leipzig: Renger.
Wolters, Gideon. 1980. *Basis und Deduktion. Studien zur Entstehung und Bedeutung der Theorie der axiomatischen Methode bei J. H. Lambert (1728-1777).* Berlin/New York: De Gruyter.
Zimmermann, Robert. 1879. *Lambert, der Vorgänger Kants. Ein Beitrag zur Vorgeschichte der Kritik der reinen Vernunft.* Wien: K. Gerold's Sohn.

Part II
The Systematicity of Special Sciences

3 Kant's Early Cosmology, Systematicity, and Changes in the Standpoint of the Observer[1]

Fabian Burt and Thomas Sturm

3.1 Introduction

The most fully developed and comprehensive scientific theory in Kant's work is his early empirical cosmology and cosmogony: his theory of the structure as well as the genesis of the Universe, as presented especially in his *Universal Natural History and Theory of the Heavens* of 1755 – *UNH* for short. Here, Kant applies Newtonian mechanics to the explanation of the motions of celestial objects, their densities and masses, the distances between these objects, and the historical formation of these and other phenomena. He does so at the level of the Solar System, the Milky Way, and even the Universe as a whole.[2] *UNH* already shows Kant, then only 31 years old, to be a highly creative and innovative thinker. It contains ideas that he became rightly famous for – such as the nebular hypothesis concerning the formation of the Solar System (and other celestial systems), the claim that distant nebulae are galaxies just like the Milky Way, that there are many more such galaxies, and that the Universe must therefore be much larger and structurally more complex than commonly assumed in his time. Even if he had not written the *Critique of Pure Reason* or other major works, Kant might be recognized for these claims about the system of the Universe up until today.[3]

Scholars have mostly studied either the relation between *UNH* and Kant's later views on metaphysics and, specifically, rational cosmology, or its place in the history of natural science.[4] While the latter approach is obviously legitimate, the former can also be justified. For even though the work is predominantly empirical, it contains assumptions about the spatiotemporal nature of the Universe that are related to rational cosmology and, moreover to a physico-theology that is metaphysical by the standards of the critical philosophy. Here, however, we will focus on another topic, namely Kant's extensive use of the language of systematicity in *UNH*. This language, which becomes so important in many ways throughout his oeuvre, is born here. Of course, there are major differences in the meanings and roles of relevant expressions as they are used later on, in the critical period. But there are also, beyond the terminological similarity, deeper connections. They pertain to important aspects of his thinking about fundamental questions of scientific knowledge. We shall show how Kant's concern with systematicity was firmly rooted in specific issues of scientific observation and reasoning before, in the critical phase, moving up to philosophical levels of reflection about the very nature of science.

In Sections 3.2 and 3.3, we will explain the two main ideas of systematicity contained in *UNH*, doctrinal and ontic. In Section 3.4, it will be shown how Kant connects the doctrinal and ontic meanings through an epistemology of forms of observation and reasoning that are fruitful for progress in cosmology and cosmogony. We will make clear that he not only uses principles of Newtonian mechanics, but also a technique of changing imagined standpoints of observation, connected to rules for legitimate reasoning about hypotheses that he makes explicit in logic lectures since the 1770s. In Section 3.5, we shall turn to Kant's thinking about these issues in his critical phase. We will clarify his views on the epistemological character and function of such standpoint changes. Moreover, we will show how this notion helps to recognize that Kant's concept of systematicity is not only about the definition of 'science' but also about the dynamic and diachronic progress of scientific research – a feature that has been neglected by scholars.

3.2 The Doctrinal Notion of Systematicity in *UNH*

The term 'system' (*System*) and related expressions (e.g. the adjective 'systematic' or the noun 'the systematic'; *systematisch, das Systematische*) occur 168 times within *UNH*, making them prominent technical terms in this work. However, in different contexts Kant uses them in different ways. Two meanings stand out: a doctrinal one, related to the theories scientists accept relative to the evidence they possess and view as relevant; and an ontic one, which concerns what the theories state about reality. While these meanings need to be distinguished, they are also in an important way connected.

As a doctrinal term, Kant uses 'system' synonymously with German expressions like *Lehre, Lehrverfassung,* or *Lehrbegriff,* all of which are translated as either 'doctrine' or 'theory.' Although Kant often uses this meaning, he never explains it. Thus, *UNH* does not discuss what turns a set of cognitions into a system of knowledge. However, in 1755 Kant has no need to give such an explanation, since he is not breaking any new terminological ground in this regard.[5] On the contrary, the doctrinal notion can be traced back even to the Stoics, and while it was lost in the Middle Ages, it reappeared in Early Modernity. After the publication of Copernicus' groundbreaking *On the Revolutions of the Heavenly Spheres* in 1543,

Kant's Cosmology, Systematicity, and Standpoint Changes 71

it became popular to use the Latin 'systema mundi' not only in the traditional way, namely to signify the sum total of relations between the Sun, the planets, and the fixed stars – perhaps most influentially, Newton devotes the whole Book III of his *Principia* (1687) to "The system of the world" in this sense. The terminology also began to be used to refer to theories about the world – specifically, *competing* theories, as witnessed by Galileo's *Dialogue Concerning the Two Chief World Systems* of 1632.[6] In the same vein, Kant speaks of the systems of different authors, as when he writes that he "cannot determine exactly the borders between the system of Herr Wright and my own and in what ways I have merely imitated his model or have explained it further" (I:232). Depending on context, he accordingly uses 'system' both for his own cosmological theory as well as for those of his predecessors, and context always makes clear how the term is to be understood. In sum, Kant had good reasons to assume that his readers would be familiar with this general doctrinal meaning of the term. A system in this sense is, of course, a combination of a plurality of scientific cognitions or propositions of some form or another, and in this context clearly something that aims at explaining a set of phenomena. That much is implicitly assumed. Given that the expression is used in this way without an explicit definition or explanation, and without considering what makes such systems possible and/or reasonable, the question of whether or how a body of knowledge is *scientific* due to its systematicity does not yet come up. This is an issue which would only become important for the critical Kant.

3.3 The Ontic Notion of Systematicity in *UNH*

In its ontic use, predominant in *UNH*, the meaning of 'system' is explicitly clarified. The central features are presented in an early introductory section titled "Short summary of the most essential basic concepts of Newtonian science, which are necessary for understanding what follows" (I:243). Kant closes this section as follows:

> In the treatise, I shall frequently use the expression of a systematic constitution of the universe. So that there will be no difficulty in understanding what is meant by this, I shall explain it briefly. Actually, all the planets and comets that belong to our universe constitute a system simply because they orbit around a common central body. But I take this term in a narrower meaning in that I consider the more precise relationships that have made their connection to one another regular and uniform. The orbits of the planets relate as closely as possible to a common plane, namely to the extended equatorial plane of the Sun; the deviation from this rule occurs only at the outermost border of the system, where all motions gradually

cease. If, therefore, a certain number of heavenly bodies that are arranged around a common central point and move around this, are simultaneously restricted to a certain plane in such a way that they have the freedom to deviate from it to either side only as little as possible; ... then, I say, that these bodies are related to each other in a *systematic constitution*.

(I:246)

Kant here not only states that he uses the terminology repeatedly and wishes to define it, but also that he takes "this term in a narrower meaning." He is refining what he views as the broader or perhaps more common meaning ("all the planets and comets that belong to our universe constitute a system simply because they orbit around a common central body"). According to his refined understanding of a "systematic constitution of the universe," at least three features are central:

1) A gravitational center must exist,
2) around which all other bodies orbit
3) as closely as possible to a common plane.

Kant does not claim that this list is complete, and several other passages suggest that, for him, a fourth feature is essential too. This feature is introduced in the first chapter of the second part, where Kant develops his cosmogony, i.e. a mechanical history of the original formation of the Solar System. This cosmogony is meant to outline the causes of the present constitution of the Solar System. But why is such a history needed to give a complete explanation of its current constitution? Kant argues that Newton's account given in the *Principia* is of limited explanatory value: it only explains *how* it is possible that the planets orbit a gravitational center on stable elliptical paths. As Kant summarizes Newton (I:243–6), the stability and the elliptical form of the orbits can be explained by means of the two necessary mechanical forces, namely the "*shooting force*" (Kant's name for inertia) and "the sinking, the centripetal force or also gravity" (I:243). While the former force keeps bodies in rectilinear motion, the second force is responsible for the curvature of planetary orbits. The shape of the orbits depends on the intensity of the two forces in relation to each other. A perfect equilibrium of the two forces results in a perfect circle. That none of the existing orbits describes a circle, but that all of them are elliptical, is due to slight deviations from this equilibrium.

Although Newton was able to explain the stability and shape of the orbits, there are observable regularities in the Solar System that he could not account for. Two of them are that almost all orbits around the Sun approximate a common plane, and that they do so in the very same direction as the Sun's axial rotation (namely, counterclockwise when considered from its north pole). While absent from Kant's initial determination of systematicity, the latter observation, i.e. the joint direction of motion of the orbiting bodies, is central to his cosmogony. We find it in the first paragraph of Chapter 1, which he begins with a description of the Solar System, concluding that there must be a joint cause of all the regularities that comprise its systematic form:

> If, on the one hand, we consider that six planets with ten satellites describe orbits around the Sun as their centre and all of them move towards one side, namely that side to which the Sun itself turns, ... if, as I say, we consider all these connections: then we are moved to believe that one cause, whatever it may be, has had a pervasive influence in the entire space of the system, and that the *unity in the direction and position of the planetary orbits* is a consequence of the agreement they all must have had with the material cause by which they were set in motion.
>
> (I:261–2, emphasis added)

Kant's cosmogony, then, traces the unity in the direction and relative position of the planetary orbits back to the beginning of the Universe. Shortly after the divine creation of all matter, gravity and repulsion forced the particles that later should make up the bodies of our Solar System in the positions we can still observe them in today and impressed motion on them. Since this history of the Solar System is seen by Kant as the cause for its systematic constitution, the joint direction of motion can be regarded as the fourth essential feature of this constitution. Thus, we can sum up Kant's ontic notion of systematicity as follows: *a system of celestial bodies is given when a gravitational center exists around which the majority of bodies orbit in elliptical lines as close as possible to a common plane and in the very same direction.*[7]

Kant's main example of a system in this sense is the Solar System. But it is more than just an example. Kant uses it as a model for all systems of celestial bodies, their properties, and relations. It serves this function both synchronically and diachronically: *first*, the *structure* of the Solar System provides the model for the structure of the Milky Way and ultimately of the Universe as a whole, only at larger scales. *Second*, the *evolution* of the Solar System provides also a model for the evolution of the Universe as a systematic whole.

But if Kant thinks that these are the real structures of the Solar System, the Milky Way, indeed the whole Universe, how does he know all these things about the world? How can he justify all these bold claims about reality? In the present case (as in others), this is particularly urgent because what Kant's theory states to be the true structure of the world deviates quite dramatically from our ordinary, earthbound observations. Thus, Kant needs to give arguments for why his own theory or *system* ought to be viewed as the most acceptable one. We could not know the ontic systematicity of the world independently of an explanatory theory and appropriate modes of scientific reasoning that connect principles and phenomena in a legitimate way.

3.4 Connecting the Doctrinal and Ontic Notions of Systematicity: Standpoint Changes and Reasoning with Hypotheses

As we have already seen, Kant's own theory is embedded in a specific historical context. For instance, the structural analogies he presents between celestial systems of different scales, and between the past and the present of the Universe, all hold because they are derived from theoretical assumptions of Newtonian mechanics as Kant understands it. Not surprisingly, the context is still broader than this, and it needs to be understood if we are to grasp how Kant thought that his ontic assumptions could be justified.

Consider the first of the two respects just mentioned, the scale-indifference of the constitution of celestial systems. This aspect of the theory is developed in the first part of the book, entitled "Concerning the systematic constitution among the fixed stars."[8] This part is particularly interesting, since here the ontic notion of systematicity becomes connected to a special methodological device: namely, analogies that are made plausible by inviting the reader to consider a change in the "standpoint" (*Standpunkt*) of the observer. Kant's idea can be summed up as follows. To conceive of the Universe exhibiting a certain systematic structure B instead of systematic structure A presupposes a change in the standpoint of the cognizing subject from standpoint A* to standpoint B*. The new standpoint makes possible a more simplified and, moreover, a more comprehensive account of our earthbound observations of the behavior of celestial bodies.[9] This methodological device, thus, connects the two notions of systematicity, since the change in standpoint of the observer is an essential feature of new cosmological theories. Essential not for the content of the theory, but essential for arriving, in a reasonable manner, at the theory. That Kant's cosmology entails such a change in standpoint is going to be demonstrated now, by showing how he arrives at his thesis that the Universe as a whole constitutes a system.

We start with the major shift in perspective that accompanies the establishment of the heliocentric model by Copernicus. According to the Ptolemaic geocentric model, the Earth rests at the center of the Universe with all other visible bodies orbiting it. Here, the relations between celestial bodies and consequently the standpoint of the subject within the system are derived from the way these relations appear to an observer on Earth. Seen from Earth, all bodies visible in the sky, i.e. the Moon, the planets, the Sun, and the fixed stars, seem to circle around Earth once a day. Over a longer period, a second motion can be observed: all the neighboring bodies in the Solar System describe different patterns of motion against the background of the fixed stars. The planets seem to switch between prograde and retrograde movements, orbiting Earth in loops. The relative position of the fixed stars, on the other hand, seems to remain stable (hence the ancient concept for them), creating an unchanging background image for the motion of the other bodies. These observations amount to the Ptolemaic model in which the Earth is resting at the center of the Universe, while the Moon, the planets, and the Sun orbit around it. The stability of the positions of the fixed stars is explained by means of a solid sphere rotating around the Earth's axis on which the fixed stars are assumed to be pinned. The heliocentric model, then, is made possible only by a fundamental change in the subject's standpoint – not literally, but in thought or imagination. We thereby consider how we would perceive systems of celestial bodies and their motions if we were located, as it were, sufficiently far outside of these systems while also capable of viewing them from more or less all of their sides. In that case, the Earth only *appears* to be at the center, while in fact it is orbiting the Sun together with other planets and comets. Note that another theoretical assumption is connected to this shifted perspective: the fixed stars remain pinned to the sphere, but Copernicus suggests that it is not the sphere's rotation that accounts for their apparent motion, but the Earth's rotation around its own axis (Schönecker, Schulting, and Strobach 2011). This rotation also explains why the Moon, the planets, and the Sun seem to orbit earth every 24 hours. Finally, the Earth's orbital motion explains that the planets perform loops when observed over a long period, although they are actually moving in orbits parallel to that of Earth itself.

Now, what makes this model better than the old Ptolemaic one?[10] The advantage of the heliocentric model was not that it could give empirically adequate explanations of *more* celestial phenomena than the geocentric one, thus being more comprehensive (a typical desideratum of good scientific theories or explanations). Ptolemy's refined system was in that regard already quite a success.[11] What made the heliocentric system preferable was its ability to give a truly unified and empirically adequate explanation of those phenomena *by minimizing the number of hypotheses necessary for doing so*. This parsimony or, as Kant comes to call it in lectures on logic since the 1770s, "unity" of hypotheses is best demonstrated by a brief sketch of Copernicus' explanation of the retrograde motion of the planets – the most notorious celestial phenomena with which astronomers had been struggling for centuries. The tool of choice to explain retrograde motion within the geocentric framework was epicycles: a second circular motion overlapping the revolution of each planet around the Earth. While this model was, after a long history of geometrical adjustments and refinements, ultimately successful in giving an empirically adequate explanation of the motion of the planets, its biggest disadvantage was that each epicycle has to be considered as what Kant, again in his lectures on logic, calls a "subsidiary hypothesis": i.e., a hypothesis that only serves to help another hypothesis to correspond to the phenomena in an adequate way but, because it is also hypothetical, diminishes the *probability and credibility* of the whole theory.[12] This is a problem that Kant also diagnoses in Tycho Brahe's model, which still has the Earth at the center of the Universe, with the Sun revolving around the Earth, but with all other planets revolving around the Sun. According to Kant, Tycho is forced to develop numerous additional hypotheses about planetary motion, the combination of which reduces the probability and credibility of his general model (cf. XXIV:220, 223, 888–9; IX:85–6). In Kant's view, the parsimony of theoretical hypotheses is therefore not so much an economic or instrumental requirement but a genuinely epistemological one.[13] The heliocentric model only needs two further assumptions from which the observable phenomena necessarily flow: the assumed axial rotation of the Earth and its assumed revolution around the Sun. From this, every retrograde motion can be explained in the same way, as the scheme in Figure 3.1 shows.

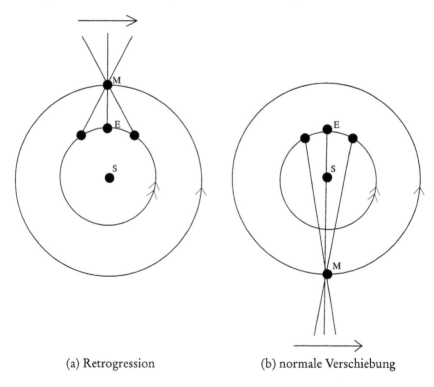

(a) Retrogression   (b) normale Verschiebung

Figure 3.1 Retrograde motion (from: Martin Carrier 2001, Nikolaus Kopernikus, München: Beck, 91; reprinted with permission of the publisher.)

Here, the Earth (E) is shown orbiting the Sun (S) closer to and faster than Mars (M). The diagram on the left pictures the moment in which the Earth is overtaking Mars. For an observer on Earth, Mars will appear to move backwards during this period, describing a loop on the night sky. However, when Mars and Earth oppose each other on their orbits, as is shown in the second diagram, Mars is seen from Earth as moving in the regular direction on the night sky, just like all other visible stellar objects. The observable change in the pattern of Mars' motion is thus only a result of its position and orbital velocity in relation to that of Earth and does not correspond to any type of real change in motion at all. Thus, the heliocentric model is, in principle, simpler or more unified than the geocentric one. However, this requires a methodology of imagining changes of observational standpoints.

Let's turn to Kant's use of this methodological device that Copernicus started to implement in cosmology – the change of the standpoint of the observer. We will see Kant expanding it to celestial structures beyond our Solar System, finally arriving at an all-encompassing cosmological model that explains the phenomena in a similarly elegant, unified way like Copernicus' heliocentric model had done 200 years earlier. Kant starts the first part of *UNH* with a summary of the cosmological knowledge of his time. The sphere of the fixed stars, which Copernicus still assumed in continuity with the old geocentric model, had vanished by the 18th century. Actually, already since Giordano Bruno (whom Kant does not mention) some astronomers assumed that the fixed stars form planetary systems just as our Solar System does and that these systems of fixed stars are scattered all over endless space.[14] In this model, the observer on Earth is located at a random spot within a disordered mass of planetary systems. Motion is restricted to bodies within a planetary system.[15] The apparent motion of the fixed stars is explained by the motion of the Earth within the Solar System and its rotation around its own axis. Kant, however, is not satisfied with a disordered and stagnant Universe. Thus, he claims that the fixed stars make up a system in the same way that its components, the planetary systems, display a systematic constitution:

> According to this representation, the system of the fixed stars may be described approximately by the planetary one, if the latter is extended infinitely. Because if, instead of the six planets with their ten satellites, we assume as many thousands of them and instead of the twenty-eight or thirty comets that have been observed, we assume a hundred or thousand times as many, if we think of these very bodies as self-illuminating, then to the eye of an observer looking from the Earth, they would create the appearance as of the fixed stars of the Milky Way.
> (I:250–1)

This assumption entails a change in the standpoint of the cognitive subject, the latter being addressed by Kant speaking of "the eye of an observer looking from the Earth." The Solar System, containing the earthbound observer, is now orbiting a gravitational center instead of ever remaining at the exact same spot in space. Because the system of fixed stars is constructed in analogy to the Solar System, these stars are no longer viewed as fixed, but as orbiting that same center. Also, all orbits approximate a common plane. The last assumption is linked by Kant to an observation in the night sky that could not be coherently accounted for from the old standpoint. The assumption that the Solar System orbits around a gravitational center on a common plane together with the fixed stars explains why we see them form a circle shaped band in the night sky: "This light band will extend in the direction of a largest circle because *the position of the observer is in the plane itself*" (I:249, emphasis added). If the fixed stars were really scattered across the Universe, we would not perceive them in this shape. That the stars constituting the light band on the night sky appear as fixed, although Kant is ascribing motion to them, is explained by means of the vast distances between the bodies in the Milky Way (I:251–3).

What are the epistemological characteristics of Kant's mode of scientific reasoning here? Two related features are most important. First, there is an increased complexity in the cognitive activity of changing observer standpoints. Like Copernicus, Kant is not merely using the imagination to place an observer outside of the system in order to describe, as objectively as possible, the structure of that system. He also considers how the system will appear to an observer with certain cognitive capacities located at a certain place *within that very system*. To use ideas of recent discussions on scientific objectivity, Kant is not attempting to create a model of the Universe that is objective in the sense of representing only mind-independent objects and their relations; he also incorporates the – mind-dependent – perspective of human observers from their contingent and limited position into that framework.[16] By doing so, his model itself seems to receive a certain piece of confirmation: it fits with the observations that we make of the starry heavens above us. Second, however, this reasoning is open to a familiar worry. Kant uses what we, since Charles Sanders Peirce, call 'abductive reasoning': from a certain phenomenon we infer its cause or perhaps, as is here the case, a complex causal model that would show why we perceive things the way we do. Logically speaking, however, this mode of reasoning is a case of the *affirming the consequent*:

1) If the fixed stars are orbiting the center of the Milky Way, and all orbits approximate a common plane, then we perceive the fixed stars as forming a circle shaped band in the night sky.
2) We perceive the fixed stars as forming a circle shaped band in the night sky.
3) Therefore, the fixed stars are orbiting the center of the Milky Way, and all orbits approximate a common plane.

This argument has the form of 'If p, then q; q; therefore p'; which is non-deductively valid and does not guarantee the truth of the conclusion. However, we do not therefore have to discard Kant's mode of reasoning. It is widespread in science: we devise theoretical models that can explain the phenomena.[17] Once again, no later than by the early 1770s we see Kant as being aware of the possible fallaciousness of such an inference, which he relativizes by embedding it in a broader picture of scientific reasoning:

> But since this mode of inference ... yields a sufficient criterion of truth and can lead to apodeictic certainty only when *all possible* consequences of an assumed ground are true, it is clear from this that since we can never determine all possible consequences, hypotheses always remain hypotheses, that is, presuppositions, whose complete certainty we can never attain. In spite of this, the probability of a hypothesis can grow and rise to an *analogue* of certainty, namely, when all the consequences *that have as yet occurred to us* can be explained from the presupposed ground.
> (IX:85; see also IX:51–2; cf. e.g. XXIV:221)

So, any hypothetical explanation must face the test of as many empirical consequences as possible. Kant surely thinks his cosmology does pretty well in this regard. Still, his model of the galaxy only provides a possible explanation for the relevant explananda. Others are conceivable, just as Kant also admits that Copernicus's heliocentrism could conceivably be falsified (but simply has not been; XXIV:888). Two points are important here. On the one hand, Kant declares several times in *UNH* that many of his theoretical assumptions are hypotheses, even partly quite speculative ones. On the other hand, he also emphasizes that these assumptions are not entirely arbitrary but embedded in his general mechanical framework that is well supported: the universal forces of attraction and repulsion. That makes the analogy and the expansion of the model of the Solar System to one of a vastly larger scale more plausible.[18] Kant is trying to tease out consequences of this underlying assumptions and see whether they fit with the data.[19] If anyone wants to argue that his explanation of the phenomena does not provide an "inference to the best explanation" (Lipton 1991), they have to provide a better one. It is true, however, that something needs to be said to justify the assumptions that there are just those two basic forces, attraction and repulsion and that they are at work everywhere in the Universe in the same way. In *UNH*, Kant does not provide anything that would be of help here.

So, Kant radically expands the idea of a change in standpoint introduced by Copernicus. Just as the heliocentric model revealed how the geocentric model misleadingly took the apparent motions of the Moon, the planet, and the Sun at face value, Kant's thesis of the systematic constitution among the fixed stars demonstrates how their apparent fixity is misleadingly inferred from taking an inadequate epistemic standpoint.

But Kant does not even stop here. At the end of the first part of *UNH*, he famously claims that the Milky Way might only be a subsystem of a much larger system of galaxies, just as the Solar System is only one of many thousand subsystems (cf. I:240) within our galaxy. Once again, this change in standpoint uses an analogy and is explicitly meant to account for specific observations in the night sky:

> [I]f such a world of fixed stars is viewed at such an immeasurable distance from the eye of the observer which is outside it, then it will appear under a small angle as a minute space illuminated by a weak light, the shape of which will be ... elliptical when it is seen from the side.
>
> (I:253–4)

At the time, what Kant describes here as "a minute space illuminated by a weak light" was subject to a lively astronomical debate. The debate concerned the nature of the so-called "nebulous stars" (I:254) that differ from the normal fixed stars in that they possess a particular elliptical shape and emit noticeably less light.[20] Kant claims that these nebulous stars are galaxies just like the Milky way is, i.e. they are "systems of fixed stars."[21] And again, he does not want to set an arbitrary boundary to systematicity, but suggests that the nebulous stars, together with the Milky way, form a large system of galaxies:

> One could also speculate that these higher orders of worlds are not without connection to one another and that, through this mutual relationship, they constitute in turn an even more immeasurable system. Indeed, it can be seen that the elliptical figures of this type of nebulous star ... are very closely related to the plane of the Milky Way.
>
> (I:255)

Regarding the system of galaxies, the standpoint of the observer is important with regard to two observations. First, it is central that the observer on Earth is completely outside of the other galaxies and that these galaxies, in turn, are not part of our galaxy, the Milky Way. These assumptions make sense of the elliptical shape and the weak light of the nebulous stars. The intensity of their light is so low because, as distant galaxies, they are even farther away than the fixed stars within our galaxy (on the scale of cosmic distances in *UNH* cf. Burt 2025). Their shape is explained by the assumption that they are stellar systems themselves, i.e. they contain thousands of shining stars revolving around a gravitational center in elliptical orbits. The fixed stars do not possess such a shape because they are orbited by planets and planets themselves do not emit light. Second, the assumption that the nebulous stars form a system together with our galaxy is linked to the observation that most of them "are very closely related to the plane of the Milky Way." Kant asserts that the plane on which the galaxies orbit around a common center can only be thought of as the extension of the Milky Way. Therefore, the nebulous stars must appear close to or even in the same plane as the Milky Way on our night sky.

This last observation already indicates the form of the Universe as a systematic whole: a plane, consisting of a hierarchy of systems in infinite extension. That Kant indeed conceives of the Universe as spatially infinite becomes obvious by the end of the first part, where he states:

> We see the first members of a progressive relationship of worlds and systems, and the first part of this infinite progression already gives us to understand what we can suppose about the whole. There is no end here but rather an abyss of a true immeasurability into which all capacity of human concepts sinks.
>
> (I:256)[22]

To sum up, Kant distinguishes a doctrinal and an ontic meaning of the terminology of 'systems' but also closely connects them. Scientific reasoning about observable celestial objects, their formation, their properties, relations, and structures requires that one integrates them into a unified theory or set of explanatory assumptions. Again, arguing for these theories against alternative options requires transcending one's ordinary observational standpoints and to imagine models of celestial systems as if one could view them from the outside. At the same time, the scientist has to reinterpret one's ordinary observations of the starry heavens in the light of the new theories. The Universe forms a system that can be known only by reasoning, in as systematic a manner as possible, about theories or 'systems' in the doctrinal sense.

3.5 Standpoints and Systematicity in the Critical Phase

We have shown how Kant achieves important progress in scientific reasoning in his early cosmology by changing the standpoint of the observer. Any attentive reader of Kant will already have been thinking about his famous claim that metaphysics, in order to overcome its *aporia* and to reach the "secure path of a science" (Bvii), requires a "revolution in the mode of thought." Kant compares this required revolution with Copernicus:

> [L]et us once try whether we do not get farther with the problems of metaphysics by assuming that the objects must conform to our cognition, which would agree better with the requested possibility of an *a priori* cognition of them, which is to establish something about objects before they are given to us. This would be just like the first thought of Copernicus, who, when he did not make good progress in the explanation of the celestial motions if he assumed that the entire celestial host revolves around the observer, tried to see if he might not have greater success if he made the observer revolve and left the stars at rest.
>
> (Bxvi)

Metaphysics can reach the "secure path of a science" only if it makes a change in its standpoint, just like cosmology was brought on the path of a secure science by Copernicus imagining a change in the standpoint of the observer – and, we may speculate, Kant also thought of himself as having advanced cosmology still further by more complex and more far-reaching changes in the standpoint of an observer. Here, we shall not engage in the dispute concerning what Kant means by a revolution of metaphysics, nor do consider the general notion of a 'standpoint' in his philosophy.[23] Instead, we focus on the special concept of the 'standpoint of an observer.' What, according to Kant's views in the critical phase, is the epistemological character of such standpoints and changes thereof, and what role do they play for the systematicity of science? And how reasonable is the use of such standpoint changes?

We have seen that this concept functions in Kant's empirical cosmology within a successive series of analogical arguments. He supports these with the assumption that the same basic forces, viz. attraction and repulsion, work in the same way throughout the whole Universe. By means of the imagined changes of standpoints, the fundamental forces become applied in iterated ways in models of celestial systems at different scales. It is best to view the changes of standpoints as a kind of necessary methodological, specifically heuristic device; no more, but also no less. No more, because such changes are not themselves any explanatory tools; this is left to the system of physical forces and the concrete models of celestial systems. No less, because without the changes in standpoints we could not deal appropriately with many observations. We could not properly distinguish between real and apparent motions, sizes, or distances, for instances. We would not be able to make the step from our common observational knowledge to a more thorough scientific knowledge of the Universe; nor could we make progress in cosmological knowledge. Let us elaborate these points a bit.

Start with mundane changes of viewpoint, where one moves from one location to another in order to look at the same thing from different angles or distances. These can already lead to different and interesting results. In his *Anthropology*, Kant declares that to view a landscape from a bird's eye view rather than from the ground level can "prompt a completely different judgment of the region" (VII:216). Such a standpoint change is, in principle, not impossible and can be an activity of real observation, as when we climb up mountains, use a hot air balloon (which Kant knew as "Aërostat"; VII:247n) or, today, take flights.[24]

Changing the standpoint of the observer in cosmology differs dramatically from such mundane changes. In Kant's time at least, they always require an act of imagination, a kind of thought experiment. In addition, Kant is inviting us to perform a dual task, as already indicated: to ask ourselves (a) how the Milky Way would look if we had the ability to view it from a position outside of it – something not even our most advanced technologies today can achieve, and (b) if a cosmological theory like Kant's is true, then how the Milky Way would look, given that we are located inside of it. If the Milky Way has the shape of a flat rotating disk, then if you yourself are very small relative to it, and located somewhere in it, you will experience the Milky Way as a band stretched across the sky. (If one observes another galaxy outside of ours, it will look like a small, elliptic disk.) But both (a) and (b) have their problems. While (a) remains different from any actual observational practices, (b) leads to the problem that our actual, mundane observations of the starry heavens cannot immediately or fully justify the theory – rather, the theory is used to explain the phenomenal character of our ordinary observations. It is theory-laden observation.

Kant himself agrees that one can go too far in imagining different standpoints, up to the point that one reaches a stage of "vesania" (*Aberwitz*) or "positive unreason," a mental derangement or illness which implies not merely a deviation from standard rules of reason but the invention of new rules – a disease which he calls, interestingly, the "systematic" one (as opposed to others, classified as either "tumultuous," "methodical," or "fragmentary"; VII:215–6). He does not consider whether he might, in imagining radically different standpoints in his own cosmology, have himself deviated from legitimate rules of scientific reasoning. In *UNH*, he admits that he sometimes engages in hypotheses the observational support for which is still far out. But even in his critical works, Kant never denies the legitimacy of the core of his cosmological and cosmogonical theories. He does in no way commit them to the flames where dogmatic metaphysics belongs. In 1791 he even legitimizes the publication of an extract of

UNH, edited and published by his former student, the Königsberg mathematician Johann Friedrich Gensichen.[25] The critical Kant must still think that there are legitimate rules for taking certain standpoints in science, even including certain imagined ones. Without the willingness to engage in imagined standpoints, progress in cosmology may be difficult if not impossible.

But while changing the standpoint of the observer makes it possible to progressively expand and correct our current observational horizon, it cannot do that by itself alone. It must be thoroughly embedded in other elements and forms of legitimate scientific theorizing and reasoning. Kant has several plausible ideas here. First, his emphasis on the explanatory role of the fundamental Newtonian forces and laws must not be forgotten, nor his reliance on other parts of well-established scientific knowledge, partly related to Newton's physics, such as Kepler's laws of planetary motion. Second, imagined changes of standpoints of observation should cohere with existing observations. That does not mean that we take the latter for granted. On the contrary, as Kant's complex example of observing the Milky Way from different perspectives shows, our imaginative view of our galaxy from the outside and our parochial perspective on it from Earth must be brought together by theoretical interpretations. That, in turn, contributes to the growth and increased systematicity of knowledge. Thirdly, changes in standpoints must also be controlled by standards of hypothetical reasoning, one of which we have seen: hypotheses are permissible, but they must be kept small in number if one does not wish to reduce the probability and credibility of a whole theory. Kant adds two further requirements, namely that all consequences of a hypothesis must follow from the hypothesis in such a way that the data confirm the hypothesis and that each hypothesis must be such that it is certain that the object assumed in the hypothesis is possible in a real sense: it must be coherent with the system of categories and principles of the pure understanding that make any cognition whatsoever of nature possible (A769–70/B797–8; IX:85–6),[26] or – more pertinent to a natural science like cosmology – with the specific system of metaphysical concepts and principles of physical objects that Kant develops later in the *Metaphysical Foundations of Natural Science* (1786). Only under such constraints can we perform, in imagination, changes of observational standpoints that are reasonable or legitimate. What is more, only then can standpoint changes and reasoning with hypotheses be fruitful for a progressive integration of numerous different cognitions, and the discovery of new knowledge, within a unified systematic framework.

This helps, at least to some extent, to explain under what conditions or constraints we could and should engage in the activity of changing observational standpoints. Other passages in Kant's writings reinforce our claim that the notion is an important methodological device for his scientific and philosophical thinking. However, they also lead to new problems. Here are two passages.

First, in the first *Critique*'s fourth Antinomy (concerning the existence of an absolutely necessary being as cause of the world), Kant explains his idea of a compatibility of apparently incompatible metaphysical standpoints by referring to another astronomical issue:

> M. de Mairan took the controversy between two famous astronomers, arising from a similar difficulty in the choice of a standpoint, to be a sufficiently strange phenomenon that he wrote a special treatise about it. One inferred, namely, that the moon turns on its axis because it constantly turns the same side toward the earth; the other, that the moon does not turn on an axis, just because it constantly turns the same side toward the earth. Both inferences were right, depending on the standpoint taken when observing the moon's motion.
>
> (A461/B489)[27]

So, from Earth one sees only the bright, and never the dark side of the Moon. One astronomer claimed that, because we always see the same side, the Moon cannot be rotating. The other said, however, that the Moon turns on its axis and that explains why we see always the same side (you have to add in that the Moon's axial rotation takes the same time as its rotation around Earth). According to the French physicist and member of the Paris Academy Jean Jacques d'Ortous de Mairan (1678–1771) mentioned here by Kant, the former view was held by Kepler, among others, while the latter position was represented first by Giovanni Domenico Cassini and later on, for instance, by Newton.[28] Against this, Kant asserts that the *prima facie* existing incompatibilities of the two theses about the Moon's axial rotation might be dissolved by introducing different standpoints from which the observation of the Moon's bright side can be explained – and that then both "inferences" are "right" (*richtig*). However, already in 1754 Kant had argued for what is still accepted as the correct view, namely that the Moon's axial rotation is synchronized with its rotation around Earth (I:190–1).[29] So, the Moon *does* rotate around its axis and there is no compatibility of the different claims: one is true, while the other one is false. In addition, another problematic feature of Kant's example is that he tries to explain that the astronomers reached their conclusion through a "mode of inference" (*Schlußart*) and that he claims that such inferences can be affected by taking standpoints. By becoming aware that one is taking different standpoints, Kant argues, one can show that "Both inferences were right" (*Beide Schlüsse waren richtig*). But this ascribes more power to the taking of standpoints than we have argued they have. They are heuristic devices, and do not by themselves prove the truth of a claim of the legitimacy of an inference. Therefore, thinking that taking a standpoint proves that an inference is right is a nonstarter.[30]

Second, in the *Contest of the Faculties*, when discussing the question of whether humanity progresses, Kant again refers to retrograde motion in astronomy:

> If the course of human affairs seems so senseless to us, perhaps it lies in a poor choice of position [*Standpunkt*] from which we regard it. Viewed from the earth, the planets sometimes move backwards, sometimes forward, and sometimes not at all. But if the standpoint selected is the sun, an act which only reason can perform, according to the Copernican hypothesis they move constantly in their regular course. Some people, however, who in other respects are not stupid, like to persist obstinately in their way of explaining the phenomena and in the point of view which they have once adopted, even if they should thereby entangle themselves to the point of absurdity in Tychonic cycles and epicycles.
>
> (VIII:83)

This example clearly differs fundamentally from the first one: it is not aimed at making apparently incompatible explanations of phenomena compatible but at deciding which one is correct. Accordingly, the correct one will *replace* the incorrect one. Retrograde motion, says Kant, seems "senseless" (*widersinnisch*), and so does the course of human affairs.[31] So, apparently changes in standpoint can achieve quite different things; but how is that possible? Kant does not explain this in general. In addition, Kant here says that the change in standpoint can only be achieved by "reason." This might remind one of the already cited passage from the B-Preface to the first *Critique*, since both passages refer to retrograde motion. However, there Kant said that Copernicus "did not make good progress in the explanation of the celestial motions if he assumed that the entire celestial host revolves around the observer, tried to see if he might not have greater success if he made the observer revolve and left the stars at rest" (Bxvi). The observer is invoked twice here, so whatever faculty brings about the change, it is still meant to be a change in the standpoint of the observer. If it is not imagination but reason, then which concept of reason does Kant appeal to? A practical, perhaps specifically moral use of reason? Or does he point toward his views on the regulative use of ideas and maxims of reason from the Appendix to the Dialectic of the first *Critique*?[32] That Kant considers himself to be bringing sense into the chaos of human action might speak for the former, i.e. *practical* use of reason.

But this concept differs fundamentally from the methodological notion of changes in observational standpoints that is used in the decision between empirical theories – and it is therefore not clear, at least not without further investigation, what Kant really means here.

So, there are unresolved issues in Kant's concept of the standpoint of an observer. He invokes the concept for opposite solutions to fundamental conflicts of opinions, but does not explain how each type of solution is possible; he shifts too easily between saying that observer standpoint changes are produced by imagination and that they are produced by reason; and he does not clarify which notion of reason he has in mind.

However, these problems should not obscure the positive results we have reached concerning the relation of the concept to Kant's notion of scientific systematicity. The most important lesson here is that we must not understand the latter notion as the mere demand for ordering and integrating aggregates of cognitive judgments after the fact, as it were, into theories in a purely logical manner. As the discussion of meanings of 'system,' and the role of standpoint changes and reasoning with hypotheses in cosmology has shown, systematicity is very much about enabling science to progress *forward* in dynamical and diachronic ways. Kant claims that scientific systematicity demands that a scientific system may "grow from within (*per intus susceptionem*), but not by external addition (*per appositionem*)" (A832/B860). Heliocentrism and Newton's mechanics are the basis; but they can and should be expanded and refined further. Kant claims that, in order to be acceptable, a theory must conform to ideally *all* empirical consequences, that is to all observations that are already known but also those that are yet to be made (IX:84–6). We do not only deal with old observational knowledge but can also become confronted with new data – in astronomy, for instance, through telescopes, that might or might not come to confirm Kant's claims about the nebulae. The methodological device of standpoint changes often allows for new theoretical interpretations of observations, and it thereby is highly important to increase scientific systematicity, too.

3.6 Conclusion

Kant's work on cosmology and cosmogony reveals interesting features of his practices and ideals of legitimate, progressive, and fruitful scientific reasoning. It moreover shows that his notion of systematicity of the sciences is more complex than commonly assumed. *First*, it possesses ontic and doctrinal meanings which, in the practice of scientific reasoning, must be combined to achieve novel and acceptable results. As we have argued, the ability to change standpoints of observation can be essential for this. *Second*, in Kant's critical thought, systematicity becomes the *defining* feature of science; but systematicity also remains an unavoidable *normative demand* for the research practice of scientists. Far from being just an afterthought or a pedantic demand for combining cognitions in a merely logical manner, or organized by the categories and principles of the understanding, the continuous systematization of cognitions and theories is a *conditio sine qua non* for scientific progress. Systematicity, in the case of cosmology (and likely, in other disciplines too), enables us to "suitably determin[e] the greatest possible use of the understanding in experience in regard to its objects" (A516/B544). We systematically generate new hypotheses, new observations, new interpretations of our data, and thereby new knowledge. Systematicity, then, is not merely a product-notion, relating to a completed and ordered body of knowledge, but also a process-notion, relating fundamentally to the heuristics and methods for attaining such systems of knowledge (more generally on processes of research in Kant's account of science cf. Sturm and Wallner 2025).

We do not wish to deny or downplay the more familiar product-notion of scientific systematicity in Kant's thought; but the process-notion is important for him too. Stated less anachronistically, Kant does not follow d'Alembert's famous critique of the "spirit of systems" (*esprit de système*), allegedly to be replaced by the "systematic spirit" (*esprit systematique*; d'Alembert 1751). Instead, both must be connected. Neither do we think that it is correct to claim that, for Kant, "speculative system-building and scientific work are fully distinct" (Butts 1961, 159). Kant did not endorse such distinctions, and his best and most concrete scientific reasoning supports the view that he was right here.

Notes

1 For comments and criticisms, we would like to thank Sabrina Bauer, Carl Hoefer, Stephen Howard, Werner Stark, and participants at workshops in Frankfurt and Barcelona.
2 The book was published anonymously, and it is not clear why. Scholarly explanations for this include, for instance, the hypothesis that Kant thought that a naturalistic, mechanistic account of the genesis and structure of the Universe where God would no longer intervene might cause troubles with the religious zealots of Pietism, who were a strong social force in Prussia and who fought against Newton's *Principia* (1687) up until 1750 (Schoenfeld 2006, 49). Another possibility might be that Kant still felt the public criticism and even ridicule of his first publication, the *Thoughts on the True Estimation of the Living Forces* (1749). But, in either case, why did Kant then clearly and publicly announce a forthcoming book on cosmology and cosmogony less than a year before? (cf. I:191; Sturm and Burt 2023, 673–5)
3 As already argued by Adickes (1925, 310–5), this positive assessment is not undermined by the – broadly correct – criticism that Kant's scientific works reveal a lack of mathematical sophistication (e.g. Jaki 1981; Waschkies 1987; but see Palmquist 1987).

4 Cf., e.g. de Bianchi (2013) and Falkenburg (2000) for relations to Kant's discussions of rational cosmology and Waschkies (1987) for Kant's early physicotheology. A still important contribution to the second tradition is Adickes (1925). For an overview of research on *UNH* see Howard (2023, 247–9). Also, Kant's early cosmology may not be central to but is still frequently included within research that focuses on the importance of the antinomy theme for the genesis of his critical philosophy (see Hinske 1970 for such an approach).
5 For the following see Hager and Strub (1998).
6 In the course of the 17th century, the notion gained popularity in all kinds of academic contexts, as for instance in legal and political theory and, since Leibniz, in rationalist metaphysics.
7 This notion, however, should not be taken as a definition in the strict sense, since it allows for counterexamples. Kant was well aware of the threat the motions of the comets pose to this notion of systematicity, since some of them were believed to revolve around the Sun in the opposite direction to that of the planets, while others even seemed to trespass the boundaries of the Solar System. Hence, for Kant, especially the last type of comets "sets the last borders to the systematic constitution" (I:281). Due to the unsystematic motion of the comets, it seems reasonable to say that, for Kant, in some cases only the "majority of bodies" in a certain region of space comprise a systematic constitution. The same point could be applied to the (much more recent) discovery that some planets, including Venus, show a retrograde (clockwise) axial rotation.
8 Kant's views on this matter were heavily influenced by Thomas Wright's *An Original Theory or New Hypothesis of the Universe* (1750), as he admits on several occasions (cf. I:231–4; I:248). However, Kant knew of Wright's work only through a review (Anonymous 1751). The review gives only a very limited and, in parts, inadequate picture of Wright's theory; see Adickes (1925, 227–35) and Hoskin (1971) for a comparison of Kant's and Wright's views and an evaluation of the quality of the review.
9 As to comprehensiveness, this is, speaking from a strictly historical point of view, only achieved after further elaboration of heliocentrism by later researchers – but note that we said that it makes such comprehensiveness possible, not that it already achieves it.
10 Cf. the depiction of Carrier (2001, 89–94), for the following explanation.
11 While it has often been claimed, following e.g. Kuhn's account (1957, 1962/1970), that Ptolemy's theory had run into severe difficulties or even a crisis (among other things, that it could not do full justice to observations), this has also been doubted (Gingerich 1975; but see Griffiths 1988).
12 To make it worse, to account for all individual properties of the observable patterns of planetary motion the epicycles necessarily differed in a whole range of parameters, for instance in scale and velocity. Therefore, the explanation of a single observable phenomenon required a complex set of assumptions to fit with the underlying cosmological model of the Universe.
13 At this stage of our interpretation of what is going on in *UNH*, we are using texts that were produced 15 years or more after *UNH*. Is that illegitimate? We do not think so. First, we do not causally explain why Kant developed his cosmological models but identify which unspoken reasons might have stood in the background; second, it is quite significant that the Copernican theory is used, time and again, as the primary example in these later texts. Possibly, it was by reflecting on epistemological issues in his own cosmology and its

broader historical context that Kant developed his later views on the criteria for legitimacy of scientific hypotheses, one of which is that of the "unity" of a hypothesis. Thus, there is little if any historiographical or interpretive harm done by using these later texts – certainly less harm than trying to rationally reconstruct Kant's cosmology by ideas of later philosophy of science, say.

14 Besides Bruno, who developed this idea in his *De l'Infinito, universo e mondi* from 1584, Bernard le Bovier de Fontenelle (in his *Entretiens sur la pluralité des mondes* from 1686) and Christiaan Huygens (in his *Cosmotheoros* from 1698) were famous for this conception. Kant was especially familiar with the work of Fontenelle, whom he mentions in UNH (I:353) and whose *Entretiens* he quotes in his 1754 essay on *The question whether the earth is ageing* (I:197).

15 In the decades before the publication of UNH, however, occasional reports of the observation of motion among the fixed stars became published. Kant refers to a report by French astronomer Philippe de La Hire (1693/1748; see I:253), but he does not mention Edmund Halley's observation (see Halley 1718).

16 We here draw on the difference between the views on objectivity expressed by Thomas Nagel (1986) and Bernard Williams (1978, ch. 1). Only the latter considers incorporating the mind-dependent perspective into an "absolute conception of the world." However, Kant does this in the course of scientific, empirically testable investigations, not to develop a kind of metaphysics, as Williams does. There is, then, still the conceptual and argumentative room for Kant to develop his later transcendental idealism; which, as he claims, is perfectly compatible with empirical (or scientific) realism and even implies it. In our view, UNH instantiates this realism.

17 Kant also refers to the analogical character of his reasoning: "In this context, we may use the analogy of what has been observed in the orbits of our solar system" (I:250). Hence, perhaps he views his non-deductive reasoning as supported by the analogical nature of the arguments. To consider this suggestion is, however, beyond the scope of this paper, since we would need to address the problem which precise kind of analogy Kant employs in UNH; see Howard (2023) for an extensive discussion of the latter question.

18 Cf. I:234–6 and Chapter 8 of the second part in I:331–47.

19 Admittedly, the observational data Kant relied on in UNH was, as Jones (1971) has argued, rather thin.

20 See the long footnote in the preface of UNH at I:232–3 for some of the interpretations of the nebulous stars that Kant was aware of.

21 Again, Kant was not the first to come up with this assumption. As early as 1675, Christopher Wren had already sympathized with the idea that the nebulous stars are systems of stars just like the Milky Way. See Whitrow (1967) for a brief history of the observation and interpretation of the nebulous stars until UNH.

22 No wonder, thus, that at the end of the second part, UNH slides into a speculative chapter called "On Creation in the entire extent of its infinity both in space and in time" dealing with questions known rather from rational than empirical cosmology.

23 See Palmquist (1993/1998) and Longuenesse (2005) for such an approach.

24 In the 1780s, the first balloon flights made this birds-eye perspective possible for humans. They also led to a novel way of depicting landscape, namely in horizonless, aerial view images. See Thomas Baldwin's *Airopaidia* (1786) for the first images of this kind.

25 This text has so far not been included in the Academy edition but will be so in volume IX of the revised edition. See Gensichen (1791/2025); Burt and Sturm (2025).
26 On Kant on hypotheses in science, see Butts (1961, 1962). In fact, it is possible that the last condition requires not only the conformity of hypotheses with the *a priori* categories and principles, but also the regulative use of ideas of reason, as in taxonomies of objects, their properties, and relations within special sciences. It is probably no accident that Kant illustrates this use of ideas, for instance, by the system of kinds of motion in astronomy (circular, elliptical, and parabolic; cf. A662–3/B690–1).
27 In the last sentence, the Cambridge Edition translates the German term *richtig* as "correct." We try to be more literal here. In deductive logic, inferences may be judged by being merely formally correct or, in addition, having true premises and then being what is usually called 'sound.' Kant's *richtig* is more likely meant in the second sense. There is, of course, another option: viewing the "inference" not as a deductive but an abductive one: an inference toward what might *explain* the observation that we always see the same side of the Moon (this is the reading of Meer 2019, 47). In this case, *richtig* cannot mean 'sound' and not even 'deductively valid.' It would only mean that the inferences are toward possible or perhaps plausible explanations. Meer (2019, 47) claims that the inferences are drawn "in a valid manner" (*in gültiger Weise*), but we disagree. Kant knows that good or convincing explanatory inferences require more than just to explain the phenomena – precisely because one can always explain phenomena in more than one way, as this very example illustrates.
28 The "special treatise" Kant refers to is Mairan (1747) (cf. McLaughlin 1990, 115), not another text by Mairan (as Proops 2021, 332, asserts). We did not find a German translation of this article. Possibly, Kant learned about Mairan's analysis from other sources.
29 McLaughlin (1990, 115) also views a passage in *Metaphysical Foundations of Naturals Science* (1786) as implying that this rotation is a "real" motion (IV:557).
30 Scholars have primarily discussed whether taking standpoints for explaining, on opposite grounds, that we never see the dark side of the Moon's is illuminating for a metaphysical dispute such as the Fourth Antinomy. McLaughlin claims "that even if there is a 'real' rotation of the moon (measurable by the centrifugal forces), there can still be a standpoint (e.g. the earth) in relation to which the moon is at rest" (McLaughlin 1990, 115), and he argues against Michael Wolff that therefore the analogy fully works. If our concerns are correct, McLaughlin's defense of Kant is not convincing. The conflict about the Moon's axial rotation is empirically decidable, even if it involves theoretical arguments. It is categorically different from metaphysical conflicts such as the dynamical antinomies, and from solving them by invoking the transcendental distinction between things in themselves and appearances. In that regard, Proops (2021, 332–3) might be closer to the truth by claiming that the Moon example fails to illustrate the use of this distinction in the resolution of metaphysical conflicts. The whole issue is more complex than we can discuss here, however.
31 *Widersinnisch* might also be translated as 'contrary to the senses.' However, this option would refer only to sensory knowledge, whereas the German *widersinnisch* also has the broader meaning of 'absurd,' which Kant needs for the connection to the issue at hand, the apparent irregularity of human free actions.

32 Kant indeed connects the regulative use to standpoint changes, as when he writes: "Systematic unity under the three logical principles can be made palpable in the following way. One can regard every concept as a point, which, as the standpoint of an observer, has its horizon, i.e., a multiplicity of things that can be represented and surveyed, as it were, from it" (A659/B687). For the issue at stake this does not help, since he is here using the concept of a standpoint as a metaphor for explaining the functioning of conceptual taxonomies in science; he is not telling us here anything about how to make things that appear absurd or senseless intelligible.

Bibliography

Adickes, Erich. 1925. *Kant als Naturforscher*. Vol. II. Berlin: De Gruyter.
Anonymous 1751. "Review of Thomas Wright. An Original Theory or New Hypothesis of the Universe." *Freye Urtheile und Nachrichten zum Aufnehmen der Wissenschaften und Historie überhaupt*. Achtes Jahr. I. Stück, 1. Januar, 1–5; II. Stück, 5. Januar, 9–14; III. Stück, 8. Januar, 17–22.
Baldwin, Thomas. 1786. *Airopaidia: Containing the Narrative of a Balloon Excursion from Chester, the Eighth of September, 1785 [...]*. London: J. Fletcher.
de Bianchi, Silvia, ed. 2013. *The Harmony of the Sphere*. Newcastle: Cambridge Scholars Publishing.
Bruno, Giordano. 1584. *De l'infinito, vniverso et Mondi*. Venice.
Burt, Fabian. (Forthcoming). "Kant on the Age of the World." *Annals of Science*.
Burt, Fabian and Thomas Sturm. 2025 (forthcoming). "Editorischer Bericht, Sacherläuterungen und Literatur zu: Johann Friedrich Gensichen. Authentischer Auszug aus Immanuel Kants Allgemeiner Naturgeschichte und Theorie des Himmels (1791)." In *Kant's Gesammelte Schriften*. Academy edition. Vol. IX. Berlin/Boston: De Gruyter.
Butts, Robert E. 1961. "Hypotheses and Explanation in Kant's Philosophy of Science." *Archiv für Geschichte der Philosophie* 43: 153–70.
Butts, Robert E. 1962. "Kant on Hypotheses in the 'Doctrine of Method' and the Logik." *Archiv für Geschichte der Philosophie* 44: 185–203.
Carrier, Martin. 2001. *Nikolaus Kopernikus*. München: Beck.
d'Alembert, Jean Le Rond. 1751/2009. *Discours préliminaire de l'Encyclopédie*. Introduced and annotated by Michel Malherbe. Paris: Vrin.
Falkenburg, Brigitte. 2000. *Kants Kosmologie*. Frankfurt: Klostermann.
de Fontenelle, Bernard le Bovier. 1686. *Entretiens sur la Pluralité des Mondes*. Paris: Blageart.
Galilei, Galileo. 1632. *dialogo di Galileo Galilei Linceo Matematico sopraordinario dello Stvdio di Pisa. E filosofo, e Matematico Primario del Serenissimo Gr. Dvca di Toscana. Duoe ne i Congressi di Quattro Giornate si discorre sopra i due Massimi sistemi del Mondo Tolemaico e Copernicano; Proponendo Indeterminatamente le Ragioni Filosofiche, e Naturali Tanto per l'vna, Quanto per l'altra Parte*. Florence: Giovanni Batista Landini.
Gensichen, Johann Friedrich. 1791 (forthcoming 2025). "Authentischer Auszug aus Immanuel Kants Allgemeiner Naturgeschichte und Theorie des Himmels." In *Kant's Gesammelte Schriften*, edited by Fabian Burt and Thomas Sturm. Academy edition, vol. IX. Berlin/Boston: De Gruyter.
Gingerich, Owen. 1975. ""Crisis" versus Aesthetic in the Copernican Revolution." *Vistas in Astronomy* 17: 85–94.

Griffiths, Robert. 1988. "Was There a Crisis before the Copernican Revolution? A Reappraisal of Gingerich's Criticisms of Kuhn." *PSA: Proceedings of the Biennial Meeting of the Philosophy of Science Association, Vol. 1988, Vol. 1,* Contributed Papers: 127–32.

Hager, Fritz-Peter and Christian Strub. 1998. "System." In: *Historisches Wörterbuch der Philosophie,* edited by Joachim Ritter, Karlfried Gründer, and Gottfried Gabriel. Vol. 10, 824–56. Basel/Stuttgart: Schwabe.

Halley, Edmond. 1718. "Considerations on the Change of the Latitudes of Some of the Principal Fixt Stars." *Philosophical Transactions* 30: 736–8.

Hinske, Norbert. 1970. *Kants Weg zur Transzendentalphilosophie: Der dreißigjährige Kant.* Stuttgart: Kohlhammer.

Hoskin, Michael A. 1971. "Introduction." In *Thomas Wright of Durham. Original Theory or New Hypothesis of the Universe. 1750. A Facsimile Reprint together with the First Publication of 'A theory of the universe'. 1734,* edited by Michael A. Hoskin, ix–xxxiii. London: Macdonald/American Elsevier.

Howard, Stephen. 2023. "Kant's Universal Natural History and Analogical Reasoning in Cosmology." In *Between Leibniz, Newton, and Kant. Philosophy and Science in the Eighteenth Century.* 2nd ed., edited by Wolfgang Lefèvre, 247–70. Cham: Springer Nature Switzerland.

Huygens, Christiaan. 1698. *Christiani Hugenii κοσμοθεωροσ [Cosmotheoros], sive de Terris Coelestibus, earumque ornatu, conjecturae. Ad Constantinum Hugenium, Fratrem: Gulielmo III. Magnae Britannieae Regi, a secretis.* Den Haag: Moetjens.

Jaki, Stanley. 1981. "Introduction." In *Immanuel Kant. Universal Natural History and Theory of the Heavens,* edited by Stanley Jaki, 1–76. Edinburgh: Scottish Academic Press.

Jones, Kenneth Glyn. 1971. "The Observational Basis for Kant's Cosmogony: A Critical Analysis." *Journal for the History of Astronomy* 2: 29–34.

Kuhn, Thomas. 1957. *The Copernican Revolution.* Cambridge: Harvard University Press.

Kuhn, Thomas. 1962/1970. *The Structure of Scientific Revolutions.* 2nd ed. Chicago: Chicago University Press.

La Hire, Philippe de. 1693. "Observation faite a l'Observatoire Roial du passage de la lune par les Pleïades le 12 Mars au soir." *Mémoires de Mathematique et de Physique, Tirez des Regristres de l'Académie Royale des Sciences,* 37–39.

La Hire, Philippe de. 1748. "Observation des durchganges des Mondes durch das Siebengestirn, am 12ten Merz abends." In *der Königl. Akademie der Wissenschaften in Paris Physische Abhandlungen, Erster Theil, welcher die Jahre, 1692. 1693. 1699-1702 in sich hält,* translated by von Steinwehr, Wolf Balthasar Adolph, 131–4. Breslau: Korn.

Lipton, Peter. 1991. *Inference to the Best Explanation.* London: Routledge.

Longuenesse, Beatrice. 2005. *Kant on the Human Standpoint.* Cambridge: Cambridge University Press.

McLaughlin, Peter. 1990. *Kant's Critique of Teleology in Biological Explanation.* Lewiston: Edwin Mellen Press.

Mairan, Jean Jacques d'Ortous de. 1747. "Recherches sur l'équilibre de la lune dans son orbite." *Mémoires de Mathematique et de Physique, Tirez des Regristres de l'Academie Royale des Sciences,* 1–22.

Meer, Rudolf. 2019. *Der transzendentale Grundsatz der Vernunft.* Berlin: De Gruyter.

Nagel, Thomas. 1986. *The View from Nowhere.* New York: Oxford University Press.

Newton, Isaac. 1687/1714. *Philosophiae Naturalis Principia Mathematica. Neueste, verbesserte Auflage.* Amsterdam: Compagnie des libraires.

Palmquist, Stephen. 1987. "Kant's Cosmogony Re-Evaluated." *Studies in History and Philosophy of Science* 18: 255–69.

Palmquist, Stephen. 1993/1998. *Kant's System of Perspectives.* Lanham: University Press of America/Hong Kong: Philopsychy Press.

Proops, Ian. 2021. *The Fiery Test of Critique.* Oxford: Oxford University Press.

Schönecker, Dieter, Dennis Schulting, and Nico Strobach. 2011. "Kants kopernikanisch-newtonische Analogie." *Deutsche Zeitschrift für Philosophie* 4: 497–518.

Schoenfeld, Martin. 2006. "Kant's Early Cosmology." In *A Companion to Kant*, edited by Graham Bird, 47–62. Malden: Blackwell.

Sturm, Thomas. 2009. *Kant und die Wissenschaften vom Menschen.* Paderborn: Mentis.

Sturm, Thomas, and Fabian Burt. 2023. "Editorischer Bericht, Sacherläuterungen und Literatur zu: Immanuel Kant. Allgemeine Naturgeschichte und Theorie des Himmels (1755)." In *Kant's Gesammelte Schriften.* Academy edition. Vol. I, 672–822. Berlin/Boston: De Gruyter.

Sturm, Thomas, and Larissa Wallner. 2025 (forthcoming). "Kant's Empirical Account of Science: Psychological, Social, and Historical Conditions of Research." *Annals of Science.*

Williams, Bernard. 1978. *Descartes: The Project of Pure Inquiry.* Harmondsworth: Penguin.

Waschkies, Hans-Joachim. 1987. *Physik und Physikotheologie des jungen Kant.* Amsterdam: Grüner.

Whitrow, Gerald J. 1967. "Kant and the Extragalactic Nebulae." *Quarterly Journal of the Royal Astronomical Society* 8: 48–56.

Wright, Thomas. 1750. *An Original Theory or New Hypothesis of the Universe, Founded upon the Laws of Nature, and Solving by Mathematical Principles the General Phaenomena of the Visible Creation; and Particularly the Via Lactea.* London: Chapelle.

4 Kant and the Idea of a System of Logic

Clinton Tolley

4.1 Introduction: Logic and System in Kant

4.1.1 A "Faculty-first" Conception of the System of Logic

Though the idea of a "system of logic" is perhaps nowadays most often associated with post-Kantian philosophy of logic – as in J.S. Mill's influential work from 1843, for example, which claims to provide "A System of Logic" – the phrase itself does in fact occur in Kant's own works (albeit in passing; cf. VII:243; cf. IV:306). In any case, the idea that logic itself is in a special way both concerned with and exemplary of systematicity has a definite Kantian pedigree, and the goal of what follows is to explore some of the more significant relations that Kant takes to obtain logic, systematicity, and being a system.

One fairly straightforward commitment that Kant has in this regard is that logic itself, qua *science*, is a system – namely, a system of *cognitions* of its subject-matter. A second, less straightforward, commitment that I will also try to establish is that Kant takes logic qua science to *be about* a system. More specifically, I will argue that Kant takes the "object" or *subject-matter* of logic – what he identifies as the faculty of "the understanding in general [Verstand überhaupt]" and its activity – to itself be a system and to thereby provide the objective correlate to the system of cognitions that comprise the science of logic. In this sense, the first system of logic qua science represents a second system of logical activity "in" the faculty itself.

Finally, and perhaps most significantly, I will argue that Kant takes the first system of logic qua science to depend on the second system of activity "in" the faculty of the understanding. This dependence relation can be seen from the fact that Kant takes the latter system to be already present in the faculty prior to its being represented (cognized) as such in a science. This constitutes Kant's partial acceptance of the priority of what in the early modern tradition is called a "*logica naturalis,*" in contrast to the "*logica artificialis*" which consists, paradigmatically, in the scientific cognition of this more "natural logic."

By emphasizing Kant's commitment to the priority, in the order of explanation, of the system of activities present in the faculty of the understanding, over and against the systematic cognition of this activity-system, the following can be seen as a contribution to what might be called a "faculty-first" interpretation of Kant's conception of logic, and so in broad sympathy with the general faculty-first interpretive strategy of Kant's critical philosophy more generally.[1] In the case of logic in particular, however, certain distinctive problems arise for any attempt to give the faculty of the understanding priority in relation to the systematic cognition of this faculty in the science of logic – due, most importantly, to the fact that systematic cognition (science) "in general" is itself one of the activities which constitutes the primitive system of acts within the faculty itself. In the latter part of this essay, I explore some of the complications that arise due to the fact that science itself belongs to the subject-matter of logic qua science.

4.1.2 A Summary

I will proceed as follows. In Section 4.2 I will provide some of the background for my interpretive theses. I will introduce Kant's conception of *science* in general, as consisting in a "system" of representations, and then distinguish science as a representational system from the "domain" or *subject-matter* of the science, in order to introduce the idea of a further system being present in the domain (objective correlate) of the science itself, over and against the science itself as a system of representations.

In Section 4.3 I will then show how this general distinction between science and subject-matter also finds its application, for Kant, in the specific case of logic, and thereby draw out the manner in which logic itself also comprises two "systems": the system of cognitions (representations) which make up the science and a second system which obtains in the domain or subject-matter of logic, i.e., in the faculty of the understanding and its activity. Here I also spell out the elements and order of the system "in" the faculty and present evidence for the further interpretive thesis that Kant takes logic qua science to be a system that is itself grounded in, and explained by, the system of activities already present in the faculty of the understanding, prior to this system being "systematically represented" in the science of logic.

In the concluding Section 4.4 I take up several problems that would seem to arise for Kant, on the foregoing interpretation. First, I explore the complications which arise due to the fact that one of the elements of the original system of acts present in the faculty is itself "method," which seems to comprise the activity of forming a system of cognitions – i.e., the activity of systematic cognition via the formation of a science. The worry here is that there will be simply no way for the system of acts of the understanding to be specified in abstraction from the "idea" of a system of representations (cognitions) in particular. Second, I consider further worries concerning the seeming independence of the system in the faculty from not just specific "ideas" about the whole of some science or other, but from all actual representations whatsoever.

4.2 Two Kinds of "Systems" with Respect to Science

4.2.1 *System in General and Science as a System of Cognitions in Particular*

As a preliminary step toward drawing, on Kant's behalf, the distinction between the two kinds of systems that I have distinguished above (system of cognitions, system in the subject-matter of cognitions), we first need to show that Kant recognizes a broad enough sense of "system" that can apply to more than just to a whole of cognitions.[2] An important indication that Kant means to allow for this broader conception of system comes in the third *Critique*'s account of the manner in which the power of judgment represents nature itself "as a system," operating "in accordance with laws" and a "principle [Princip]" (§23, V:246). Nor is this the only place where Kant uses the phrase "system of nature" with the (seeming) intent of assigning the property of being a system to nature itself, and not to a body of cognitions of nature (cf. VI:218, 434; IX:160; XX:217–218; XXII:269; XXVIII:1307). What is more, these more generic uses also give some indication as to what pertains to the more abstract concept of being a system as such: being a whole which is ordered according to a "principle" and determined by "laws."[3]

That Kant means to be working with a more generic conception of "system" is also indicated by notes for his lectures on "philosophical encyclopedia" from around 1780. Here Kant begins by defining a "system" as consisting in a "whole" whose "idea" precedes that of the "parts," in contrast to an "aggregate," which consists in that kind of whole which "arises" due to "the parts of the whole preceding" the whole itself (cf. XXIX:5). Kant then goes on from this general definition of "system" to consider the specific case of "a system of cognitions [Kenntnissen]," which he takes to "constitute a science" (ibid.). That Kant marks a step here would again seem to imply that the first definition of "system" is intended to be more generic than the case of science, understood as a system comprised specifically of a "whole" of cognitions.

4.2.2 *System of Cognitions vs. System in the Domain of Cognitions*

Even if one grants that Kant is working with a broader conception of "system," one might nevertheless wonder whether he makes use of it to draw the even more specific distinction I sketched in the opening section between two senses of "system" that pertain to a science itself – i.e., to mark the distinction between the *science itself* as a system of *representations*, and then the separate system that obtains in the *domain* of the *objects* (subject-matter) of the science. Granting that this distinction is one that is more thoroughly explored later in the 19th century in the aftermath of Kant's writings,[4] do we find any evidence of this more specific idea in Kant's own texts?

We have already seen from Kant's lectures that a science itself is a "whole" which is "constituted" by being a kind of "system." In the first Critique's "Doctrine of Method," Kant claims that the transition from "ordinary cognition" to "science" is achieved precisely when cognition is brought into the form of a "systematic unity," such that "making" this cognition "into science" is identified with "making a system out of a mere aggregate" of cognitions (B860). In Kant's discussions of science throughout his lectures on logic, we find similar expositions of science as a "body of cognitions" about some subject-matter (a "domain") which are connected together in the form of a "system." Kant again describes the connection of cognitions into a system as being distinguished from other "wholes" that manifolds of cognitions can form (e.g., an "aggregate"), due to the cognitions being ordered in relation to a basic "principle" (or principles) concerning the domain of the science, which itself helps to specify the basic "idea" of this domain and its own structure and order (cf. IX:72; XXIV:879, 813). This sort of specification is also found elsewhere in Kant's published texts: in the *Metaphysical Foundations of Natural Science*, Kant writes that each science consists in "a system, i.e., a whole of cognition ordered according to principles" (IV:467; cf. IV:468).[5]

In the first *Critique*, however, science qua "system" is said also to consist, not just in "the unity of the manifold cognitions under one idea," but in a unity with respect to which "the domain [Umfang] of the manifold as well as the position of the parts with respect to each other are determined apriori" (B860). This reference to the ordered "domain" of the cognitions of a science suggests that, beyond taking each science to itself consist in a "system" of *cognitions* (i.e., a whole of cognitions ordered according to "principles"), Kant also takes each science to represent the "domain" of *objects* as *itself* forming a system (i.e., a whole of objects ordered according to an "idea"). In other words, while a science qua system consists in the ordered body of cognitions – such that a science is itself something that obtains or exists at the level of representations (comprising elementary concepts, "basic propositions [Grundsätze]," derived or demonstrated propositions ("theorems") and so on) – there is a further, second system which the science "represents" – i.e., a system which corresponds to the system of cognitions, but which itself obtains in the science's domain of objects. To connect this way of speaking with our earlier discussion: while the system of the cognitions of nature will constitute the *science* of nature, what it cognizes or comprehends is the system "in" *nature* itself, the "system of nature" qua organized domain.[6]

4.2.3 Science as "Logical" System vs. "Real" System in Its Subject-matter

As a further way of referring to these two different systems, we can introduce a terminological distinction between the "logical" system of representations (concepts, propositions, derivations) which constitutes a science, and the "real" system which obtains among the subject-matter ("things") of the science and which serves as the correlate to the science qua the system of representations.[7]

This distinction is meant to echo the distinction Kant himself draws between two "uses" of the understanding and reason – namely, the "logical" vs. "real" use of these faculties. One especially familiar example of this distinction occurs in the outset of the Dialectic (B355f). The logical use of reason, in its lowest degree, occurs when, instead of cognizing something "immediately," the truth of a judgment is "inferred" (B359). Beyond one-off inferences, reason in its logical use "seeks to bring the greatest manifold of cognition of the understanding to the smallest number of principles (universal conditions), and thereby to effect the highest unity of that manifold," i.e., to achieve a systematic unity of the manifold of cognitions (B361). Over and against this logical activity of reason, however, which "gives to given cognitions a certain form," in which some cognitions are "subordinated" to others, Kant also raises the possibility of a "real" use of reason which would not merely order cognitions but would "relate to objects" themselves in a distinctive manner (B362f).

In the course of drawing out this contrast, Kant also floats the possibility that, though the logical use of reason is governed by "a demand [Forderung] of reason" (i.e., to bring the "manifold" of cognitions under a "unity of principles"), this "demand" might turn out to be "merely a subjective law of economy for the provision of our understanding," and that in any case the governance of the logical use of reason by this demand in relation to our cognitions does not on its own "justify us in demanding of objects themselves any such unanimity," or allow us to "give objective validity to its maxims" (i.e., the "maxims" which guide the logical use of reason to respond to reason's demand in relation to cognitions (B362–63)).

Something more is required to demonstrate "the objective correctness" of those principles of reason – i.e., to show that the principle which is, in the first instance, directed merely at bringing a "completeness" into an ordering of cognitions is not "only a logical prescription ... to bring the highest possible unity into our cognition," but also can yield "objectively valid propositions" which posit the relevant "completeness" to be "in the objects themselves" (B365–66).[8] To align this with a related terminological distinction: while the "logical" use of reason is especially concerned with the formation of a "system" out of the body of given cognitions, at times all it can achieve is the formation of a "doctrine," rather than a "science" properly so-called; it is only in its successful "real" use that reason can achieve the cognition and demonstration that a correlative system is present in the things themselves.[9]

Kant's recognition of the possibility of a gap here – between a subjective order of cognitions (representations) and the objective order of the objects of cognitions (things) – is what is meant to be captured by my appeal to this distinction in "uses" as allowing for a way of reframing the foregoing analysis between two types of "systems." For a given science, we can say that the "logical" system which obtains among its "manifold" of cognitions would be one in which the cognitions themselves are ordered by reason according to what among the relevant set of representations functions as principle (ground) and consequence. In the "logical" system, both the relata of these connections (i.e., the individual cognitions) and the ordering relations themselves will belong to the domain of representations, and the connections between these cognitions will themselves be "logical" connections (e.g., the logical relation of ground to consequence which connects judgments; cf. B98). By contrast, the "real" system that is represented as obtaining in the domain of the objects of the science (its subject-matter) would instead have these very objects or things as its relata (e.g., the particular substances, forces or powers, etc. of its domain), and its own "order" would consist in the "real," or at least objective, connections that obtain among these things, due to the (again "real," rather than merely logical) "principles" of this domain.[10]

4.2.4 Application of This Distinction to "Abstract" Sciences (viz. Mathematics)

Using terms in this way, we can say that each science itself qua "body of cognitions" is a "logical" system that obtains among these cognitions, but is also a system which itself represents, and is distinct from, the "real" system that obtains among the objects in the domain of this science. To be sure, in the case of *logic itself*, the very idea of the application of this distinction will raise subtle questions, and for two reasons. First, what serves as the "real" system for logic (i.e., the faculty of the understanding) will have a claim to itself be "logical," at least to the extent that the particular real system in question just is the domain of the science of logic in particular. Second, insofar as logic is often thought of as having a subject-matter that is not itself "real," or at least not anything concrete, but is instead very abstract – especially in relation to what Kant characterizes as its most elementary, "pure," "general," "formal" parts – it might seem problematic to nevertheless speak of logic's domain as consisting in anything that merits the name of being a "real" system. To work our way toward this application in the next section, we can take up *mathematics* as an instructive case which in some ways parallels that of logic – especially insofar as its subject-matter is also often thought to be something abstract, and, furthermore, as something Kant takes to be "ideal," rather than "real." This will let us begin to head off misunderstandings about what is, and what is not, at issue in using the above distinctions to articulate the case of logic itself.

The main point at issue for our purposes is that even in "pure" mathematics a distinction can be drawn between, on the one hand, the representations that constitute the logical system of the science of mathematics and then, on the other, the system of mathematical objects which constitutes the domain of mathematics. For Kant as for many philosophers, the "objects" at issue in geometry and arithmetic (viz. shapes, numbers) are not themselves real things, at least in the sense of being substances with causal powers, or things that have "existence [Dasein]" in the physical sense (cf. IV:467n). Even so, it is also true that, for Kant, these objects are not identical with the representations that constitute the science which represents (cognizes, comprehends) them. This follows from Kant's conception of the standing of mathematical cognition itself as essentially "synthetic" rather than "analytic." The concept (representation) <number three>, for example, is a constituent of the science of arithmetic, but is not identical with the number three itself. Nor is the concept <right triangle> identical with any right triangle qua object (viz. of intuition). If these were identical, then the mere analysis of what is contained in these concepts should be able to disclose both all of the properties which are true of the objects of these concepts, and also the truth and falsity of judgments about these objects made on the basis of such analysis. Against this, Kant insists that there are some mathematical predicates which cannot be derived from (viz. defined wholly on the basis of) the mere analysis of mathematical concepts, and also that the truth and falsity of some judgments (viz. theorems) cannot be demonstrated on the basis of mere conceptual analysis (cf. B744).[11]

The gap between concept and property, and between analytic and synthetic judgments, thus implies that while concepts, along with the "basic propositions" formulated through them, are among the constituents of the respective sciences, mathematical concepts themselves do not belong to the "system" of the objects of the sciences. Rather, the properties, order, and relations among numbers and shapes are what can be represented via concepts and propositions (in combination with intuition, in construction). Conversely, the same gap also implies that mathematical objects – whether geometrical (the shapes and figures, the systematic relations that obtain among certain shapes and others) or arithmetical (the numbers, the systematic relations that obtain between certain numbers and others) – do not belong among the representations which constitute either science. Geometrical and arithmetical objects and relations are instead that *about which* cognition is obtained or demonstrated (i.e., they belong to and populate the "real" system of mathematics), rather than being *constituents of* the science of mathematics itself, qua ordered body of cognitions (its "logical" system).

4.2.5 The "Innateness" of the System "in" Sensibility

We can set the stage even more directly for the case of logic if we reflect on the related fact that, for Kant, the real metaphysical "principles" of the mathematical order and relations among numbers and shapes qua objects lie outside not just of concepts and propositions, but outside of representations all together, including intuitions, insofar as they lie instead in the "formal constitution" of the faculty of sensibility (B41). Though, for Kant, this non-conceptual "principle" of mathematics turns out to be a form of a faculty of representing (viz. sensibility), this "real" form as a constituent of the faculty must be distinguished from any and all representations of this form, including the "pure intuition" of this form. Kant's insistence on this distinction – between the form of a faculty and the representations which might arise from it – can be seen in his controversy with Eberhard, where he makes clear that, though he rejects the idea of any "innate [angeborne] representations" (and instead considers them "one and all" to be "acquired [erworben]"), he nevertheless accepts that the "ground" for at least some of the representations which we do end up acquiring must itself be "innate" in the subject prior to all representations (cf. VIII:221–223). One of Kant's two main examples here of "innate grounds" of this sort – i.e., grounds which are in the subject already, are not themselves representations, but which give rise to representations – is precisely the form of "receptivity," which gives rise to "the formal intuition called space" (VIII:222–223). Whereas the intuition of space is therefore something "acquired," the form of sensibility serves as the "ground" (viz. principle) of this intuition and is itself "innate."

This in turn complements and extends the more familiar point raised above about the analyticity of mathematics, which might otherwise seem to entail only that the real "principle" of mathematics must be distinguished from any representations of the form of sensibility by the understanding alone, via concepts and judgments. Here we see that the metaphysical ("real") principle of mathematics and of the systematicity in its objects should also not be identified with any representation (viz. intuition) drawn from sensibility either.[12]

4.3 Two Kinds of "System" in Logic

4.3.1 Logic Qua Science and Its "Domain"

With this discussion in mind, we can now turn to the question of the applicability of the general distinction between two kinds of systems that pertain to science in general, to the special case of the science of logic – and to the question, in particular, of whether the understanding itself can be rightly thought of along the lines of being or containing a "real" system which constitutes the domain of logic.

That Kant thinks logic qua science is itself a system is perhaps not especially controversial. Insofar as Kant's lectures on logic and published writings all make it clear that Kant takes logic to rise to the level, not just of a "doctrine [Lehre]," but of a "science [Wissenschaft]" (IX:13–16; cf. B76),[13] then logic itself, qua science, will comprise a "system" – namely, a system of a manifold of cognitions pertaining to its subject-matter, one which is organized around an "idea" of this subject-matter, which "determines apriori" the order and structure of its "domain." In other words, as with every other science, so too with logic: qua science, "logic is itself a system" (XXIV:697).

What might seem less clear is the applicability of the second sense of "system" – i.e., the idea that the domain of logic is itself something which can or should be taken to form a system, and in particular a "real" system, a "domain" of objects over and against the representations (concepts, propositions, etc.) which constitute the science of logic. One preliminary concern here might have to do with the very idea that logic, for Kant, should be thought to have an object or domain at all. Kant's most common way of specifying the subject-matter of logic, however, is to say that it is the science "of" a particular "faculty [Vermögen]" of the human mind or soul: logic is the science of the "understanding [Verstand] in general" and its "rules" (cf. B76). In his lectures, Kant takes this as equivalent to saying that logic is a science which has this faculty – and the more specific faculties of "the understanding, the power of judgment, and reason" contained within it (B169) – as its "object" (IX:14, 19; XXIV:695, 704; XXIV:792). This is also entailed by Kant's claim in his lectures that in logic the understanding and reason achieve "self-cognition" (IX:14), i.e., a scientific cognition of themselves as its object.

Still, a further worry might arise due to the fact that, at least with respect to the part of elementary logic which concerns only the "form" of the activities of the understanding and reason (the "mere form of thinking" (B78)), there could be something inapt in calling this "form" (or set of forms) an "object" or "subject-matter."[14] To be sure, it is true that a key part of the way that Kant characterizes the "forms" which are at issue in the resulting "formal logic" is precisely by reference to an abstraction that is made in relation to the activity of the understanding, which brackets out any and every object or kind of object to which the activities of the understanding might be oriented ("without regard to the difference of the objects to which it may be directed" (B76)). Even so, the fact that elementary formal logic considers only the "form" of the activity of the understanding in no way entails that the representations *in logic* do not themselves have an object. Indeed, the characterization of logic as "about" these forms is fully compatible with (and made within the same pages as) Kant's claims that the way logic treats the faculty of the understanding (and its acts) is to abstract from its relation to *further* objects (i.e., its content) and focus solely on the "form" of its activity (cf. B78–79). To be sure, the sense in which the understanding and its acts (treated as to their "form") itself serves as the "object" for logic and its cognitions need not entail many further (viz substantive) commitments as to the metaphysical status of this faculty itself. Even so, it does entail that the forms of the activity of the understanding do in fact serve as the subject of the various predications in the judgments expressed in the science of logic and also serve as the ground for the truth of these judgments.

4.3.2 The "Innateness" of the Domain of Logic, the "Acquiredness" of Science of Logic

Further reason for keeping track of the distinction between the scientific representation of the faculty of the understanding, and the faculty itself, can be drawn from Kant's various remarks on what is correct, and what is misleading, about the traditional distinction between what is "innate" and what is "acquired," with respect to the understanding – and the related distinction between two kinds of "logic," one that is already present in our nature (*logica naturalis*) and another that instead must be produced "artificially" through learning (*logica artificialis*).

We have already anticipated Kant's appeal to this distinction in his controversy with Eberhard, and also the facts that, first, Kant rejects the idea of any "innate [angeborne] representations" and considers them "one and all" to be "acquired [erworben]," and second, that Kant nevertheless accepts that the "ground" for at least some of the representations which we do end up acquiring must itself be "innate" in the subject prior to all representations (cf. VIII:221–223). We have already considered the form of receptivity as one of Kant's two examples here of "innate grounds" which give rise to acquired representations (here: the original intuition of space). The second example is even more apt for our own purposes, since it is nothing other than "the subjective conditions of the spontaneity of thought (in conformity with the unity of apperception)": these "conditions" are themselves affirmed as "innate," even as they eventually give rise to "universal transcendental concepts of the understanding," which are themselves thus "acquired" (VIII:222–223). The elementary concepts (representations) which arise in the understanding thus rest on something else which is not itself a representation, and which is also not itself acquired in any sense, but which lies a priori in the subjective conditions for thinking in general, i.e., in faculty of the understanding itself.

This idea that the understanding itself contains "grounds" which are not themselves representations and does so "innately" also figures into Kant's otherwise critical discussions of the related idea from the tradition that there is already a kind of "natural logic" present in the intellect or understanding prior to anyone (e.g., Aristotle) coming to represent this logic systematically or scientifically. In the textbook used by Kant for his lectures on logic, for example, Georg Meier distinguishes between, on the one hand, "the innate natural doctrine of reason [angebohrne natürliche Vernunftlehre]," for which he also gives the name "*logica naturalis connata*," and then, on the other, "the artificial [künstliche] doctrine of reason," which he associates with "*logica artificialis*" (Meier 1752, §533).[15]

Now, as we would expect from Kant's rejection of innate representations in general, Kant takes this use of the term "logic" here to be misleading if it is supposed to refer to a particular set of representations (viz. "doctrine [Lehre]") that are somehow present "innately."[16] Kant accepts that the core sense of the term "logic" is the sense in which it consists in a body of representations or cognitions, but he then insists that the only "logic" there is, in this sense, is one which is acquired (like all representations) and so in this sense an "*logica artificialis*" (cf. XXIV:696f, 792f; IX:17). Still, we have also seen that Kant himself accepts the related idea that the understanding is structured by nature ("innately") so as to contain specific conditions or "grounds" for specific representations (concepts), and so to this extent he does not reject one dimension of the traditional thought behind the positing of a "*logica naturalis.*"

4.3.3 The "Real" and "Logical" Systems That Pertain to Logic

The foregoing should suffice for making it at least prima facie plausible that, in the case of logic, too, we can apply the basic distinction between the representations (concepts, basic propositions, etc.) which constitute

106 *Kant and the Systematicity of the Sciences*

the science of logic and the object or subject-matter which is represented through these concepts and propositions – with the latter being supplied by the elementary forms of activity of the understanding in general. We should now consider whether this can also license an application of the distinction between the "logical" and the "real" system in the case of logic itself.

4.3.3.1 The "Real" System of the Acts of the Understanding

The first order of business in this respect is to identify more concretely the elementary activity which would serve as the "real" system of logic – and to bring out, further, the manner in which Kant takes it to already function as a *system*, i.e., whole of a plurality ordered in relation to an idea which sets out the principle of its unity. A second step will then be to draw more sharply the contrast between these "real" activities and the "logical" system of concepts and propositions which represent them, and so which function as constituents of the science of logic.

As we have anticipated above, the signature activity of the faculty of the understanding "in general" is identified as "*thinking*" (cf. B74–79; IX:13–16). In the introductory sections of many of his lectures on logic, Kant distinguishes different moments in the activity of thinking, and then sets them into an order of progression or development. In the Vienna Logic, for example, Kant follows tradition (the "ancients") in presenting thinking in its "beginning [Anfang]" as starting from the "apprehension" of a "*concept*," and then "combining" concepts in "*judgment*," and then "combining" judgments in a way that "*inferences* arise therefrom" (XXIV:904; cf. XXIV:701, 762–763). These three different act-types then provide the subject-matter for each of the three main parts of logic's own "universal doctrine of the elements" (cf. IX:91; IX:101; IX:114). In the final part of the lectures, Kant then singles out a fourth act of thinking, i.e., that of methodically ordering all of the previous acts (cf. IX:139). This series of acts and their ordering are what Kant alludes to in the A-Preface, when he describes "common logic" as having "fully and systematically enumerated" the "simple acts of reason" (Axiv).

For present purposes, what is of further interest about the final type of act is not only that it is this act which is involved in connecting the previous "manifold" of acts of thinking into a "science," but also that it is what brings the activity of the understanding as a whole to completion or "perfection" (cf. IX:139–140). In his lectures Kant further specifies the "perfection" of the activity of the understanding along several further dimensions: concerning the scope or domain ("extension") of the thinking (cf. IX:40f); concerning the "truth" of the thinking (IX:49ff); concerning the "clarity" and "distinctness" of the contents of the thinking (IX:58ff); and concerning the "certainty" that pertains to the thinking (IX:65ff). The highest perfection of thinking across all of these dimensions (comprising all objects, true, clear and distinct, certain) is itself associated with what Kant calls "absolute comprehension" – though he also claims that this is impossible for humans (while allowing that we can achieve "relative" comprehension in relation to this or that domain, cf. XXIX:731; IX:65).

If the foregoing helps to spell out the manner in which Kant can be seen to specify a "real" system in logic – one which is constituted by a manifold (i.e., of types of "acts" of thinking) that is unified according to an ordering and orientation toward an "idea" of the perfection or completion of this activity (in "science," and ideally, "absolute comprehension," viz. omniscience) – one thing which would seem to be still outstanding, however, is the real "principle" or "principles" which give rise to the manifold of activities and perhaps even to the general order or orientation itself.[17] To be sure, Kant himself refers to "the universal and necessary laws [Gesetzen] of thinking" as the "principles [Principien]" which pertain to logic, "from which all of its rules [Regeln] are derived and demonstrated" (IX:15). Importantly for our purposes, however, these "principles" are not to be taken as mere representations – in the sense that, e.g., a "basic proposition [Grundsatz]," might also be called a "principle" (as in "the principle of contradiction [Satz des Widerspruches]," IX:52)). Rather, these "principles" are themselves laws which belong to the real essence or nature of the faculty of the understanding itself, such that "agreement" with these principles is in effect "agreement" of the understanding "with itself" (cf. XXIV:823–824; IX:51).

As we saw above in our discussion of Kant's reply to Eberhard, Kant accepts that the faculty of the understanding is itself something constituted by an essence or nature, which is given and present "innately" in the soul, prior to any representation (and so, any "Satz") being "acquired" by its exercise. The "universal and necessary laws of thinking" which function as the real "principles" of the activity which issues forth from this nature or essence cannot therefore be identified with any propositional expression or any other representation. To be sure, there is reason to think that precisely the "basic propositions [Grundsätze]" that constitute the most elementary judgments in the science (textbook) of logic will be those which *represent* exactly these "universal and necessary laws of thinking." Nevertheless, the ("logical") representations of these laws, as constituents of the science of logic, should be distinguished from the ("real") laws of the faculty of the understanding.[18] What is at issue in the science of logic is never "merely" the clarification, ordering, etc. of some batch of representations, but always the "real" cognition of the system of (forms of) acts present in the faculty of the understanding itself. As Kant puts it at the outset of the Transcendental Logic's "Analytic of Concepts," what is most at issue in logic is not "analyzing the content of concepts that present themselves and bringing them to distinctness, but rather the ... *analysis of the faculty of the understanding* itself" (B90).

4.3.3.2 The "Logical" System of Concepts and Propositions about This Activity

With this last point, we can shift our focus from the "real" system in domain of the *activity* of the understanding ("*logica naturalis*," in one sense of the term) to the "logical" system of the *representations* of this activity in the science of logic itself (viz. "*logica artificialis*"). This system will be comprised, not by the activity itself, but by the concepts and propositions which are about this activity – i.e., the kinds of thoughts (judgments, arguments, demonstrations) which are about the understanding, and which are themselves what are expressed and ordered in the sentences, paragraphs, and sections written out in a textbook of logic. The understanding itself, its essence, and the "forms" of its activities are what is referred to by means of these thoughts and the correlative sentences which are contained in a logic textbook – and, importantly, what makes these thoughts (and sentences) true.

Here we can be much briefer, since we can simply reiterate the fact that the structure of Kant's own notes and lectures on logic, along with the "handbook" prepared by Jäsche, are themselves visibly organized entirely according to the very same system (ordered series) of acts of the understanding, oriented toward its "perfection," which we have met with above. After an introductory overview of philosophy itself as a system, and the place of logic within it, it is the systematic ordering in the subject-matter itself – i.e., in the activity of the understanding, ordered from conceiving, to judging, to inferring, to "method" – which is what drives the order of exposition in the textbook presentation of the science of logic qua system of cognition of the understanding. What is more, the various "principles" which are identified in the course of this exposition (viz. the principle of contradiction, of excluded middle, etc.) are themselves expressions or representations of those laws which "really" govern and determine the elementary cases of these activities more directly, i.e., without requiring the mediation of being represented by the understanding, or being referred to or appealed to with consciousness in order for the understanding to engage in the relevant activities. To be sure, the achievement of the kind of "self-cognition" of the understanding that is manifest in the science of logic itself does require the representation of these basic laws, the consciousness of these representations and the cognition of their objects by means of them, the ordering of these cognitions into a system of cognitions, and so on. But the formation of the science of logic is an "artificial" achievement, something over and above the system of the elementary activity of the understanding itself.

4.4 Some Difficulties Concerning the Priority Thesis

In the foregoing I have argued that, just as in the case of the science of *nature*, there is not only the general distinction between the objects which exist in nature and the cognitions (representations) of these objects, but also at least in principle a distinction between the system which obtains among the objects in nature and the system which obtains among the cognitions of the objects of nature, so too in the case of logic, as the science of the *understanding*, there is both a general distinction between the faculty of the understanding itself and the cognitions (representations) which are about the understanding or have it as their object, and then also the distinction between the system which obtains within the (forms of) acts of the understanding and the system which obtains among the cognitions whose object is this same activity and faculty. In conclusion I want to draw out and critically assess two consequences of the further commitment I have ascribed to Kant – namely, that the real system of forms of activity of the understanding has an explanatory priority over the system of cognitions that constitute the science of logic – and in this way raise and partially respond to some challenges to any "faculty-first" interpretation of Kant's views on the idea of a "system of logic."

First, it seems to be a consequence of the foregoing position that the "real" system of activities in the understanding can in some sense be *present* ("innately") even without any representation, or perhaps even any consciousness – let alone cognition or (in particular) systematic, scientific cognition – of these acts, of their order, of their real principles, let alone of the real system itself qua system.

In fact, at the beginning of several of the student transcripts of his lectures on logic, Kant seems to more or less straightforwardly endorse this position, insofar as he himself explicitly draws the parallel between the way in which laws in nature (e.g., the laws of the planets, gravity, etc.) determine the elements in their domain and the way in which the laws ("rules") of the understanding determine the elements in their domain. In the Vienna Logic, for example, Kant is recorded as saying: "Everything in the world happens according to rules[;] as we perceive this in the corporeal world, so do we find it even in the employment of our own powers, although we are not immediately conscious of the rules at all" (XXIV:790; cf. XXIV:693). Jäsche's edition of Kant's logic follows suit: "Everything in nature, both in the lifeless and in the living world, takes place according to rules, although we are not always acquainted with these rules … The exercise of our powers also takes place according to certain rules that we follow, unconscious of them at first" (IV:11). While this might not be a welcome consequence for those who want to ascribe to Kant a more cognitively demanding conception of what laws (principles, standards, norms) must be present to consciousness (in some sense or other) for thinking to take place at all (perhaps especially if it is to be an act of reason or of rational freedom),[19] it does seem neatly continuous with the general flow of the texts noted above.

To be sure, this now raises a second challenge of concretely and differentially characterizing the higher forms of thinking, such as inferring and systematically ordering or achieving science, as activities which are performed in accordance with the specific laws or principles which are disclosed in the science of logic as governing inference and method, but not (or at least not necessarily) performed in light of any involvement of the consciousness, cognition, or knowledge of these principles. Concerning inference, it is unclear how we might characterize what is distinctive about the activity of coming to form a syllogism or the activity of inference without making an appeal to some at least minimal consciousness of the logical forms which help to constitute the elements of the syllogism: e.g., consciousness that the predicate-term (concept) in the judgment (proposition) which serves as the major premise is the same as the subject-term (concept) in the judgment (proposition) which serves as the minor premise. Concerning method, or the achievement of science itself, it is even more unclear what it would mean for Kant to be open to the possibility that, with respect to some "body" of cognitions, all of the logical forms of the given "manifold" (of thinkable contents, their inter-relations) might be left entirely "implicit" (unconscious), and yet there would still be sufficient reason to count the cognitive relation to this "body" as rising to the level of scientific comprehension.[20]

On the other hand, in the Prolegomena, Kant also describes the very same forms of thinking which he characterizes in his debate with Eberhard as the innate ground of the acquisition of the pure concepts qua representations, as the forms which are themselves the subject-matter of "the work of the logicians" (IV:323).[21] Even if the concepts which the logicians fashion to represent and ultimately comprehend these forms have to wait to be "made" artificially (as it were), the logical forms themselves, as an original system of logic, "lie ready" in the human understanding (cf. B91).[22]

Notes

1 For the more general idea of a faculty-first interpretation of Kant's method, compare Schafer (2020); for its development in relation to the interpretation of Kant's method in the first *Critique* and the *Prolegomena*, see Tolley (2021).
2 This more generic concept of "system" is neglected, for example, by Garrett Bredeson in his entry on "System" for the *Cambridge Kant Lexicon*, which is written as if the only kind of system, for Kant, is a unity of cognitions (cf. Bredeson 2020). Daniel Dahlstrom gives more hints as to a broader sense of "system" at work in Kant, in the course of his contribution to the De Gruyter *Kant-Lexikon*, though he too does not present this broader sense directly (cf. Dahlstrom 2015).

3 Beyond the properties of being a whole whose parts are determined by (the idea of) the whole, according to a "principle," further material for the articulation of this more generic conception of "system" is given in the course of Kant's derivation of the three "concepts of reason" in the Dialectic. Here Kant describes the concepts of reason in general as expressing concepts of "totalities" of "relations" between "conditions" and "conditioned things" and then considers in particular the totality which is constituted by the relations ("synthesis") of "parts in a *system*" as mirroring the exclusive and exhaustive predication-relations that obtain across the propositions which form a "disjunctive" syllogism (B379). For a recent report of the corresponding dearth of discussions of "system" and "systematicity" in the abstract (among recent philosophers as well), see Hoyningen-Huene (2013).
4 Careful, sustained expressions of this distinction can be found, for example, in the influential attempts to provide a comprehensive "doctrine of science [Wissenschaftslehre]," such as the 1837 work of this name by Bolzano, and the later (Bolzano-inspired) *Logical Investigations* of Husserl. Compare Bolzano (1837) and Husserl (1900-1).
5 For more on the relation between science, system, and systematicity, compare Blomme (2015), de Jong and Betti (2010), Gava (2014), Hoyningen-Huene (2013), Sturm (2009), and Van den Berg 2021.
6 This second "system" is not always theorized as such, though it is common to use "the solar system" as an example; cf. Friedman (2001).
7 In this spirit, Rudolf Eisler, for example, draws the distinction between a "unified thoroughgoing ordering of a manifold of cognitions according to a principle, in the form of a whole of knowledge," on the one hand, and then "the real system of things," on the other, with the former (ideally) consisting in "the closest possible correlate" to the latter; cf. Eisler (1910, 1478).
8 For more on this distinction between the logical and real "uses," and also the cases in which they might come apart, compare Watkins (this volume).
9 For more on the distinction between mere "doctrine" and "science" proper, see Gava (2014), Brittan (1978), and Plaass (1965). Putting matters this way would also let us say that, in cases of mere "doctrines," though reason in its logical use has achieved some success in the formation of an at least somewhat or partially systematized body of cognitions, reason has not yet been able to achieve the requisite success via its "real" use, i.e., to demonstrate that the relevant systematic connections and unity that it has identified as obtaining among a body of cognitions has its correlate systematic unity "in" the domain of the objects of the cognitions themselves.
10 For this reason, the "principle" of the "real system" must be understood as a "real" (metaphysical) ground or cause, lying "in" the objects or things themselves, in their own properties, natures, essences, etc., and so not as a "basic proposition" or anything consisting merely of representations (concepts, judgments, etc.). More on this below.
11 Compare Anderson (2015) and Sutherland (2021). For important questions concerning the precise content of mathematical concepts, considered per se, in abstraction from the corresponding intuition and process of construction (i.e., as to the "real definition" of the mathematical object), see Heis (2014).
12 It is an important (and much-disputed) question whether, in a footnote at B160, Kant means to be claiming that activities of the understanding are in fact required for even the most elementary representations of these forms of intuition as objects (viz. as in an original "intuition" of them), or whether they are required only for the representation of these forms in cognition and especially

in scientific cognition (viz. as "in geometry"). (For the latter sort of interpretation, see Tolley 2016.) Either sort of interpretation, however, can be made compatible with the more basic distinction being appealed to here – namely, that the form of sensibility itself, and with it, the objects of geometry and arithmetic, which eventually come to be represented via whichever activities are required to represent them, that all of these *objects* are themselves still not identical with any of the resulting *representations* (whether these representations are "conceptual" or merely sensible).

13 For discussion of the distinction between (mere) "doctrine" and "science," compare again Plaass (1965).
14 For some concerns about the idea that, for Kant, logic has any "object" at all (due especially to its "formality"), see MacFarlane (2002). For discussion of the different sense in which 'transcendental' logic focuses on 'objects in general', see Tolley (2012).
15 Compare Lu-Adler (2018).
16 In the section quoted above, Meier suggests they are present as "indistinct [undeutliche] cognition" (Meier 1752, §533).
17 Famously the German Idealists, and especially Hegel, complained about Kant's failure to even pretend to offer an actual derivation, from a principle, not only of the forms of judgment he famously put on a "Table" in the first *Critique*, but also of the fourfold distinction he draws within thinking more generally, between conceiving, judging, inferring, and comprehending. For more recent attempts to reconstruct, on Kant's behalf, a kind of derivation from a principle of at least the fragment of the "system" which would comprise the forms of judgment, see Reich (1932), Krüger (1968), Brandt (1991), Wolff (1995), and Hoeppner (2021).
18 The possibility of maintaining this kind of distinction in general on Kant's behalf, with respect to principles (and laws), has not always been foregrounded in recent discussions of Kant on laws; for an attempt to give Kant a "univocal" conception of law which aims to stay neutral on whether laws are always (or always involve) representations or not, see Watkins (2019).
19 See, e.g., Land (2021).
20 Brandom (1994) remains perhaps the most sustained among recent explorations of the possibility of a characterization of cognitive activity which itself is in fact logical and even sophisticatedly so, albeit only "implicitly," as it has not yet made this fact about itself "explicit" to itself.
21 Though Kant sometimes might seem to claim that the pure concepts are "nothing but" the logical forms ("functions") themselves (cf. IV:324), we have seen that elsewhere he insists on the logical forms and functions of thinking as instead the "ground" of the pure concepts.
22 Thanks to Colin McLear, Karl Schafer, Thomas Sturm, and Eric Watkins for helpful comments on earlier drafts.

Bibliography

Anderson, RL. 2015. *The Poverty of Conceptual Truth*. Oxford: Oxford University Press.
Blomme, H. 2015. "La notion de 'système' chez Wolff, Lambert et Kant." *Estudos Kantianos* 3(1): 105–26.
Bolzano, B. 1837. *Wissenschaftslehre*. Sulzbach: J. E. v. Seidel.
Brandom, R. 1994. *Making It Explicit*. Cambridge: Harvard University Press.

Brandt, R. 1991. *Die Urteilstafel*. Hamburg: Meiner.
Bredeson, G. 2020. "System." In *Cambridge Kant Lexicon*, edited by J. Wuerth, 434–45. Cambridge: Cambridge University Press.
Brittan, G. 1978. *Kant's Theory of Science*. Princeton: Princeton University Press.
Dahlstrom, D. 2015. "System." In *Kant Lexicon*, edited by M. Willaschek, 2238–42. Berlin: De Gruyter.
de Jong, Willem, and Arianna Betti. 2010. "The Classical Model of Science: a Millennia-Old Model of Scientific Rationality." *Synthese* 174(2): 185–203.
Friedman, M. 2001. "Matter and Motion in the Metaphysical Foundations of Natural Science." In *Kant and the Sciences*, edited by E. Watkins, 53–69. Oxford: Oxford University Press.
Gava, G. 2014. "Kant's Definition of Science in the Architectonic of Pure Reason and the Essential Ends of Reason." *Kant-Studien* 105: 372–93.
Heis, J. 2014. "Kant (vs. Leibniz, Wolff, and Lambert) on Real Definitions in Geometry." *Canadian Journal of Philosophy* 44: 605–30.
Hoeppner, T. 2021. *Urteil und Anschauung*. Berlin: De Gruyter.
Hoyningen-Huene, P. 2013. *Systematicity*. Oxford: Oxford University Press.
Husserl, E. 1900-1. *Logische Untersuchungen*. Leipzig: Veit.
Krüger, L. 1968. "Wollte Kant die Vollständigkeit seiner Urteilstafel beweisen?" *Kant-Studien* 59: 333–56.
Land, T. 2021. "Epistemic Agency and the Self-Knowledge of Reason." *Synthese* 198(13): 3137–54.
Lu-Adler, H. 2018. *Kant and the Science of Logic*. Oxford: Oxford University Press.
MacFarlane, J. 2002. "Frege, Kant, and the Logic in Logicism." *Philosophical Review* 111(1): 25–65.
Meier, G. 1752. *auszug aus der Vernunftlehre*. Halle.
Plaass, P. 1965/1994. *Kant's Theory of Natural Science*. Dordrecht: Springer.
Reich, K. 1932. *Die Vollständigkeit der kantischen Urteilstafel*. Rostock.
Schafer, K. 2020. "Transcendental Philosophy as Capacities-First Philosophy." *Philosophy and Phenomenological Research* 103(3): 661–86.
Sutherland, D. 2021. *Kant's Mathematical World*. Cambridge: Cambridge University Press.
Sturm, T. 2009. *Kant und die Wissenschaft vom Menschen*. Paderborn: Mentis.
Tolley, C. 2012. "The Generality of Kant's Transcendental Logic." *Journal of the History of Philosophy* 50(3): 417–46.
Tolley, C. 2016. "The Difference between Original, Metaphysical, and Geometrical Representations of Space." In *Kantian Nonconceptualism*, edited by D. Schulting, 257–85. London: Palgrave.
Tolley, C. 2021. "From 'facts' of Rational Cognition to Their Condition: Metaphysics and the 'analytic' Method." In *Kant's Prolegomena: A Critical Guide*, edited by P. Thielke, 48–70. Cambridge: Cambridge University Press.
Van den Berg, Hein. 2021. "Kant's Ideal of Systematicity in Historical Context." *Kantian Review* 26(2): 261–86.
Watkins, E. 2019. *Kant on Laws*. Cambridge: Cambridge University Press.
Wolff, M. 1995. *Die Vollständigkeit der kantischen Urteilstafel*. Frankfurt: Klostermann.

5 Mathematics: Systematic Unity and Construction in the Theory of Conic Sections

Katherine Dunlop

5.1 Introduction

In Kant's theory of knowledge, mathematics features as an important illustration of how much our rational faculty can achieve. Interpreters have been most concerned with how the example of mathematics proves the possibility of synthetic *a priori* judgment and illustrates its ampliative character, as Kant contends in the second-edition Introduction to the *Critique of Pure Reason* and the chapter "Discipline of Pure Reason." Indeed, in these contexts Kant's overriding purpose is to set boundaries on the rational faculty's legitimate use, so it is important for him to determine whether the conditions under which such cognition is possible are met when reason is used "discursively," merely "from concepts." (His conclusion is that mathematical cognition is made possible by the construction of concepts in pure intuition; without this resource, concepts can be connected in synthetic *a priori* judgment only by taking account of the "synthesis of possible intuitions," in possible experience.[1]) Judgment is, however, a function of the understanding, and I will argue in Section 5.2 that Kant places still more importance on the way mathematics exemplifies reason's distinctive powers.

Kant distinguishes between the understanding as "a faculty of unity of appearances by means of rules" and reason as "the faculty of the unity of the rules of understanding under principles" (A302/B359). On his account, the unity sought by reason is "higher" or more encompassing than that sought by understanding (see, e.g., A298/B355; A310–311/B367). In a penetrating study of Kant's view of "the cognitive resources available to us for doing mathematics," Alison Laywine brings out how Kant uses one particular mathematical theory – the geometrical theory of conic sections – to illustrate the surpassing unity that reason seeks. This chapter's Sections 5.2–5.4 aim to build on Laywine's treatment by explaining in detail how this theory serves that illustrative role. In Section 5.2, I explain in general terms how particular examples of systematically unified cognition fulfill the need for a "schematization" of reason's principles of unity.

Mathematics: Systematic Unity, Construction, and Conic Sections 115

Kant indicates that mathematics serves specifically to "guide reason's insight into nature" (A464/B492); to show how the theory of conic sections plays this role, I consider its important applications in Newton's theory of motion (in Section 5.3) and in astronomy (in Section 5.4). I argue that these applications are made possible by the unity that the theory manifests, as a piece of pure mathematics.

Section 5.5 takes up a tension that Laywine finds in Kant's account, between his overall view that mathematics is distinguished by its constructive method, and his use of the theory of conic sections to illustrate the systematic unity sought by reason. Laywine argues that because Kant understands the theory of conic sections as fundamentally constructive, the theory as he understands it lacks the unity and power of algebraic treatments; his example thus seems inapt, and his view seems to leave no room for "the distinctive and remarkable ways that reason ... was achieving higher degrees of systematic unity in the theory of conic sections" (2014, 746). I show (in Section 5.5) that Laywine's point holds for the constructive approach of Apollonius as compared with algebraic treatment. The projective treatment of conic sections represents an alternative and (in its "synthetic" version) more purely geometrical approach, which powerfully unifies the theory of conic sections, as I explain in Section 5.6 But I think Laywine is correct to argue that Kant's writings do not show familiarity with projective techniques, so we cannot take this to be the constructive approach he has in mind.

Despite Kant's apparent disregard of projective geometry, I think we can turn back the charge that Kant's adherence to a constructive understanding of mathematics puts him out of step with mathematical progress. For the passage on which Laywine's interpretation is based, which discusses Apollonius, also alludes to Newton's mathematical work; and while the discussion of Apollonius can be explained by the dialectical context, Kant takes the initiative in bringing Newton to the fore. After I lay out the textual evidence for Newton's relevance (in Section 5.7), I explain (in Section 5.8) how Newton's treatment of the conic sections in *Principia Mathematica* displays the power and generality of a construction-based approach. Newton's treatment relies on a particular method of construction (which he calls the "organic description" of conic sections), which, as I show in Section 5.9, can be taken as a hypothetical or fictive manner of origin. I argue that this way of taking it makes Newton's treatment especially suitable to illustrate the role of ideas in schematizing reason's principle of systematic unity.

5.2 Mathematics' Role in "Guiding Reason's Insight"

Mathematics' exemplification of the systematic unity sought by reason appears even more important to Kant than the proof it supplies of the possibility of synthetic *a priori* judgment in general. This higher importance becomes apparent in a remark on the four "natural and unavoidable problems of reason" posed in the Antinomies. Kant first asserts that "the mathematician would gladly give up his entire science for" the solution to these questions, since mathematics gives no "satisfaction in regard to the highest and most important ends of humanity." He then indicates what worth mathematics has:

> Even the proper dignity [*Würde*] of mathematics (that pride of human reason) rests on this, that it guides reason's insight into nature in the great as well as the small in its order and regularity, and in the admirable unity of the forces moving nature, far beyond every expectation of any philosophy built on common experience, and thereby gives occasion and encouragement to the use of reason which extends beyond all experience, just as it provides to the philosophy concerned with it [*die damit beschäftigte Weltweisheit*] the most excellent materials for supporting its inquiries, as far as their character allows, with appropriate intuitions.
> (A464/B492, translation modified[2])

Kant's characterization of mathematics' "proper dignity" as its role in "guiding reason's insight into nature" is striking. For the investigation of nature is presumably beyond the scope of mathematics itself, given that Kant closely associates the notions of nature and of existence, and insists that mathematics does not pertain to either.[3] We would thus expect that, on Kant's view, mathematics most clearly fulfills the function of "guiding reason's insight" where it contributes most directly to natural science.

A further conclusion we can draw from Kant's specification of the direction in which mathematics guides reason's insight – namely, toward nature – is that the experience-transcending "use of reason" which it prompts is a theoretical rather than practical use.[4] We thus know that this use is narrowly circumscribed, insofar as it issues in genuine cognition. Kant makes clear that in its theoretical use, reason "never relates directly" to anything outside the understanding, but rather "has as its object only the understanding and its purposive application" (A643-4/B671-2). Thus the use in question cannot be to cognize special objects lying beyond all possible experience, but rather to facilitate the "empirical use" of the understanding (A680/B708, quoted below). Our question is then, how exactly does mathematics function to support the inquiries of natural science and stimulate this use of reason?

The answer seems to lie in in Kant's claim, in the Introduction to the first *Critique*, that mathematics "gives us a splendid [*glänzend*] example of how far we can go with *a priori* cognition independently of experience" (A4/B8). When Kant reiterates this point, at the start of the first

Mathematics: Systematic Unity, Construction, and Conic Sections 117

section of the "Discipline of Pure Reason" chapter, he indicates that "we" refers to the use of our reason: mathematics "gives the most resplendent [*glänzendst*] example of pure reason happily expanding itself without assistance from experience" (A712/B740).

One function that an example can serve is that of increasing confidence or emboldening. Kant indeed speaks of mathematics as giving "encouragement" [*Aufmunterung*] to the use of reason beyond all experience (A464/B492) and as providing "proof of the power of reason" (A5/B8). But reason does not seem to need such encouragement, since in general it does not shrink from extending itself beyond experience. Kant identifies the "wholly isolated speculative cognition of reason that elevates itself entirely above all instruction from experience" with metaphysics (Bxiv), and he makes clear that reason is inexorably driven to pursue metaphysical inquiry (B22), despite the dismal progress of metaphysics to date (Aviii-x). But since our concern is not this general tendency of reason, but rather its use to facilitate empirical cognition, it may be easier to see how there can be room for encouragement.

I think it is easier to see how the "splendid example" of mathematics can foster the use of reason beyond experience if we consider another way that examples can function, namely providing a template to follow. However, this point is a delicate one. Where Kant describes mathematics in these terms, his concern is to explain why the "transcendental" use of reason (in philosophy) cannot follow mathematics' way of proceeding. In the Introduction, he notes straightaway that mathematics "is occupied with objects and cognitions only so far as these can be exhibited in intuition" (A4/B8), and suggests that without such a constraint, philosophy can "make no headway" (A5/B9). In the "Discipline" chapter, Kant explains in detail that, while the expectation that philosophy can follow mathematical method is "entirely natural" (in light of the "great good fortune that reason enjoys by means of mathematics"; A724/B752), nonetheless it cannot be fulfilled. For mathematics "brings all of its concepts to intuitions that it can give *a priori* and by means of which, so to speak, it becomes master over nature," but intuitions cannot be given *a priori* for the philosopher's "pure and even transcendental concepts" (A725/B753; cf. A720–721/B748–749). Although these passages do not specifically concern mathematics' role in the study of nature, there too mathematics proceeds in this distinctive way; it provides "appropriate intuitions" for natural science (A464/B492).

Although reason cannot follow mathematics' manner of proceeding, mathematics can still illustrate or give specific form to the systematic unity that reason demands. Though it cannot just provide "appropriate intuitions" for philosophy (as it does for natural science; A464/B492), it can supply particular examples of cognition that meet reason's standards, specifically in regard to the interrelationships between concepts. I will close this section by explaining, in general terms, how such examples remedy a difficulty in applying reason's principle, which arises from its indeterminacy.

When Kant sets out to "place the entire result of the Transcendental Dialectic clearly before our eyes," he begins by reminding us that pure reason "is in fact concerned with nothing but itself," rather than with objects. Thus its unity, which is that of a system,[5] "does not serve reason objectively as a principle, extending it over objects, but subjectively as a maxim, in order to extend it over all possible empirical cognition of objects" (A680/B708). So that he should not be understood to deny all objective application[6] to reason's principle of systematic unity, Kant then reiterates the point (made at A663/B691) that the principle is "objective but in an indeterminate way" (A680/B708). His remark that that the principle functions as "a *merely regulative* principle and maxim for furthering and strengthening the empirical use of reason" (A680/B708, emphasis added) offers a way to understand its objectivity, namely by contrast with "constitutive" principles, which "determine something in regard to [their] direct object" (A680/B708; cf. A671/B699).

Part of what Kant means in describing the principle's objectivity as "indeterminate" is that it does not express the requirement of systematic unity with any specificity or detail. As Rachel Zuckert observes, one reason that the principle is a "*mere* guide to investigation" is that it does not "articulate any characteristics an object might have" (2017, 94). Hence, the principle lacks a schema (which we may understand as a rule specifying the feature(s) in virtue of which an object falls under it). Kant brings to the fore this facet of the constitutive/regulative contrast by recalling that the understanding's dynamical principles are "constitutive in regard to *experience*" (A664/B692) despite being merely regulative in regard to intuition. This means at least that the corresponding ("dynamical"; cf. B110) categories apply to appearances by means of "rule[s] in accordance with which unity of experience is to arise from perceptions" (A180/B222) (although they are not directly instantiated by individual appearances, as are the "mathematical" categories; see Section 5.4). The principles of pure reason, "on the contrary, cannot be constitutive even in regard to empirical *concepts*, because for them no corresponding schema of sensibility can be given, and therefore they can have no object *in concreto*" (A664/B692). Kant then directly addresses the question of how such principles can nevertheless have "a regulative use, and with this some objective validity," arguing that their application requires something analogous to the schemata of sensibility.

> The understanding constitutes an object for reason, just as sensibility does for the understanding. To make systematic the unity of all possible empirical actions of the understanding is a business of reason, just

Mathematics: Systematic Unity, Construction, and Conic Sections 119

as the understanding connects the manifold of appearances through concepts ... The actions of the understanding, however, apart from the schemata of sensibility, are *undetermined*; likewise the *unity of reason* is also in itself *undetermined* in regard to the conditions under which, and the degree to which, the understanding should combine its concepts systematically.

(A664–665/B692–693)

Since Kant characterizes the schemata of pure concepts of the understanding as "*a priori* formal conditions of sensibility that contain the general condition under which alone the category can be applied to any object" (A140/B179), the analogy between the understanding and reason suggests that the latter's activity of unifying requires specification of the "conditions of understanding," as it were, under which its principle applies. What is needed here seems to be an indication of *what*, exactly, to look for: *which* relationships among the understanding's cognitions (concepts, judgments, and principles) satisfy reason's demand. Kant then contends that this need is met by "the idea of reason," insofar as it expresses "the *maximum* of division and unification of the understanding's cognition in one principle" (A665/B693).

We will return to the view that an idea of reason fulfills the role of a schema in Section 5.9. For now, let us stay with Kant's account of how the fulfillment of this need gives reason's principle "some objective validity":

Thus the idea of reason is an analogue of a schema of sensibility, but with this difference, that the application of concepts of the understanding to the schema of reason is not likewise a cognition of the object itself (as in the application of the categories to their sensible schemata), but only a rule or principle of the systematic unity of all use of the understanding. Now since every principle that establishes for the understanding a thoroughgoing unity of its use *a priori* is also valid, albeit only indirectly, for the object of experience, the principles of pure reason will also have objective reality in regard to this object, yet not so as to *determine* something in it, but only to indicate the procedure in accordance with which the empirical and determinate use of the understanding can be brought into thoroughgoing agreement with itself, by bringing it *as far as possible* into connection with the principle of thoroughgoing unity ...

(A665/B693)

So, just as the schema of a concept is "a general procedure of the imagination for providing a concept with its image" (A140/B179–180; here the concept in question is a mathematical or "pure sensible" one), the schema of a principle of reason is, or at any rate yields, a procedure for bringing unity to the understanding's cognition. Although we still lack a full explanation of how this effort of unification makes reason's principles objectively valid, we can at least see that they have application to objects of experience insofar as the concepts, judgments, and principles that they unify have such application. My aim in Sections 5.3 and 5.4 is to show how the theory of conic sections exhibits systematic interconnection among concepts, of the kind reason seeks; thus the relation of these concepts to objects (through construction in pure intuition[7]) mediately relates reason's systematic unity to objects.

5.3 The Theory of Conic Sections in Relation to Laws of Motion

We may now begin to consider how the theory of conic sections, in particular, fulfills the role we have characterized: an example of pure reason "happily expanding itself" (A712/B740), which we need in order to conceive of systematic unity in any specificity or detail. We have seen that Kant thinks mathematics most clearly fulfills the function of "guiding reason's insight" where it contributes most directly to natural science. Indeed, the theory of conic sections has an important role in demonstrating the "admirable unity" of moving forces in nature (cf. A464/B492). Laywine observes that, of four discussions of conic sections in Kant's published works, two explicitly concern laws of motion (2014, 736). We will consider these passages here, and a third passage (not specifically discussed by Laywine) which deals with astronomy in Section 5.4.

In the *Only Possible Proof of the Existence of God*, Kant is mainly concerned to show how the circle manifests "order and ... complete unity in the manifold" (II:94), of a kind that is unexpectedly complex. He chooses two examples to illustrate this point. The first is a pair of results from Euclid's *Elements*. Proposition 35 of Book III states that for any two chords that intersect within a circle, the rectangle on the segments (created by the intersection point) of the first chord – that is, the area that results from carrying out the geometrical operation of multiplication on these two segments – equals the rectangle of the segments thus created on the second chord. The following Proposition 36 states that for a segment drawn from a point outside a circle to the circle's opposite side, the "rectangle contained by the whole of [this segment] and the straight line intercepted on it outside between the point and the convex circumference" (Heath 1926/2014, vol. 2, 74) equals the square on the tangent drawn from this point. Kant's second example is that free fall through any chord which meets the circle's vertical diameter will take the same time, "regardless of whether [the chord extends] from the point at the top or at the bottom"

Mathematics: Systematic Unity, Construction, and Conic Sections 121

(II:94). Galileo proves this result as Proposition VI and its first Corollary in the "Third Day" of the *Two New Sciences*, making use of *Elements* III.35.[8]

The circle is a limiting case of a conic section. So by themselves, these results do not show that the theory of conic sections is suitable to illustrate the systematic unity found in mathematics. But the circle interests Kant precisely as one of the "various species of curved line" whose "affinities" are revealed by "higher geometry." In this context, his point is that these affinities comprise some of the "immeasurably" many "harmonious relations which inhere in the properties of space in general," whose existence can thus be inferred from the "infinitely many" surprising properties which hold of "the figure of the circle alone" (II:95). In *Prolegomena to Any Future Metaphysics*, Kant describes these affinities more precisely.

> If we ... pursue further the unity of manifold properties of geometrical figures under common laws, and consider the circle as a conic section, which thus stands with other conic sections under the same fundamental conditions of construction, we find that all chords which intersect inside the latter, inside the ellipse, the parabola, and the hyperbola, always do so in such a way that the rectangles constructed with their segments, though not equal, always stand in equal relations to each other.
>
> (IV:321)

This common property of the conic sections can be stated a bit more formally: for any two chords AB and FG of a conic section that are parallel to one another, and any other chord DE which intersects them (at points called C and H, respectively), the rectangle on AB created by their intersection (AC × CB) stands to the rectangle created on DE (DC × CE) as the rectangle on FG created by their intersection (FH × HG) stands to the rectangle created on DE (DH × HE).

I take this property as the clearest and most important example of unity among diverse cases within the theory of conic sections. The importance of this instance of unity, as an illustration of the systematic unity sought by reason, is suggested by its relevance for natural science, to which Kant alludes in the following sentence: "If we go still further, to the fundamental doctrines of physical astronomy, we find a physical law of reciprocal attraction extending over the whole material nature," namely the law that attraction is proportional to the inverse square of the distance between attracting points (IV:321). Of course, Newton is credited with establishing the inverse-square law of universal gravitation, in *Principia*. So Kant's contention that we reach the law by proceeding "still further" from the theory of conic sections invites us to consider how Newton uses this theory to prove the law.

One important use that Newton makes of the theory is showing that a body will describe a trajectory that is a conic section, with the center of force located at one focus, if and only if the body is acted on by an inverse-square centripetal force (Propositions 11 through 13 and Corollary 1, Prop. 13, Book I, *Principia*). Michael Friedman has noted that conic-section orbits and the inverse-square law are both linked, further, to Kepler's area rule.[9] These linkages enable Newton to infer the forces acting on satellites from features of their orbits and, thus, to settle the question of the "world-systems" in Book III of Principia.[10] Friedman also makes the important observation that Kant seems to have "something even more specific in mind" when he speaks of going "still further" to the inverse-square law: that "Newton appeals explicitly to a special case of the particular property of the conics to which §38 refers," viz. the proportionality of rectangles on intersecting chords, when he derives the inverse-square law (in Propositions 11 through 13 of Book I; Friedman 1992, 192).[11]

I think Friedman is exactly right to focus on Newton's use of this property, which, as I will explain, goes well beyond the derivation of the inverse-square law; we will see in Section 5.8 that Newton uses it to derive impressive results in pure geometry.

5.4 The Unity of the Theory of Conic Sections, and Its Application to Orbits

I have thus far argued that the theory of conic sections has the kind of application, within natural science, that serves to illustrate mathematics' role in "guiding reason's insight." What gives this theory its strongest claim to illustrate systematic unity, however, is how it itself manifests unity among diverse cases falling under a common principle. Indeed, as we shall see, on Kant's telling it is precisely this feature of the theory that makes possible its application to astronomy.

To see how the theory of conic sections manifests unity among diverse cases, we may begin with Kant's own example of the "unity of manifold properties of geometrical figures," the proportionality of the rectangles on intersecting chords. Kant's claim that the rectangles, "though not always equal," still "always stand in equal relations to one another" (IV:321) brings out how this property generalizes the property of circles proved in *Elements* III.35. In the case of the circle, for two parallel chords AB and FG that intersect a third chord DE, the rectangle created on AB by the intersection (AC × CB) equals the rectangle created on DE (DC × CE), and the rectangle created on FG by the intersection (FH × HG) equals that created on DE (DH × HE). But for conic sections in general, we have the property that the ratio (AC × CB):(DC × CE) equals the ratio (FH × HG):(DH × HE), which includes the special case in which each of these ratios is equality (that is, (AC × CB) = (DC × CE) and (FH × HG) = (DH × HE)). There is a further difference between this general property of conic sections and the property of circles proved in *Elements* III.35 that bears mentioning. In the case of the circle, as Laywine explains, "we do not have to define the directions of the divided lines, on whose segments the rectangles of interest are formed"; they may intersect at any angle. In the general case, however, DE and the pair of parallel chords AB and FG must each be parallel to one of a pair of intersecting segments (Laywine 2014, 740).

Kant was by no means the only eighteenth-century thinker who took this property to afford a general understanding of conic sections, with respect to both the properties of each curve and the curves' relations to one another. Like Kant, if the argument of Sections 5.7-8 is correct, these other thinkers were stimulated to understand conic sections in these terms by Newton's work. As the historians of mathematics Del Centina and Fiocca write in a recent paper, the "geometrical achievements" of Section V, Book I of *Principia* drew geometers' attention to the property, and led them to recognize its "foundational value" (2021, 7). Elsewhere, Newton extends the property to cubic curves (those whose equations involve terms raised to the third power); stimulated by this work in particular, the mathematician James Stirling published a treatment in which, he claimed, "through many corollaries of this proposition [asserting the property in question] flows the whole of conic sections" (1717, 77, quoted by Del Centina and Fiocca 2021, 8). In his 1707 *Traité analytique des sections coniques*, the Marquis de L'Hospital departs from his "analytical" (algebraic) approach to prove the property "synthetically" and "in the solid," that is, by geometrical reasoning on three-dimensional shapes. L'Hospital subsequently claims his "design" is "to show what can be the usefulness of passing into space in order to prove, everything at once and without any calculation, the properties on which depend all the others" (quoted and translated in Del Centina and Fiocca 2021, 25–26).

Leaving aside, for now, the unity that the theory of conic sections manifests as a piece of pure mathematics, let us return to the theory's role in guiding insight into nature. Kant's account of how the circle manifests "order" and "complete unity in the manifold" (II:94) concludes by emphasizing the "unity alongside the highest complexity" found in the "necessary properties" studied by geometers (II:95). In his discussion of the proportionality of rectangles on intersecting chords, Kant likewise stresses the "unity of manifold properties" (IV:321).[12] In the *Critique of Pure Reason*, Kant explicitly claims that "the affinity of the manifold, without detriment to its variety, under a principle of unity, concerns not merely the things, but even more the mere properties and powers of things" (A662/B690). His immediate purpose is to explain the order of application of the three principles of systematic unity that he has distinguished, namely manifoldness, affinity, and unity,[13] and he takes conic-section orbits to illustrate the principle of affinity, which "offers a continuous transition from every species to every other through a graduated increase of varieties" (A657–8/B685–6). In this case, the "things" whose properties the principle especially concerns are celestial bodies.

> Hence if, *e.g.*, the course of the planets is given to us as circular through a (still not fully corrected) experience, and we find variations, then we suppose these variations to consist in an orbit that can deviate from the circle through each of an infinity of intermediate degrees according to constant laws; *i.e.*, we suppose that the movements of the planets that are not a circle will more or less approximate to [the circle's] properties, and then we come upon the ellipse. The comets show an even greater variety in their paths, ... yet we guess at a parabolic course for them, since it is still akin to the ellipse and, if the major axis of the latter is very long, it cannot be distinguished from it in all our observations.

Kant then indicates what "property," and what "power," are shown to have unity in this example: "Thus under the guidance of those principles we come to a unity of genera *in the forms of these paths*, but thereby also further to unity in *the cause of all the laws of this motion* (gravitation)" (A662/B690, emphasis added).

What Kant calls the "unity of genera" in the curves – specifically, the idea that one can "deviate from" another, as the ellipse deviates from the circle, "through each of an infinity of intermediate degrees according to constant laws" – is a different kind of unification that the common characterization of the curves in terms of the intersecting-chord property, which seems more suited to exemplify the first principle (by giving content to its assertion of "sameness of kind" among diverse species). So this appeal to the theory of conic sections brings out another way in which it is apt to illustrate the systematic unity sought by reason, insofar as the principles each specify an importantly different aspect of this unity. The possibility of systematically transforming the curves into one another has powerful applications in pure mathematics, as we will see;[14] and it played a role in the theory of motion since its modern beginnings. Thus, in a passage immediately preceding his proof that a body moving in an ellipse is subject to an inverse-square centripetal force (Proposition 11, Book I, *Principia*), Newton attributes to Galileo the theorem that if "the center of the ellipse goes off to infinity, so that the ellipse turns into a parabola, the body will move in this parabola, and the force, now tending toward an infinitely distant center, will prove to be uniform" (Newton 1999, 460). Newton proceeds to discuss the transformation of the parabola into the hyperbola

Mathematics: Systematic Unity, Construction, and Conic Sections 125

and to explain how a known property of the circle and ellipse generalizes to "all figures universally" (1999, 461).

We have now seen two ways in which the theory of conic sections manifests the unity of diverse cases falling under a common principle: in the common property of proportionality of the areas created by intersecting chords and in the existence of systematic transformations between the curves. Taking this theory as example, we can better understand how mathematics provides for the schematization of reason's principle of systematic unity. At the end of Section 5.2, we saw that because the principle of systematic unity does not "determine something in the object," its application requires a specification of the "conditions" under which the understanding should "combine its concepts systematically" (A664–665/B692–693). This need is met by indicating what kind of unity to look for in these combinations (just as the analogous need for the understanding's pure concepts is met by indicating how the objects falling under them are determined in time[15]). Since the unity sought by reason pertains, in the first instance, not to objects (of experience or intuition) but to concepts of the understanding, reason needs an example of unity that "concerns not merely the things, but even more the mere properties and powers of things," as Kant says (A662/B690).

The move from the "unity of genera in the forms of" conic sections to the "unity in the cause of the laws of all these motions" (A662/B690) illustrates how a pattern of unification displayed by mathematical objects can also be applied to concepts, even concepts that cannot be exhibited in intuition as directly as mathematical concepts. Here the dual status of the categories of relation, and their associated principles, becomes relevant. In contrast to reason's ideas and principles, they are constitutive in regard to experience; but, in contrast to the "mathematical" categories and principles, they are merely regulative in regard to "the way in which [objects are] apprehended in appearance." (For only the latter concepts and principles "determine *a priori*" that a certain rule of synthesis "at the same time yields this intuition *a priori* in every empirical example" (A178/B221), so that they can be understood as rules for a kind of construction within the apprehension of appearances.[16]) In Kant's example, we bring unity to the concepts of celestial bodies and their properties by seeking, first, a common power (universal gravitation) to posit as the cause of conic-section orbits, and, second, force laws that describe the systematic relationships between the curves in terms of physical factors. The relationships to which we are thereby led are among causal factors, and the concept of causality is of course a "dynamical" category and thus cannot be constructed (in this sense). So, in this example, the way that mathematics guides reason's insight into nature – by providing an example of systematically interrelated objects and properties – also provides for a manifestation of systematic unity that is distinctively conceptual, in the sense that it cannot be constructed in intuition.

5.5 The Theory of Conic Sections in Apollonius's *Conica* and Descartes's *Géométrie*

Laywine finds "a problem associated with Kant's account of conic sections" (2014, 719), namely that "it is just not obvious how" geometrical constructions "yield overarching, systematic unity." Kant's "deep commitment to geometrical constructions" within the theory (729–730) thus appears in tension with his use of the theory to illustrate the systematic unity sought by reason.

Laywine first raises this problem by contrasting Apollonius's treatment of conic sections, which exemplifies the unifying power of the "classical" (2014, 724) approach based on geometrical construction, with the algebraic treatment originating in the work of Descartes. What shows Apollonius's *Conica* and Cartesian algebraic geometry to be the appropriate historical points of reference, according to Laywine, is her reading of Kant's "On a Discovery," composed in reply to the Wolffian J.A. Eberhard. For, on the one hand, Kant explicitly adverts to the algebraic equation for the parabola ($ax = y^2$; VIII:192), but on the other hand, Kant claims that the algebraic treatment involves awareness of "the pure, merely schematic construction" of the figure (VIII:192), which Laywine takes to refer to Apollonius's construction. (We will consider Laywine's argument for this interpretation in Section 5.7.) Laywine further contrasts the Apollonian treatment that she takes to be Kant's point of reference with the development of projective geometry, as we will see in Section 5.6. After setting out the terms of this threefold contrast, I will argue (in Sections 5.7–5.8) that Newton also belongs on the side of the classical approach, and the generality and power of Newton's treatment make evident that geometrical construction can yield systematic unity, *pace* Laywine.

What makes Apollonius's treatment suitable to illustrate the unifying power of a construction-based approach is how it differs from earlier Greek work. As Laywine explains, in earlier treatises (by Euclid and Aristaeus) "conic sections were always generated as sections of a right cone" – that is, one generated by rotating a right triangle around one of the sides containing the right angle – by "cutting" the cone along a plane that meets the hypotenuse of the axial triangle at right angles. Hence, "conic sections had to be classified in terms of the vertical angle of the axial triangle." Indeed, before Apollonius the curves now called "parabola," "ellipse," and "hyperbola" were referred to as "section of a right-angled cone," "section of an acute-angled cone," and "section of an obtuse-angled cone," respectively (2014, 730). Apollonius's construction involves fixing a point that lies on a different plane than the circle and "carrying" a straight line from this point "around the circumference of the circle until it returns to the place from whence it started" (Laywine 2014, 726, quoting *Conica*, Book I), so that the line joining the fixed point to the circle's center (the axis) is "a sort of tent pole" (Laywine 2014). Unlike the axis of a Euclidean or Aristaean right cone, the "tent pole" can list, so this procedure can generate the surface of an oblique cone. In that case, the figure thus generated will lack the radial symmetry of a right cone.[17] Furthermore, Apollonius does not require the plane along which the conical surface is cut to be perpendicular to a side of the right triangle: the parabola, ellipse, and hyperbola are generated by cutting the surface along a plane parallel to one side of the axial triangle, intersecting one side of the axial triangle, or intersecting two sides of the axial triangle, respectively (as Laywine explains; 2014, 730–1).

Laywine uses the example of the parabola to illustrate how Apollonius's approach achieves greater generality. Apollonius characterizes each section in terms of its "symptom," a geometrically expressed property satisfied by each point on the curve. This is, specifically, a relationship between segments, called "ordinates" and "abscissas," that the point determines in relation to a "diameter" of the curve (conic section). Apollonius defines the diameter as "that straight line ... which, drawn from the curved line, bisects all straight lines drawn to this curved line parallel to some straight line."[18] The ordinate is the segment which is drawn, parallel to a certain line determined by the section's construction,[19] from the given point on the curve to its diameter. The abscissa is the segment on the diameter bounded at one end by the intersection with the ordinate, and at the other by the intersection with the curve. The characteristic property (symptom) of the parabola is that the square on the ordinate is equal to the rectangle on the abscissa and another segment, called the "latus rectum" or "parameter." When the parabola is generated from a right cone (as by Apollonius's predecessors), the radial symmetry of the cone ensures that the latus rectum equals twice the line from the vertex angle of the cone to the intersection of the diameter with the conic section.[20] But this does not hold for parabolas in general. By characterizing the latus rectum as having "the same ratio to the straight line between the [vertex] angle of the cone and the [intersection of the diameter with the section] as the square on the base of the axial triangle has to the rectangle contained by the remaining two sides of the triangle,"[21] Apollonius formulates the symptom in a manner valid for every parabola, whether generated from a right or an oblique cone.

Still more importantly, Apollonius's approach extends to the hyperbola and the ellipse, as Laywine explains. As the characteristic property (or symptom) of the parabola is a relationship to the rectangle on the latus rectum and the abscissa, so the characteristic property of the hyperbola is a relationship to an area that is the *sum* of the rectangle on the latus rectum whose width is equal to the abscissa and a rectangle similar to a given rectangle, and the characteristic property of the ellipse is a relationship to an area that is the *difference* between this rectangle on the latus rectum and a rectangle similar to a given rectangle. These relationships are expressed in his new names for the curves, the ones we still use.[22] Apollonius is thus "able to characterize the three conic sections in general terms, but also in such a way as to show, in general terms, how they relate to one another – thereby achieving still greater systematic unity" (Laywine 2014, 731).

Apollonius's characterization of the conic sections enables him to achieve systematic unity in a further respect: it conduces to establishing analogues, for the conic sections, of the circle's properties. Indeed, a standard textbook on history of mathematics suggests that "among [Apollonius's] motivations was the desire to generalize various theorems on circles, including those proved in *Elements*, Book III" (Katz 2009/2018, 120). Laywine focuses on how, "in a cluster of thematically unified propositions" (Propositions 16–29; Laywine 2014, 743) of Book III, Apollonius develops the common property of the conic sections stated by Kant in *Prolegomena* §38 (that the rectangles created on parallel chords by their intersection with a common chord stand in the same ratio to the rectangle created on the intersecting chord, subject to the condition on the chords' directions).

Laywine adds, further, that "the systematic unity of these propositions can also be seen in the work they do at the very end" of *Conica*'s Book III, which gives solutions to the converse of the "Pappus problem," discussed below. But because Laywine is setting up a contrast between Apollonius and Cartesian algebraic treatments, her aim is to show how this body of propositions allows us to "taste … the 'systematic unity' that Kant says is the ideal of mathematics and indeed all other sciences." She stresses that they give *only* a taste, "because they may be understood to whet our intellectual appetite, without properly satisfying it" (2014, 742). Different propositions of the group III.16–29 treat different cases: for instance, Propositions 18 and 19 extend the results of Propositions 16 and 18 to the double-branched hyperbola, and Propositions 23 through 26 extend earlier results to conjugate hyperbolas. Hence, to establish the general property of conic sections that Kant mentions, "we have to trudge through 13 propositions and consider separately cases for which Kant would surely predict we would desire a single, overarching treatment – and this comes after a trudge, more dogged still, through the first 15 propositions of Book Three that … serve as lemmas for Propositions 16–29." Laywine concludes that while the example of Apollonius (which she takes Kant to have in mind in the reply to Eberhard) "gives us such a distinctive taste of what we are looking for" when we demand systematic unity, "it would also just

Mathematics: Systematic Unity, Construction, and Conic Sections 129

as naturally have produced a sense of let-down in mathematically minded readers" (2014, 743). Laywine gives Book Four of *Conica* as a further example in which Apollonius separately treats many cases – in this case, intersections of the conic sections with one another and with circles – that Cartesian algebra handles uniformly (specifically by solving equations to find the coordinates of intersection points).

Another respect in which Cartesian algebraic treatment appears more unified than Apollonius's treatment is one that Descartes himself brought to prominence: its application to the "Pappus problem." Briefly stated, the problem is to find a plane curve that satisfies certain conditions, which are set by the properties of a certain number of lines, the size of as many angles, and a ratio.[23] Laywine relates Apollonius's announcement, in a letter "that serves as a general preface to the *Conica* as a whole," that he has a solution, which uses propositions from Book III, to the problem for the case of three lines and that of four lines. The solution itself is not given in *Conica*, but has been reconstructed by modern scholars, using Propositions of 16–17 of Book III; and the final propositions of Book III, which appeal to Propositions 16, 18, and 20, give solutions to the converse of the problem (that is, they "show that the central conic sections and the circle are loci [for the cases of] three or four lines"; Laywine 2014, 742). In his *Géométrie*, Descartes takes the Pappus problem as an illustration of how "the ancient mathematicians … put so much labor into writing so many books in which the very sequence of propositions shows that they did not have a sure method of finding all, but rather gathered together those propositions on which they had happened by accident" (1637/1954, 304). In contrast, Descartes shows how his own "sure method of finding all" not only solves the problem for the case of three and four lines, but easily generalizes to cases of $n > 4$. Descartes's solution to the problem is, in fact, the first application of his algebraic method, and not only does it play "the role of a paradigm example of how to solve all geometrical problems," but its sweeping generality underwrites his "claim to a unique mathematical achievement" (Mancosu 1996, 66).

5.6 Projective Geometry and Kant's Apparent Disregard of It

To "sharpen the thought" that appealing to Apollonius's treatment "would have produced a sense of let-down" in Kant's readers, Laywine introduces a further contrast with "the development of projective geometry." Projective geometry, as she puts it, appears to achieve "the systematic unity that eluded Apollonius," because it "shows us that circle and conic section are the same thing by finding a characteristic property of conic sections that is also a property of the circle" in a much more direct manner (2014, 743). Indeed, when Jean-Victor Poncelet recreated projective geometry in the nineteenth century (after it had fallen into obscurity), his stated motivation was to eliminate the need for multiple proofs within geometry of results that could be proved by a single argument within algebra (Barrow-Green, Gray, and Wilson 2019, vol. 2, 416).

Projective geometry is, in the first instance, the study of geometrical figures under projection, which is the drawing of lines or rays from a point to points or lines (Morehead, Jr., 1955, 2). The most familiar example of projection is a "perspectivity," as in a perspective drawing: lines or rays are drawn from an origin, corresponding to the location of an observer, through a figure or scene onto a certain plane (the "picture plane"). Each point of the figure or scene is thus mapped to a corresponding point in the picture plane, which together comprise a second figure or scene, the "perspective image" of the first. The projections treated in projective geometry are sequences of perspectivities. What distinguishes projective geometry from the study of perspective is, first and foremost, that it takes as its object those properties of figures which remain invariant under projection. (Starting at the end of the nineteenth century, projections came to be considered as just one class of transformations (functional mappings between geometrical objects), and geometers focused their study on those properties which remain invariant under transformations of various kinds.) Second, projective geometry extends consideration to "ideal" elements to the extent needed in order to understand a figure and its image as the same, that is, to understand both as characterized by these properties. For example, projection can map a pair of intersecting lines onto a pair of parallel lines or vice versa. Thus the projective plane (the analogue in projective geometry of the picture plane) includes a point "at infinity" for each line, so that all the lines in a sheaf of parallels coincide in their points at infinity.

Because each conic section is an image of the circle under a suitable projection, projective treatment is indeed well suited to directly characterize the circle and conic sections in terms of common properties. Though projective geometry only came into its own in the nineteenth century, in Kant's time transformations had already been used to important effect precisely in the theory of conic sections. Their use in this theory by, in particular, Johann Kepler[24] and Girard Desargues anticipated later developments in the way it involved points at infinity and, especially, in the highly general characterization it afforded. Desargues, for his part, has been credited with "the first unified treatment of the conic sections" that surpasses "the classical theory due to Apollonius" (Gray 2004, 900), which Desargues achieves by relating the conic sections to the circle by projection and reasoning in terms of ratios – in particular, the "cross-ratio" of the ratios between distances of "harmonically separated" collinear points[25] – that remain invariant under projection.

Desargues's work was known to and continued by Blaise Pascal and Philippe de la Hire, but his writings did not circulate until the nineteenth century. Laywine surveys the evidence that Kant was aware of de la Hire's work, which was in circulation, and finds it inconclusive. Kant's discussion of how a surface can be seen as either a circle or ellipse, depending on the direction from which it is viewed (I:253–254), and his reference to the "*focus imaginarius*" with respect to which all "guide lines" concur in a point (A644/B672) are "suggestive of central notions from projective geometry"; but "they are only suggestive and just as plausibly suggest a concern with more traditional problems of perspective" (2014, 745).[26] I concur with this assessment, and I would further note that Leibniz's *Theodicy* could have given Kant the idea that the conic sections are "harmoniously" ordered by familiar perspective transformations[27] (so that Kant would not have to appeal to projective transformations to account for their systematic interrelationships). Leibniz introduces this example to show that "there must always be an exact relation between ... different representations of one and the same thing."

> The projections in perspective of the conic sections of the circle show that one and the same circle may be represented by an ellipse, a parabola and a hyperbola, and even by another circle, a straight line and a point. Nothing appears so different nor dissimilar as these figures; and yet there is an exact relation between each point and every other point. Thus one must allow that each soul represents the universe to itself according to its point of view, and through a relation which is peculiar to it; but a perfect harmony always subsists therein.
> (1985, 339)

Although Leibniz was conversant with the work of Desargues, Pascal, and de la Hire, this passage makes no specific reference to their contributions.

I would not draw the same conclusion as Laywine, however, which is that Kant did not "keep up" with developments in mathematics to the extent he did in natural science.[28] It would be difficult to avoid this conclusion if, with Laywine, we understand Kant's advocacy of a constructive approach in terms of Apollonius's treatment. But I contend that it is better understood in terms of Newton's treatment; and as I will argue, Newton's treatment not only incorporates insights of projective geometry, it represents the most systematic approach to the theory of conic sections until projective geometry emerged full-fledged (in the nineteenth century). So, Kant's apparent failure to engage with de la Hire's work is beside the point.

5.7 Apollonius and Newton in Kant's Reply to Eberhard

It is now time to revisit Kant's discussion of conic sections in the reply to Eberhard. Kant was provoked by Eberhard's claim that "mathematicians have completed the delineation of entire sciences without saying a single word about the reality of their object" (VIII:190), which draws for support on a remark about Apollonius's procedure by J.A. Borelli. In his edition of the *Conica*, Borelli says "one may assume the subject as defined, so that the variety of its affections may be shown, even if the manner of describing the subject's formation has not been presupposed" (quoted by Kant at VIII:192). Kant objects to Eberhard, first, that Borelli refers only to "mechanical" construction, which teaches how to produce a physical inscription with the aid of instruments. Thus Borelli should not be taken to deny that construction in Kant's sense – the "exhibition of a concept through the (spontaneous) production of a corresponding intuition" (VIII:192n.) – is required, specifically to prove the objective reality of the constructed concept. Kant then proceeds to explain how "the modern geometers" satisfy the requirement of proving objective reality, and precisely how their approach falls short in comparison with the ancients'.

> One could rather address to the modern geometers a reproach of the following nature: not that they derive the properties of a curved line from its definition without first being assured of the possibility of its object (for they are fully conscious of this together with the pure, merely schematic construction, and they also bring in mechanical construction afterwards if it is necessary), but that they arbitrarily think for themselves such a line (*e.g.*, the parabola through the formula $ax=y^2$), and do not, according to the example of the ancient geometers, first bring it forth as given in the conic section.
> (VIII:192)

Laywine helpfully reformulates Kant's point as a twofold clarification: "unlike the historical Apollonius, the modern algebraic mathematicians proceed hypothetically, but unlike Eberhard's Apollonius, they (can and) will discharge their hypotheses ... in their treatment of conic sections – if they are challenged to do so" (2014, 734).

Where I differ from Laywine is her account of "how Kant thinks the algebraic mathematicians would" discharge their assumption of the curves' objective reality: "if pressed, they will invoke the historical Apollonius' construction" (2014, 734). On her reading, when Kant gives the algebraic equation for the parabola ($ax = y^2$), he "must take this equation to stand in some relation to Apollonius' geometrical characterization of the parabola in terms of its principal property." For, she argues, Kant has not only "deliberately picked the example of a parabola to make a certain comparative point about geometry and algebra," but has himself "explicitly alluded to Apollonius' characterization a few lines earlier" (2014, 733).

Here Laywine refers to Kant's claim that Apollonius "extracts a concept of the relation in which" the ordinates of a curve "stand to the parameter" (VIII:191), which fits Apollonius's formulation of the parabola's "symptom" (that the square on the ordinate is equal to the rectangle on the abscissa and the parameter). As Laywine points out, "there is indeed a relation" between this formulation and the equation $ax = y^2$, since we "can take the y-value in the algebraic equation for the parabola to express the length of the ordinates, the a-value to express the length of the [parameter] and the x-value to express the length of the abscissae" (2014, 733). In this context it is plausible, as Laywine suggests, that Kant takes the "algebraic mathematician" to rely on the correspondence between the algebraic formula and Apollonius's characterization:

> [T]he algebraic mathematician knows how to back translate the relevant equation into the language of the geometers, and hence she always knows what she is talking about. This does not mean that the 'modern' algebraic geometers are above reproach ... For the algebraic mathematician's awareness that, say, the equation for the parabola can be back translated into the language of the geometers is not all by itself a proof that the equation has 'objective reality,' as Kant understands it ... [This means that] on Kant's view, [the algebraic mathematicians] will have to do something to reassure those of us in need of reassurance that conic sections are not a matter of mere speculation. The only reassurance they can give is to appeal to Apollonius' techniques for constructing and sectioning the cone. If Kant has a complaint about the algebraic mathematicians, it is that they do not voluntarily do this.
>
> (2014, 734–5)

This interpretation has Kant putting forward Apollonius's geometrical characterization, and relating it to the algebraic formula, on his own initiative. Alternatively, the emphasis on Apollonius could be taken to carry over from Eberhard's attack. Kant is, after all, responding to Eberhard's claim that "Apollonius and his interpreters built up the whole theory of conic sections without anywhere explaining how the ordinates are applied to the diameters of these curves," thus without securing the possibility of constructing the curves (quoted in Laywine 2014, 727–8). Once we consider that Apollonius may not have been the only or primary point of reference for Kant himself, we can begin to inquire what other sources could have informed Kant's understanding of construction in the theory of conic sections. I will now argue for the relevance of Newton as a source.

As we saw, Kant "reproaches" the "modern geometers" for not "first bringing forth" the curve "as given in the conic sections." Doing so, he concludes, would be "more in accordance with the elegance of geometry, an elegance in the name of which we are often advised not to completely forsake the synthetic method of the ancients for the analytic method which is so rich in inventions" (VIII:192). This appears to echo a complaint in Newton's *Arithmetica Universalis*:

> Multiplications, divisions, and computations of that sort have recently been introduced into geometry, but the step is ill-considered and contrary to the original intentions of this science: for anyone who examines the constructions of problems by the straight line and circle devised by the first geometers will readily perceive that geometry was contrived as a means of escaping the tediousness of calculation [namely] by the ready drawing of lines. Consequently these two sciences ought not to be confused. *The Ancients* so assiduously distinguished them one from the other that they never introduced arithmetical terms into geometry; while *recent people*, by confusing both, have lost *the simplicity in which all elegance in geometry consists*.
> (Whiteside ed., 1967–1981, vol. V, 429, emphasis added)

Kant's contrast between "the synthetic method of the ancients" and the richly inventive "analytic" method is implicit in this passage, but easily seen from its broader context. Newton's topic is the admissibility of various classes of curves as auxiliary figures in geometrical construction, which received much discussion in ancient and early modern writings on geometry. It was widely agreed that simpler auxiliaries should always be used in place of more complex ones, but the advent of Cartesian analytic geometry brought new criteria for simplicity. Newton notes that "mathematicians of recent times" have "welcomed into geometry all lines which can be expressed by means of equations, and have distinguished those lines into classes ... in line with the dimensions of the equations by which they are defined," then "laid down" the "rule that it is not permissible to construct a problem by means of a curve of a superior class when it can be constructed by one of a lower" (*loc. cit.*, 425). He rejects this innovation on the grounds that "it is not the simplicity of its equation but the ease of its description which primarily indicates that a line is to be admitted into the construction of problems" (*loc. cit.*, 425.). For instance, the ancient Greek preference for circles over parabolas should be upheld (because the circle is easier to construct than the parabola), although by the Cartesian criteria the circle and parabola are equally simple (because their equations are of the same degree).

Further evidence that Kant is echoing Newton's view – namely that geometry has its own criteria for simplicity, which give the discipline an "elegance" that is lost by substituting the algebraic criteria – is that Kant too links the "elegance" of geometry to the simplicity of constructive techniques. In a footnote, he classifies constructions as either pure ("schematic") or empirical ("technical"), depending on whether they "occur through the mere imagination" or are carried out "on some kind of material." Kant then classifies "technical" constructions as either "*geometrical*, by means of compass and ruler, or ... *mechanical*, for which other instruments are necessary as, for example, the drawing of the conic sections besides the circle" (VIII:192*n*.). In privileging the use of circles over that of conic sections, Kant makes evident that he shares Newton's understanding of comparative simplicity.[29]

It is easy to see how this passage could have caught Kant's attention. It occurs in a section titled "The Linear Construction of Equations." Despite its title, *Arithmetica Universalis* is best understood as a treatise of algebra,[30] and deals in large measure with geometrical problems. It belongs to a tradition, going back at least to the fourth-century CE *Collectio* of Pappus and continued by Descartes's *Géométrie*, of solving problems in two distinct stages: analysis (or resolution) and synthesis (or composition).[31] Pappus describes analysis as proceeding from what is sought, which is "assume[d] as if it has been achieved," to something that is already established or taken as a first principle.[32] The early moderns took this stage to consist in algebraic reasoning, in which the "assumed" (unknown) quantities are at first treated on a par with known quantities, "as if [they had] been achieved," by being expressed in combination with known ones. The isolation of unknown quantities on one side of an equation was only the culmination of the analytic stage; a complete solution of the problem required, in addition, the construction of the symbolically designated quantity (which comprised the synthetic stage). Newton's inclusion of this section demonstrates his adherence to this view. Since this view informs Kant's understanding of the role of construction in algebra (as Lisa Shabel persuasively argues in (1998)), Newton's remarks on the topic might well have been a resource for Kant.

5.8 The Theory of Conic Sections in *Principia*, Book I, Section V

To show that a "synthetic" approach, based on geometrical construction, could rival the generality and power of Cartesian analytic geometry, it is not enough merely to attribute it to Newton. What shows this are the advances Newton made in the theory of conic sections.

Kant was surely in a position to appreciate Newton's achievements, for Newton not only mentions them in *Arithmetica Universalis*, but showcases them in *Principia Mathematica*. In all fairness to Laywine, we should note her remark that "the *Principia* ostentatiously engages with the classical, ancient treatments of conic sections." But she takes Newton's engagement with these treatments to exhibit only "the special kind of systematic unity achieved by reason when physics and mathematics are united in what Kant takes to be the right way." So on her account, Kant's regard for the way Newton applies the theory of conic sections in physics "does not make it any less odd that we do not find a treatment [in Kant's works] of the distinctive and remarkable ways that reason, in the eighteenth century, was achieving higher degrees of systematic unity in the theory," within pure mathematics (2014, 746). I think this pronouncement overlooks the contributions to pure mathematics contained in *Principia*, including ideas that gave the theory of conic sections what unity it achieved in the eighteenth century. These are found in Section V of Book I, which is aptly described as treating "the point-properties and the tangent-properties of conics" with "the utmost generality" (Taylor 1881, *lxvi*).

The manuscript record shows that Section V (titled "To find orbits when neither focus is given") and the preceding Section IV ("To find elliptical, parabolic, and hyperbolic orbits, given a focus") were composed as separate tracts prior to writing *Principia*.[33] Even in their new setting, the sections read as an excursus on pure mathematics. As the scholar Derek T. Whiteside comments, it is "obvious" that "Newton's overriding interest in exhaustively exploring the geometrical subtleties of the problems ... far transcends the immediate, practical aim of effectively and accurately constructing planetary and cometary orbits which is [the sections'] professed *raison d'être*" (Whiteside ed., 1967–1981, vol. VI, 245–6 *n*.23).[34] When we consider Section V in light of Newton's polemics against Cartesian analysis, we readily see that his pursuit of "the geometrical subtleties" also vindicates his preferred construction-based approach; and, as I will now argue, it points the way to powerful unifications of the theory of conic sections. Newton already treats the proportionality of rectangles on intersecting chords as a characteristic property of the curves, and his reasoning makes use of additional, important projectively invariant properties, as well as ideal elements.

The first three propositions of Book V (Lemmas 17 through 19) and their corollaries yield, as Newton claims, "not a computation but a geometrical synthesis, such as the ancients required, of the classical problem of four lines" (1999, 485). This refers to the "Pappus problem" for the case of four lines. So we can take Newton to claim that in this instance, a "synthetic" approach – one which satisfies "the ancients'" demand for an explicit constructive procedure – equals the power of Cartesian analysis, i.e., "computation" of the equation that expresses the locus. Del Centina and Fiocca's examination of Newton's solution indicates two ways in which

Mathematics: Systematic Unity, Construction, and Conic Sections 137

it conduces to a unified treatment of conic sections in general. First, the initial Lemma 17, which Newton needs for an important proportionality result in Lemma 18, "rests essentially on" the proportionality of rectangles on intersecting chords (Del Centina and Fiocca 2021, 21; cf. Guicciardini 2009, 91). Newton's solution of the four-line Pappus problem can thus be considered the first of the "extraordinary applications Newton made" of the property, which demonstrated its foundational significance (*op. cit.*, 8). Second, in a Scholium Newton considers that "one or two of the four points" at which the given lines intersect "can go off to infinity," "and in this way the sides of the figure which converge to these points can turn out to be parallel, in which case the conic will pass through the other points and will go off to infinity in the direction of the parallels" (1999, 483). Del Centina and Fiocca cite this remark as evidence that "Newton had a clear 'projective' vision of the problem" (2021, 22).

The four-line Pappus problem generalizes in a natural way to the problem of constructing a conic through five given points (because, as Newton shows in Lemma 19, a fifth point can be constructed on the curve on which lie the four points at which the given lines intersect; see Guicciardini 2009, 93). Newton solves this and several related problems in the following five propositions of Section V, as Taylor concisely summarizes.

> Then follows a theorem (Lemma 20) which may be thus stated: If *ABPC* be four fixed points on a given conic, the chords from *B* and *C* to a variable point on the curve meet the parallels through *P* to *AB* and *AC* respectively in points *T* and *R*, such that *PT* varies as *PR*, and conversely. From a limiting case of this lemma he deduces his organic construction of a conic by means of two rotating angles (Lemma 21) … By means of [these Lemmas] he shows how to describe a conic when five points on it are given, or four points and a tangent, or three points and two tangents (Propositions 22–4).
>
> (Taylor 1881, *lxvi*)

The main idea of Newton's "organic" construction is, briefly, that "if two angles of given magnitudes turn about their respective vertices (the poles) in such a way that the point of intersection of one pair of arms always lies on a straight line (the describing line), the point of intersection of the other pair of arms will describe a conic (the describend curve)" (Guicciardini 2009, 94). Newton himself placed particular importance on this construction, which has a special claim on our attention as well. I will conclude this section by discussing its importance for Newton's pursuit of a constructive approach and in the development of the theory of conic sections.

Newton includes his organic construction of a conic in *Arithmetica Universalis* (Problem 55, Whiteside ed. 1967–1981, vol. V, 309*ff.*) and in an appendix to *Optics*, entitled *Enumeratio Linearum Tertii Ordinis* (Whiteside ed. 1967–1981, vol. VII, 637), as well as in *Principia*. One reason for giving it such prominence is that it represents an alternative to Cartesian algebraic techniques. Newton wrote in one manuscript that such "descriptions are of greatest use in determining [conics] and so on," "for given merely five points[,] *without any preparatory calculation* or knowing the vertex, axis, diameters, center and species of the curve, ... you should even so be able to describe it" (quoted in Guicciardini 2009, 101, emphasis added). By "preparatory calculation," Newton appears to mean algebraic computation of the curve's equation.

Although Newton saw himself as vindicating the methods of "the ancients,"[35] in retrospect his approach can be seen to lead the way for projective geometry. In the early decades of the 1700s, an extension of Newton's method was published by William Braikenridge (and also claimed by Newton's protégé Colin MacLaurin) which eliminates the "metrical element" of the angle of fixed measure and thus qualifies as "purely projective" (Coolidge 1940, 48). More importantly, Newton's own reasoning shows insight of a projective kind. Newton asserts in Lemma 21 that the line traced by the "intersection of the other pair of arms," i.e. the "describend" curve, will be a conic; and the converse, namely that if the curve on which lie these intersection points is a conic passing through A, B, and C, and the angles at the "poles" B and C remain the same as the intersection point of the lines from B and C moves along this curve, then the intersection point of the "first" pair of arms "will lie on a straight line given in position" (Newton 1999, 487). As Taylor observes, Newton uses the result of Lemma 20, that a certain ratio (PR: PT) is constant just in case A, B, P, and C all lie on a conic, to show that the "described" curve is a conic. He deduces the fixity of this ratio from the fixity of a certain cross-ratio ((PB × NC):(NB × PC)). Commentators agree that Newton's reasoning here is "guided by an understanding of the fact that conic sections can be defined as those curves that satisfy the 'anharmonic property'" (Guicciardini 2009, 92): the property that the cross-ratio of four lines drawn to fixed points (A, B, C, D) from a variable point (P), i.e. the ratio ((PA × PB):(PC × PD)), is constant. This property is projectively invariant and takes on crucial importance in projective treatments of conic sections, beginning with the work of Michel Chasles and Jacob Steiner.

While eighteenth-century readers could not yet see the power of a projective approach, they were in a position to appreciate the unity that Newton brought to the theory of conic sections. It bears emphasizing that Newton shows not only that his construction yields a conic, but that every conic can be considered as the result of such a construction. This is important because Newton appeals to his Lemma 20 to show that the construction yields a conic, and the proof of Lemma 20 makes crucial use of the proportionality of rectangles on intersecting chords. So once we understand why the curves yielded by this procedure are conics, we are also in a position to grasp that they have this property; and once we understand that every conic can be regarded as the result of the procedure, we can immediately see that all conics must have the property. Because Newton's treatment thus makes known the proportionality of rectangles on intersecting chords without need of any further argumentation – whether algebraic calculation, as on the Cartesian approach, or consideration of separate cases, as in Apollonius' *Conica* – it gives the most powerful illustration of the unity of diverse cases falling under a common principle.

5.9 Illustrating the Role of Reason's Ideas in Schematizing Reason's Principle

I now want to bring out a further respect in which Newton's treatment of conic sections is apt to illustrate the systematic unity sought by reason. In this regard, it is noteworthy that Newton appeals to known properties of conic sections to show that the lines produced by his procedure are conics. Since Newton does not set out to deduce these properties *ab initio* using his novel constructive procedure, he at least leaves open that conics are to be constructed – in the proofs that establish these properties – by other means (such as the sectioning procedures of Apollonius).

In fact, Newton openly allows for a plurality of construction techniques in his proof of Proposition 22 of Section V, Book I, *Principia* ("To describe a trajectory through five given points"). Newton's first solution makes use of Lemma 20 to show that the intersection points of lines drawn from two of the given points lie on a conic that also passes through two other given points (and argues, further, that the fifth point is also on the conic). His second solution is, in essence, his organic construction. In a "Scholium" following Proposition 22, Newton presents a simplification of the first solution which allows "the points of the trajectory [to be] found most readily, unless you prefer to describe the curve mechanically, as in the second construction" (1999, 490). Newton here uses "mechanical" in the same way that he elsewhere uses "organic," to indicate the use of tools (such as a pair of rulers joined at a fixed angle and made to rotate around their joined ends) for tracing curves.[36] Kant uses "mechanical" with a similar meaning, as we saw in Section 5.7. Newton's usage is broader, however, in that it includes (while Kant's excludes) the use of ruler and compass; more importantly, Newton breaks from tradition in not privileging geometrical constructions over mechanical ones. In this Scholium, Newton suggests the mechanical construction can be employed at pleasure. In a passage to which Kant alludes in *Metaphysical Foundations of Natural Science*

(4:478), Newton makes the stronger claim that "mechanical" construction techniques are indispensable:

> For the description of straight lines and circles, which is the foundation of *geometry*, appertains to *mechanics*. *Geometry* does not teach how to describe these straight lines and circles, but postulates such a description. ... *Geometry* postulates the solution of these problems [to describe straight lines and circles] from *mechanics* and teaches the use of the problems thus solved. And *geometry* can boast that with so few principles obtained from other fields, it can do so much. Therefore *geometry* is founded on mechanical practice.
> (1999, 381–2)

Geometry thus presupposes operations that are "mechanical" in this sense (belonging to the discipline of mechanics) whenever it calls for the construction of a figure; and it at least permits the use of constructions that are "mechanical" in the sense of Section V, Book I (involving the use of specified tools or instruments).

Newton's pluralism with regard to construction techniques is important because it suggests that we have the latitude, but are not required, to consider any given curve as the result of any particular construction technique. In particular, we may regard the curve as a result of his organic construction if that serves the purpose of our inquiry (as, in the most obvious case, it could serve our further purposes to assure that the curve in question has the property of the proportionality of rectangles on intersecting chords). It is only a small step to suppose, further, that we can so regard the curve without committing ourselves to its actually having such an origin; that is, we can regard the curve "as if" it resulted from Newton's organic construction.

On this line of thought, the theory of conic sections, as Newton develops it in *Principia*, not only supplies a specific example of the systematic interconnection (of concepts) that reason seeks, but also illustrates the manner in which an idea of reason mediates the application of reason's principle (of systematic unity) to the understanding's cognition. As briefly mentioned in Section 5.2, Kant claims that "the idea of reason" supplies reason with the schema for its principle of systematic unity. At the beginning of the Transcendental Dialectic,[37] Kant introduces three ideas of reason (the soul, the world, and God), corresponding to the three types of Aristotelian syllogism (categorical, hypothetical, and disjunctive; A323/B379). But in his account of the ideas' role in schematizing reason's principle, the idea of God has pride of place.[38] Kant also discusses the regulative employment of the idea of God more extensively than that of the other ideas.[39] Indeed, it may not be too strong to claim, with Lawrence

Mathematics: Systematic Unity, Construction, and Conic Sections   141

Pasternack, that the overall strategy of the Appendix is "to justify the maximal application of regulative principles through the supposition of a 'highest intelligence', *i.e.* an author capable of ordering nature with maximal systematicity" (2011, 418, citing A670–671/B698–699). Of course, this reading would have to be supplemented by some further account of the other ideas' positive contribution (to theoretical cognition).

Even if Kant highlights the idea of God not to give it a unique role, but only as a perspicuous example, it still appears that the best way to understand the ideas' function (in schematizing reason's principle) is to consider this example. What the example gives us is a way to "study nature *as if* systematic and purposive unity together with the greatest possible manifoldness were to be encountered everywhere to infinity," as "the regulative law of systematic unity" demands (A700/B728, emphasis original). This way is, namely, to

> consider everything that might ever belong to the context of possible experience *as if* this experience constituted an absolute unity, ... yet at the same time *as if* the sum total of all appearances (the world of sense itself) had a single supreme and all-sufficient ground outside its range, namely an independent, original, and creative reason, as it were, in relation to which we direct every empirical use of *our* reason in its greatest extension *as if* the objects themselves had arisen from that original image of all reason.
> (A672–673/B700–701, emphasis original)

It is important for Kant that we must consider "everything that might belong to ... possible experience" only *as if* it had such an origin (not as actually having such an origin), on pain of giving the idea of reason direct application to objects, in contravention of his strictures concerning reason's theoretical use. What the example of Newton's "organic" construction shows is that this "as if" mode of consideration already has a legitimate and fruitful use in science. To this extent, Kant's account of the role of reason's ideas in schematizing its principle (of systematic unity) corresponds to actual scientific practice. This gives plausibility to his view that reason's regulative use is apt to further natural-scientific cognition.

5.10   Conclusion

In this chapter, I have tried to show how reading Kant's remarks on conic sections as pertaining to Newton's treatment yields a more satisfactory understanding of Kant's thought. First and most obviously, Kant was in position to appreciate that Newton set out to surpass Descartes' "analytic" treatment through a "synthetic," construction-based approach. I think

Kant's allusion to *Arithmetica Universalis*, in the reply to Eberhard, shows that he did appreciate Newton's endeavor. And, the subsequent development of the theory of conic sections – first basing it on the proportionality of intersecting chords, and ultimately in projective terms – demonstrates the surpassing power and generality of Newton's approach. So Kant's view of mathematics as fundamentally constructive is no bar to understanding this theory's development toward greater elegance and unification. It thus becomes easier to see how Kant is entitled to take this theory as his illustration of the systematic unity sought by reason. Second, consideration of Newton's "organic" construction brings out a deep and subtle way in which Newton's treatment is apt to illustrate the schematization of reason's principle (of systematic unity). The theory of conic sections not only serves to exemplify the unity that reason seeks, but shows that we can achieve such unification precisely by attributing a hypothetical or fictive origin to the objects of investigation. It thus illustrates the way that ideas of reason function in the schematization of reason's principle.

Notes

1. A719/B747; cf. A301/B357, A477–479/B505–507.
2. The phrase which I have supplied in the original is translated by Guyer and Wood "the philosophy concerned with nature." However, *damit* could refer either to the use of reason beyond all experience (or even the encouragement of this) or to nature. Kemp Smith's rendering of *damit* as "with it" preserves this ambiguity. One consideration favoring Guyer and Wood's reading is that it is much harder to see how mathematics could supply intuitions to support reason's use beyond all experience than to see how it could supply them to natural philosophy.
3. For instance, Kant denies that geometrical objects have "natures," "since in their concept nothing is thought that would express an existence" (IV:467*n*.) Elsewhere, he insists that mathematical problems are not concerned with existence "at all" (A719/B747). And Kant explicitly ties the notion of nature to that of existence when he glosses "nature" as "the *existence* of things, insofar as the latter is determined according to universal laws" (IV:294).
4. Kant characterizes the objects of natural science as "appearances ... given to us independently of our concepts, to which, therefore, the key lies not in us and in our pure thinking, but outside us," in contrast to the subject-matter of ethics, where "nothing can be uncertain, because the propositions are either totally nugatory and empty, or else they have to flow merely from our concepts of reason" (A480/B508). More generally, a main way in which he understands "nature" is by a contrast with determination of the will, in respect to their laws, according to which the former laws are cognized theoretically and the latter are cognized practically. See V:19–20 and A547/B575.
5. Already at A337/B394, Kant asserts that pure reason "brings all its cognitions into a system." He reiterates that reason's unity is that of a system when he makes the point that reason applies itself to the understanding, rather than directly to objects: "The understanding constitutes an object for reason, just

as sensibility does for the understanding. To make systematic the unity of all possible empirical actions of the understanding is a business of reason" (A664/B692).
6 That the subjectivity of this use by reason of its own unity does not rule out the possibility of objective application is already implicit in the terminology "principle" and "maxim." Kant explains in *Groundwork* that a maxim, as a "subjective principle of acting," "must be distinguished from the *objective* principle" (IV:421*n*.); yet a maxim can coincide with the objective principle, since it can "contain in itself its own universal validity for every rational being" (IV:438) and can thus be "raised to the universality of a law" (IV:424).
7 On Kant's view, construction in pure intuition proves the objective reality of mathematical concepts insofar as pure intuition is the form of empirical intuition (B147; A223–224/B271; A239–240/B298–299).
8 Galileo's use of this result, which is noted by Michael Friedman (1992, 186–187*n*.), comes in showing that the squares on chords AB and AC, drawn from the top of the vertical diameter, are equal to the rectangles (AD × AF) and (AE × AF), respectively, for points D and E where perpendiculars drawn from points B and C meet the diameter. To show that $AB^2 = (AD \times AF)$, we first observe that by the Pythagorean Theorem, $AB^2 = AD^2 + BD^2$. By *Elements* III.35, the rectangle on AD and DF equals the rectangle on BD and the segment that prolongs BD to the circle's opposite side. But since AF is a diameter and BD is perpendicular to it, this segment is equal to BD, hence (AD × DF) = (BD × BD). Since the rectangle on AD and AF is equal to the square on AD plus the rectangle (AD × DF), we have that $AD \times AF = AD^2 + BD^2 = AB^2$. The same reasoning applies to chord AC and rectangle (AE × AF).
9 As Friedman lucidly summarizes, "if a body moves in a conic section in such a way that it satisfies Kepler's law of areas with respect to a focus of the conic, then its total acceleration is directed to that focus and is inversely proportional to the square of its distance from that point. Indeed, the relationship here is an equivalence: if the total acceleration of a body with respect to a given point is inversely proportional to the square of the distance, then the body moves in a conic section with that point as focus and satisfies Kepler's law of areas with respect to it" (1992, 191–2).
10 See Smith (2002).
11 The special case is: "Let *PG* and *DK* be two conjugate diameters of the conic (that is, each bisects every chord parallel to the other) intersecting (and therefore bisecting) one another at C and meeting the conic at *P*, *G* and *D*, *K* respectively; and let *QV* be an ordinate of the diameter *PG* (that is, *QV* is one-half of a chord in a system of parallel chords bisected by *PG*) meeting the conic at Q (and hence the diameter *PG* at V); then $QV^2/(PV \times VG) = CD^2/PC^2$" (Friedman 1992, 192).
12 An important difference between these passages is that in *Prolegomena to Any Future Metaphysics*, Kant speaks of space as "so homogeneous and in respect of particular properties so indeterminate" (IV:321), whereas in the earlier work he spoke of geometrical properties as properties of space. For discussion, see Laywine (2014, 736–8).
13 For discussion, see Proops (2021, 440–5).
14 When Kant speaks of a "constant law" according to which the paths of the planets "can deviate from the circle through each of an infinity of intermediate degrees," thus becoming ellipses (A662/B690), it is not clear whether he has in mind laws which apply to the curves as geometrical objects, or only physical

laws which apply to them as orbits (determining how their shape varies in accordance with the direction and strength of the forces producing them). But the former would fit a pattern noted by Laywine: when Kant speaks of geometrical figures (and their properties) as having "unity among the greatest manifoldness" (quoting II:95), he takes this unity to involve a law, "in virtue of [which] unity and order are exhibited" in the manifold (Laywine 2014, 721). For instance, in this passage Kant says of the intersecting tangent and secant treated in *Elements* III.36 that while they "can assume" infinitely many "different positions ... in intersecting the circle as described," they "are nonetheless constantly subject to the same law, from which they cannot deviate" (II:94).

15 "The schemata [of the understanding's pure concepts] are therefore nothing but *a priori* time-determinations in accordance with rules" (A145/B184).
16 Kant argues that in contrast to the mathematical principles, the dynamical principles must remain merely regulative because "they are to bring the existence of appearances under rules *a priori*," but "this existence cannot be constructed" (A179/B222).
17 To say, with Laywine, that Apollonius "generates each of the conic sections from any cone whatsoever" (2014, 730) glosses over some detail. Viktor Blåsjö has observed that while Apollonius's Definition 1 expresses "a more general understanding of a cone" (in terms of the construction, just described in the main text, that carries a line from a fixed point around the circumference), Apollonius limits himself to right cones when he "sets out to actually construct conics with given properties," in Propositions 52–60 of Book I (2022, 661). I am indebted to Blåsjö for bringing this limitation to my attention.
18 *Conica*, Book I Definition 4, quoted in Barrow-Green, Gray, and Wilson (2019, 155).
19 Apollonius's construction of the parabola involves cutting the conical surface by two planes. The line in question is the line common to the relevant cutting plane and the base of the cone (for details, see Laywine 2014, 728).
20 See Laywine (2014, 727), for details and references.
21 *Conica*, Book I Definition 11, quoted in Barrow-Green, Gray, and Wilson (2019, vol. 1, 156). The authors stress "how complicated the parameter p is in Apollonius's formulation, compared with the simple distance along the cone previously." The increased complication is a "price that must be set against the benefits of the new unified theory" (2019, 157).
22 "'Parabola' comes from ... the technical term for applying a figure to a straight line segment. It designates the section characterized by a rectangle applied *exactly* to the latus rectum. 'Hyperbola' comes from [a term meaning] to exceed. It designates the section characterized by a rectangle applied to the latus rectum with an *excess*. 'Ellipse' comes from [a term which means], in the passive, to be wanting or defective. It designates the section characterized by a rectangle applied to the latus rectum that *comes up short*" (Laywine 2014, 731).
23 Here is a rephrasing of Descartes's statement of the problem: "Having three (or four) lines given in position, it is required to find the locus of points C from which drawing three (or four) lines to the three (or four) lines given in position, and making given angles with each of the given lines, the following condition holds: the rectangle of two of the lines so drawn shall bear a given ratio to the square of the third (if there be only three), or to the rectangle of the other two (if there be four)." Slightly amended from Guicciardini (2009, 54), quoting Descartes (1637/1954, 307).

Mathematics: Systematic Unity, Construction, and Conic Sections 145

24 Anticipating the definitions that are now standard (see Taylor 1881, 1, and Glaeser et al. 2016, 12–13), Kepler gave the first systematic treatment of the conic sections in terms of their foci (a term he introduced). He did so by regarding the conics as subject to a principle of "continuity," that is, as related to each other by a gapless series of intermediate forms, so that "from a pair of intersecting lines one passes through an infinite number of hyperbolas to the parabola and thence through an infinity of ellipses to the circle" (Boyer 1956/2004, 68). Kepler describes the line-pair as "the most obtuse" and the parabola as "the most acute" of the hyperbolas, and likewise describes the parabola as "the most acute" and the circle as "the most obtuse" of ellipses (Taylor 1881, lvii). The hyperbola and ellipse each have two foci, equidistant from the center, while the parabola has one focus internal to the curve. To understand the parabola as related by gapless transitions to the ellipse, on the one hand, and the hyperbola, on the other, Kepler attributes a second, "blind," focus to the parabola, "which is to be taken at infinity on the axis either *without* or *within* the curve. The parabola may therefore be regarded indifferently as a hyperbola, having (relatively to either of its branches) one external and one internal focus, or as an ellipse, having both foci within the curve" (Taylor 1881, lix).
25 For a point C lying between points A and B on a line, C is said to divide the points A, B in the ratio of distance AC to distance BC (AC:BC). There can be found point D, the "harmonic conjugate" of C with respect to A, B, which divides A and B "externally" in the same ratio in which C divides them "internally": AC:BC = –AD : BD (the negative sign is required because segment BD is measured in the opposite direction from the others). Then points A, B, C, and D are said to be in harmonic separation or in "involution." The cross-ratio (AC : BC) × (–AD : DB) is invariant when points A, B, C, D are projected on to points A′, B′, C′, D′, i.e. (AC : BC) × (–AD : DB) = (A′C ′ : C′B′) × (–A′D′ : D′B′).
26 Thus, as Laywine elaborates, "the '*focus imaginarius*' might be taken to be the vanishing point of a picture rather than the point at infinity in the projective plane" (2014, 745).
27 As Laywine notes, Kant sometimes uses the term "harmony" to refer to the kind or order or unity "within a manifold" that the conic sections exhibit (2014, 722).
28 "Kant had a serious, life-long engagement with the physical sciences: he kept up; he was fluent with the technical details ... But, in the case of mathematics, things look different ... [Kant] does not seem to have been immersed in mathematics as a living discipline" (Laywine 2014, 745).
29 Kant singles out ruler and compass constructions as basic in other texts from this period (XX:411; XX:198). For discussion, see Friedman (2000).
30 Newton begins the book by distinguishing two ways of computing: a "definite and particular approach" which proceeds "by means of numbers, as in common arithmetic"; or "advancing in an indefinite, universal way," through "general variables," i.e. algebraically. He declares that he will treat arithmetic and algebra together, as "jointly constitut[ing] a single, complete computing science" (*NMP* V, 55–57).
31 That it belongs to this tradition is evident from its full title: *Arithmetica Universalis, sive De Compositione et Resolutione Arithmetica Liber.*
32 Quoted in Guicciardini (2009, 33).
33 Cohen (1999, 136).

34 Whiteside notes, further, that the "essential incongruity" of Sections IV and V became apparent to Newton, for in a "radical restructuring of the *Principia* which he began about 1692 but never fully carried through," he planned to relocate their contents "into a separate appended 'Tractatus Geometricus' where they would occupy a more suitable place" (Whiteside ed., 1967–1981, vol. VI, 245–6 *n*.23).

35 Newton saw himself, in particular, as rediscovering the lost doctrine of geometrical "porisms," which ancient commentators described as a constructive approach that proceeds by assuming the solution is effected. See Guicciardini (2009, 82). Accordingly, Newton remarked of his own organic construction that "a resolution which proceeds by means of appropriate porisms is more suited to composing demonstrations than is common algebra. Through algebra you easily arrive at equations, but always to pass therefrom to the elegant constructions and demonstrations which usually result by means of the method of porisms is not so easy" (quoted in Guicciardini 2009, 102).

36 As Guicciardini notes, since antiquity the practice of tracing curves by means of devices, such as "moving rulers or strings," had been called "organic" (from the Greek οργανον, "instrument") (2009, 46).

37 Within the Transcendental Dialectic Kant discusses additional ideas of reason (Proops 2021, 422), at least some of which are plausibly regarded as more determinate representations of the totalities associated with the soul, the world, and God (as Marcus Willaschek argues in (2018)).

38 When he asserts that the ideas "should not be assumed in themselves, but their reality should hold only as that as a schema of the regulative principle for the systematic unity of all cognitions of nature" (A674/B702), Kant "does not use the plural 'schemata,'" as Zuckert notes; and despite taking Kant to speak of all three ideas here, Zuckert further observes that Kant "does tend to use the locution 'schema' for systematic unity more frequently for the idea of God specifically than for the other two ideas" (2017, 91*n*.4).

39 As noted by Zuckert (2017, 91*n*.4) and Hoffer (2019, 221). Noam Hoffer identifies "systematic reasons for the emphasis on the idea of God in comparison with" the others: the "scope" of the idea of the soul, as a representation of systematic unity, "is limited to the domain of a subject's ideas and unrelated to the rest of nature"; and the idea of the world does not represent "unity … as a single entity" but is only "a negative idea of indefinite extendibility." In contrast, "the idea of God is tied from the outset to the idea of a *system*, the general aim of reason" (2019, 221–2).

Bibliography

Barrow-Green, June, Jeremy Gray, and Robin Wilson. 2019. *The History of Mathematics*. 2 vols. Providence: AMS Press.

Blåsjö, Viktor. 2022. "Operationalism: An Interpretation of the Philosophy of Ancient Greek Geometry." *Foundations of Science* 27: 587–708.

Boyer, Carl. 1956/2004. *History of Analytic Geometry*. New York: Dover. First published in *Scripta Mathematica Studies*.

Cohen, I. Bernard. 1999. "A Guide to Newton's *Principia*." In *The Principia: Mathematical Principles of Natural Philosophy*, translated by I. Bernard Cohen and Anne Whitman, 1–370.

Coolidge, J.L. 1940. *A History of Geometrical Methods*. Oxford: Clarendon Press.

Del Centina, Andrea, and Alessandra Fiocca. 2021. "The Chords Theorem Recalled to Life at the Turn of the Eighteenth Century." *Historia Mathematica* 56: 6–39.
Descartes, Rene. 1637/1954. *The Geometry of Rene Descartes*, edited and translated by D.E. Smith and M.L. Latham. New York: Dover. First publication Leiden: Maire.
Friedman, Michael. 1992. *Kant and the Exact Sciences*. Cambridge: Harvard University Press.
Friedman, Michael. 2000. "Geometry, Construction, and Intuition in Kant and His Successors." In *Between Logic and Intuition*, edited by Gila Sher and Richard Tiezsen, 18–218. Cambridge: Cambridge University Press.
Glaeser, Georg, et al. 2016. *The Universe of Conics*. Berlin: Springer.
Gray, Jeremy. 2004. "Projective Geometry." In *Companion Encyclopedia of the History and Philosophy of the Mathematical Sciences*, edited by I. Grattan-Guiness, 897–907. Baltimore: Johns Hopkins University Press.
Guicciardini, Niccolò. 2009. *Isaac Newton on Mathematical Certainty and Method*. Cambridge: MIT Press.
Heath, T.E. 1926/2014. *The Thirteen Books of Euclid's Elements*. 3 vols. New York: Dover.
Hoffer, Noam. 2019. "Kant's Regulative Metaphysics of God and the Systematic Lawfulness of Nature." *The Southern Journal of Philosophy* 57: 217–39.
Katz, Victor. 2009/2018. *A History of Mathematics*. 3rd ed. New York: Pearson.
Laywine, Alison. 2014. "Kant on Conic Sections." *Canadian Journal of Philosophy* 44: 719–58.
Leibniz, G.W.L. 1985. *Theodicy*, edited by E.M. Huggard. LaSalle: Open Court.
Mancosu, Paolo. 1996. *Philosophy of Mathematics and Mathematical Practice in the Seventeenth Century*. Oxford: Oxford University Press.
Morehead, James C. Jr. 1955. "Perspective and Projective Geometries: A Comparison." *Rice Institute Pamphlet – Rice University Studies* 42: 1–25.
Newton, Isaac. 1967–1981. *The Mathematical Papers of Isaac Newton*, edited by D.T. Whiteside. 8 vols. Cambridge: Cambridge University Press.
Newton, Isaac. 1999. *The Principia: Mathematical Principles of Natural Philosophy*. Translated by I. Bernard Cohen and Anne Whitman. Berkeley and Los Angeles: University of California Press.
Pasternack, Lawrence. 2011. "Regulative Principles and the 'Wise Author of Nature'." *Religious Studies* 47: 411–29.
Proops, Ian. 2021. *The Fiery Test of Critique*. Oxford: Oxford University Press.
Shabel, Lisa. 1998. "Kant on the 'Symbolic Construction' of Mathematical Concepts." *Studies in History and Philosophy of Science* 29: 589–621.
Smith, George E. 2002. "The Methodology of the *Principia*." In *The Cambridge Companion to Newton*, edited by I.B. Cohen and G.E. Smith, 138–73. Cambridge: Cambridge University Press.
Taylor, Charles.1881. *An Introduction to the Ancient and Modern Geometry of Conics*. Cambridge: Deighton, Bell and Co.
Willaschek, Marcus. 2018. *Kant on the sources of metaphysics: The dialectic of pure reason*. Cambridge: Cambridge University Press.
Zuckert, Rachel. 2017. "Empirical Scientific Investigation and the Ideas of Reason." In *Kant on the Laws of Nature*, edited by Angela Breitenbach and Michela Massimi, 89–107. Cambridge: Cambridge University Press.

6 Kant's Conception of the *Metaphysical Foundations of Natural Science*

Subject-Matter, Method, and Aim*

Thomas Sturm

6.1 Introduction

In the Architectonic of Pure Reason, near the end of the *Critique of Pure Reason* (henceforth *CPR*) Kant declares:

> Nobody attempts to establish a science without grounding it on an idea. But in its elaboration the schema, indeed even the definition of the science which is given right at the outset, seldom corresponds to the idea; for this lies in reason like a seed, all of whose parts still lie very involuted and are hardly recognizable even under microscopic observation.
>
> (A834/B862)

Here Kant suggests that any science (*Wissenschaft*) needs to be grounded on an "idea" – a concept of reason which helps to distinguish that science from others and which makes possible the integration of a plurality of cognitions into a systematic whole. This means that natural science requires such an idea. Since Kant takes the *Metaphysical Foundations of Natural Science* (henceforth *MFNS*) to be a constitutive part of natural science, it is incumbent upon him to clarify the idea upon which it is grounded. This important task is carried out in the Preface (*Vorrede*) to *MFNS* (IV:467–479). While stating the task is easy, it took Kant great intellectual effort to complete it in his own characteristic way.

The Preface can be divided into five main parts. It begins with general considerations concerning the concepts of (1) nature (paragraph 1, IV:467) and (2) science in order to explicate the specific concept of "natural

---

*This chapter has been published first as: Thomas Sturm (2022). Kant's Conception of the Metaphysical Foundations of Natural Science: Subject Matter, Method, and Aim (pp. 13–35). In Michael Bennett McNulty (Ed.), Kant's Metaphysical Foundations of Natural Science: A Critical Guide © Cambridge University Press. Reprinted with permission.

science" (*Naturwissenschaft*).¹ Using a series of successively more restrictive requirements, Kant arrives at the conclusion that in order to qualify as a "proper natural science" (*eigentliche Naturwissenschaft*; IV:469), a discipline requires metaphysical principles or presuppositions (paragraphs 2–6, IV:467–470; see Breitenbach 2022). He then (3) casts doubts on the pretensions of empirical psychology and chemistry to claim such a status (paragraphs 7–12, IV:470–471).² Only the doctrine of matter or body, i.e. physics, fully qualifies as a natural science "properly so called." However, the argument for this claim relies not on the necessity of metaphysical but of *mathematical* presuppositions of natural science. Only physics satisfies (at least to a sufficient degree) the condition that a proper natural science requires the application of mathematics. (4) At the same time, the application of mathematics to physics itself requires an explanation, which Kant promises to provide by way of metaphysical presuppositions. For this, he introduces the table of categories of the first *Critique* "as the schema of the system of the metaphysics of corporeal nature" – that is, as providing the basis for a complete system of such metaphysical principles, and thereby the structure of *MFNS* (paragraphs 13–16, IV:472–478). That he calls the table of categories the "schema" for the system provided in *MFNS* refers back to his claim in the first *Critique* that each "idea" of a science requires a "schema" for its "execution" (A833/B861). Stated differently, in the Preface, Kant has first given us the elements by which we can distinguish the science contained in *MFNS* from others – its idea – and is now telling us how *MFNS* is systematically structured. (5) Kant concludes the Preface by comparing his project with Newton's *Philosophiae Naturalis Principia Mathematica*, thereby elucidating the book's title and emphasizing the considerable importance he ascribes to his own work (paragraphs 17–18, IV:478–479).

My aim in this chapter is not to provide a complete analysis of the Preface.³ Instead, I shall focus on a question underlying several of the text's considerations concerning the "idea" and "schema" of *MFNS*: How to understand the conception of a metaphysical foundation of natural science and the systematic structure of the doctrines developed in *MFNS*? And how are Kant's conceptions of a metaphysical foundation of natural science and of transcendental philosophy (as presented in the first, constructive half of *CPR*) related to one another? Some scholars have argued that the relation must be a very close one: that in Kant's view, the only real point of his transcendental philosophy is to provide a foundation for science, or – what is more – that transcendental philosophy requires to be completed by the *MFNS* (e.g. Friedman 1992a, 1992b, 2001b, 2013; Westphal 1995; Lyre 2006). Others insist that Kant's transcendental philosophy in no way depends on metaphysical foundation of natural science: there is a "looseness of fit" between these two projects (Buchdahl 1969,

Allison 1994). This dispute possesses a broader significance, since it relates to the important debate concerning how we should assess the positive contributions of Kant's philosophy, especially in the light of post-Kantian developments in philosophy and science.

I will proceed in three steps. In Section 6.2, I will consider the entanglement between metaphysics and the sciences in Kant's thought more generally. This gives preliminary support to the claim that *MFNS* is important for understanding the *CPR*. In Section 6.3, I will turn to the debate concerning Kant's views on the relation between "metaphysical" and "transcendental" principles of cognition. By analyzing, in Section 6.4, the defining features of the aims, methods, and object given in the Preface to *MFNS*, I defend a moderate interpretation: the *MFNS* presents important concretizations of Kant's doctrines of synthetic *a priori* conditions of empirical cognition, but is not necessary to complete transcendental philosophy.

6.2 The Entanglement of Science and Metaphysics

Since his earliest writings, Kant was deeply interested in empirical science. He worked on cosmology, geology, and physics and increasingly on medicine, anthropology, geography, and human history as well. From what we know, he did not undertake wide-ranging and systematic field studies, experiments, or measurements. Nonetheless, he produced novel and even lasting results, such as the famous nebular hypothesis concerning the formation of the universe, a correct explanation of the Earth's axial rotation, and a field theory of matter (Adickes 1924–1925; Falkenburg 2000; Schönfield 2000). As Ian Hacking (1983, 100) has remarked, Kant often picked winners, and he would have made a formidable member of decision panels for research projects.

In addition, Kant's lifelong interest in metaphysics and its reform originated, in part, in questions concerning science. His first publication, *On the True Estimation of Living Forces* (1749), his *Physical Monadology* (1756), and the inaugural dissertation, *De mundi sensibilis atque intelligibilis forma et principiis* (1770), among others, addressed metaphysical issues related to the concepts of matter, force, and motion and considered the difference between the methods of mathematics and metaphysics. These views were partly revised, sometimes radically, in his critical works.

We should not be surprised to find these entanglements between philosophy and the sciences. In Kant's day, the two areas were often conjoined; one needs only to think of names like Descartes, Leibniz, or Newton to see as much. To think that Kant pursued the critical project independently of his interests in the sciences is interpretively inadequate. Increasingly, scholars have emphasized and discussed these links, and rightly so (cf.,

Kant's Conception of Metaphysical Foundations of Natural Science

e.g. Plaass 1965; Buchdahl 1969; Gloy 1976; Brittan 1978; Kitcher 1984; Butts 1986a; Röd 1991; Friedman 1992a, b, 2001b, 2013; Watkins 1997, 1998a, b; Lefèvre 2001; Pollok 2001; Lyre 2006; Massimi 2013; Watkins and Stan 2014; Breitenbach and Massimi 2017).

Upon closer inspection, however, the entanglement between science and metaphysics is a complex one, especially in Kant's critical philosophy. One can distinguish at least three basic dimensions of this entanglement (see Sturm 2012): (1) some special sciences provide a model for his project of critically reforming metaphysics; (2) some special sciences are supported by metaphysical presuppositions; and (3) all special sciences should serve the goals of philosophy according to its "world concept" (*Weltbegriff*) (A838–839/B866–867).

Regarding (1), Kant emphasizes that logic, mathematics, and parts of natural science are all "sciences of reason" (*Vernunftwissenschaften*).[4] Reason here means not only the faculty that provides nonempirical, *a priori* knowledge and methods of justification, but also concerns the very *subject-matter* of these sciences.[5] So in a sense, reason "knows itself" in these sciences by knowing its *a priori* representations (concepts and principles) and their legitimate uses. It does so in different ways: (general formal) logic has no specific object whatsoever; it concerns "nothing but the pure form of thinking" (A54/B58) or, more specifically, the formal aspects and rules of concepts, judgments, and inferences (IX:89ff.). Mathematics, again, applies reason by constructing objects *a priori* in pure intuition; and the rational parts of natural science make possible and guide experimentation and observation by means of various principles of reason, which are partly domain-specific. But despite these different uses of reason, it is the apriority of these sciences that has allowed them to succeed (Bvii–xiv). Kant claims that Euclidean geometry and Newtonian physics provide undeniable examples of synthetic *a priori* cognition and that these sciences (or their rational parts) are here to stay: such cognition is *necessary* and therefore secure from empirical refutation (A4/B8; B14–21; B40; IV:280, 295). In addition, only through a unified and complete system of *a priori* cognition can the search for new knowledge in each science be framed and guided and its results be structured systematically – systematicity being the most fundamental requirement for a body of cognition to count as a science (A832/B860; IV:467–468). Given that metaphysics is likewise concerned with *a priori* concepts and principles of reason, it must develop its own systematic *a priori* framework of legitimate representations and principles if it is ever to become a successful science. The other sciences of reason can provide clues for how to get metaphysics on the "secure path" of a science.

I will turn to claim (2) below, since *MFNS* is Kant's outstanding and, indeed, only developed example for an *a priori* framework for a special natural science. Concerning (3), he argues that the sciences can pursue legitimate aims of their own – but, especially when we consider the fact that sciences can be used for practical aims, this ought to be done under the guidance of metaphysics:

> Mathematics, natural science, even our empirical knowledge of man, have a high value as means, for the most part, to contingent ends, but yet ultimately to necessary and essential ends of humanity, but only through the mediation of a rational cognition from mere concepts, which, call it what one will, is really nothing but metaphysics.
> (A850/B878)

As is well known, Kant uses the term 'metaphysics' in several ways. He rejects a "dogmatic" form of metaphysics, and he defends a critically sanitized version which is supposed to contribute to theory and practice in general. The meaning of 'metaphysics' in the above quotation is clearly not the dogmatic one; only a critically sanitized metaphysics can guide science's contribution to the "necessary and essential ends of humanity" and thus to the "world concept" of philosophy. Such general metaphysics,[6] again, he distinguishes from the domain-specific "special" metaphysics, which he subdivides into "the metaphysics of nature" and "the metaphysics of freedom"; these are then further divided into more special domains (A841–851/B869–879).

These complex entanglements already indicate that the *MFNS* possesses a close relation to the first *Critique*.[7] This makes it incumbent upon us to clarify the nature of that relation. As I will show next, a crucial worry concerning the relation between the metaphysical foundations of natural science and transcendental philosophy arises from claims (1) and (2). Claim (3) can be set aside in what follows (for more on this claim, see Sturm 2012, 2020).

6.3 The Entanglement of Transcendental and Metaphysical Principles

6.3.1 A Classical Objection

A familiar objection concerning Kant's transcendental philosophy starts from a reading of claim (1), i.e. his reliance upon Newtonian physics and Euclidean geometry. This claim becomes entangled with claim (2): not only does Kant take these exact sciences to be models for the development of a critically cleansed, scientific metaphysics; according to the objection, his transcendental system of forms of intuition and of concepts and principles of the pure understanding is also meant to provide a kind of permanent foundation of these sciences. Since Kant viewed Newtonian physics and Euclidean geometry as "ultimate and permanent achievements of the human mind," the argument goes, "he naturally regarded their presuppositions as absolute" (Körner 1955, 26). However, with the advent of non-Euclidean geometries during the nineteenth century,[8] and the Einsteinian revolution of the early twentieth century, the assumption of the unrevisability of Newton's physics and Euclid's geometry no longer held water. Accordingly, the synthetic *a priori* principles of Kant's first *Critique* – the transcendental principles of the understanding – came to be seen as incapable of supporting the new scientific theories: "one would hardly expect to find all of them necessary to all thinking about matters of fact" (Körner 1955, 26). Alternative presuppositions must be discovered to support the possibility of the new theories, and the status as unrevisable synthetic *a priori* principles, indeed even their truth, must be reconsidered (e.g. Reichenbach 1920; Cassirer 1921; Mittelstaedt 1994; Friedman 2001a, 2002, esp. 171–2).

A standard reply to this objection is that explicating and justifying the synthetic *a priori* presuppositions of *science* is not really the aim of Kant's transcendental philosophy in the first place; therefore, a historical change in scientific theories cannot affect it. In Peter Strawson's words, we must not give in to a merely "historical view" of Kant's transcendental conditions of the possibility of any *experience* – i.e. empirical cognition, whether scientific or not (Strawson 1966, 118–21). Especially, though not exclusively, this concerns the relation between the first *Critique*'s Analogies of Experience (A176–218/B218–265) – the synthetic *a priori* principles of the permanence of substance, causality, and interaction – and empirical laws of science. In line with this defense, the Analogies are not interpreted as grounding primarily or exclusively *scientific* laws. Their function is to explain the possibility of cognition in general, not specifically scientific cognition. They can perhaps perform the latter task as well, but only, as it were, as a bonus. In any case, the validity of the Analogies is not threatened by eventual changes in scientific theories. Such the first defense against the classical objection.

6.3.2 The Analogies of Experience and Natural Science: Four Questions

However, the issue cannot be handled so easily. It is more complex. For reasons of simplicity, let us continue to restrict ourselves to the Analogies of Experience and their relation to science, leaving aside other principles of the pure understanding of *CPR* and their counterparts in *MFNS*. The Analogies have been most often at the center of the debate just outlined, so it should help us here too. We need to distinguish at least four questions here.

First, the interpretation of the Analogies is disputed. For instance, does Kant's Second Analogy, the principle of causality, assert that each cause of a certain type brings about an effect of a certain type, or that it must be covered by a general law (in Lewis White Beck's useful terminology, the "same-cause-same-effect" rule)? Or is it the claim that all events or, in Kant's language, all "alterations" (*Veränderungen*) must have a cause (the "every-event-some-cause" rule; Beck 1978) – so that Kant's principle of causality "merely" secures that there is an irreversible time order of our experience of events?[9] The former principle is an ambitious statement of strict causal determinism, depending on unchanging laws, and highly controversial, especially in the light of post-Newtonian developments in science (e.g. Körner 1955, 87–91; Mittelstaedt 1994). The latter is a much more moderate principle of a certain causal determination of events. Second, as already indicated, Kant asserts in the Preface (IV:473–477) that he develops the systematic structure of his metaphysical principles of natural science against the backdrop of the *Critique*'s system of categories. This forms the backbone of his division of *MFNS* into four chapters, each allegedly corresponding to one set of transcendental principles. But how to understand the precise relation between the *Critique*'s Analogies and the relevant counterparts in *MFNS*'s Mechanics chapter, especially the three laws of motion (IV:541–551)? Are the latter meant to be *derived* from the former? If so, how? If not, what support does Kant provide for them, and what does that say about their epistemic status? Third, what is the relation between the laws of motion as Kant formulates them in *MFNS* and Newton's own laws? Did Kant copy Newton's versions, or was he also influenced by other authors, such as Leibniz? Fourth and finally, how does the revision or replacement of Newtonian mechanics in subsequent history of science affect the critical evaluation of Kant's views?

A full interpretation and critical assessment of his claims and arguments, even if one only considers the example of causality, is accordingly complex and far-reaching. It involves issues that cannot be discussed here (see Watkins 1997, 2005; Friedman 2001b, 2013, ch. 3; Pollok 2001, ch. 3.4). For instance, I will leave aside the third question completely[10] and only touch the fourth one. Even for the first two questions, it must suffice to make a few points in order to gain a better understanding of the relation between Kant's transcendental and metaphysical doctrines.

To begin, there are reasonable interpretations of the Second Analogy according to which Kant did *not* aim to argue for the strong "same-cause-same-effect" rule, but only the weaker "every-event-some-cause" rule (Buchdahl 1969, 651–65; Beck 1978; Allison 1994; Thöle 2004). We will see below that, in *MFNS*, Kant explicitly formulates and uses the Second Analogy as the "every-event-some-cause" rule. This provides extra

Kant's Conception of Metaphysical Foundations of Natural Science

evidence for this reading. Such an interpretation may not provide what one wishes for, but it is not nothing either.

To explain the meaning of the Second Analogy, it helps to provide a brief characterization of Kant's transcendental inquiry. This inquiry aims to state and justify what synthetic *a priori* conditions we must assume for our judgments to be determinately true or false of objects of experience.[11] Kant does not mean that the concepts and principles of the pure understanding would determine which of our cognitive judgments are 'objective' in the sense of being true but, instead, which ones are determinate enough to possess a truth-value *at all*. His main target is cognition (*Erkenntnis*) rather than knowledge (*Wissen*):[12] *Erkenntnis*, while it aims at truth, can be false, namely when it does not correspond to its object (CPR, A58/B83; an approximate German equivalent today might be *Erkenntnisanspruch*). *Wissen*, however, requires more, namely, actual truth and sufficient justification (A820–823/B848–851). In this vein, Kant characterizes the Transcendental Analytic as a "logic of truth" (*Logik der Wahrheit*; A62/B87, B170). This is also what he means when he speaks of the (real, not merely logical) "possibility" of objective cognition, or when he states that his inquiry aims to identify necessary conditions for a certain determinacy of the empirical world – so that by using appropriate criteria, we can figure out whether our cognitions are true or false. Of course, for this inquiry to succeed, the synthetic *a priori* principles of the understanding must themselves be taken to be true. To show that they are, Kant provides proofs.

In line with this, the Second Analogy states, as a minimal condition for causal judgments to possess a determinate truth-value, that an alteration is brought about by the cause that precedes it in time. This order cannot be reversed, or else we are no longer talking of a causal relation. Our knowledge of causal laws presupposes that there is a determinate temporal order of relevant causal sequences. Importantly, the reverse does not hold. Now, doubtlessly, in the *formulation* of the Second Analogy, Kant claims that every causal relation must be covered by some law (A189/B232). But this does not mean that he would *prove*, in the chapter on this Analogy, the "same-cause-same-effect" rule. If one looks at his proof, it is the claim of a determinate temporal order of causal relations that he focuses on.

What, then, is the point of the thesis of lawlikeness? Clearly, it cannot be assumed to be a mere conceptual or analytic truth. Perhaps we might read it as an epistemological claim: to *be able to know* the relation between a change and its cause, we must assume that some law exists, even if we do not know it (yet), let alone know it to be necessarily true. Alternatively, we can view it as a methodological claim: the principle of causality invites us to search for laws in order for us to be able to defend the assumption that the relation between a change and its cause possesses necessity – that the time order of the relation cannot be reversed.¹³ I have sympathy for such interpretations, though they are not without their problems and certainly stand in need of a number of qualifications. However, I shall not discuss this topic here. Let us accept that the Second Analogy does not guarantee any *knowledge* of specific causal laws, let alone its necessity, but that it directs our search for causal relation by assuming that there are such laws, and that these, when discovered, help to explain why the causal relation has the determinate time order that Kant claims it has. Beyond that, discovering causal laws is a matter of empirical inquiry – albeit an inquiry that must, in Kant's view, be guided by rational constraints of systematicity. He discusses the question of how we can discover causal laws and their necessity in other places, and his answer is based upon a "regulative" use of ideas and maxims of reason (CPR, A642–668/B670–699; *Critique of the Power of Judgment*, V:179–186).

Granted such a moderate reading of the Second Analogy, its validity is no longer necessarily threatened by fundamental changes of scientific causal laws. Insofar as there are causal relations, Kant claims, they must exhibit a determinate temporal order, and it is necessary that there is some law behind them (though, as current theoreticians of causality know, the latter is a matter of dispute; see, e.g., Krüger 2005; Hoefer 2016). The historical changes in science that are typically cited in the classical objection – for instance, the replacement of Newton's physics by Einstein's (e.g. Friedman 2002) – do not imply that we must generally abandon the theses of the temporal order and the lawlikeness of causal relations. There are indeed attempts to defend suitable forms of determinism for both special and general relativity theory and even for quantum theory.¹⁴ This, of course, does not constitute a *positive* argument for the general validity of the Kant's Second Analogy. That depends on the quality of the proof Kant provides for it in *CPR*; the same holds for all other transcendental principles.

What do these points mean for the four questions distinguished above? As noted, I will leave aside the third question here and can only superficially touch the fourth question. By favoring the moderate version of the general principle of causality (answer to question 1), we have made room for separating its validity from that of special scientific laws. Hence, we are also able to protect this principle from far-reaching revisions of our scientific knowledge of causal laws (answer to question 4). But what about the second question, concerning the fact that Kant develops the systematic structure of *MFNS* against the background of the system of categories and principles of *CPR*? If we want to maintain that transcendental principles such as the Analogies of Experience are not vulnerable to scientific change, we must clarify the relation between them and their counterparts in *MFNS*'s chapter on mechanics – since here Kant formulates (his own

Kant's Conception of Metaphysical Foundations of Natural Science 157

version) of the three Newtonian laws of motion which might be vulnerable to such change.[15]

I will return to this issue in a moment (and again in Section 6.4.3). Before that, however, it will be useful to examine a textual development in Kant's works. Consider the *Prolegomena to Any Future Metaphysics*, published between the first edition of *CPR* and *MFNS*. Here, Kant speaks of the transcendental principles of the understanding as "laws of nature" and even describes them as content of "pure natural science":

> Now we are ... in possession of a pure natural science, which ... propounds laws to which nature is subject. Here I need call to witness only that propaedeutic to the theory of nature which, under the title of universal natural science, precedes all of physics (which is founded on empirical principles). ... among the principles of this universal physics a few are found that actually have the universality we require, such as the proposition: that substance remains and persists, that everything that happens always previously is determined by a cause according to constant laws, and so on. These are truly universal laws of nature, that exist fully *a priori*. There is then in fact a pure natural science ....
> (IV:294–295; cf. IV:280, 306–307, 322, 327)

This passage refers to the first two of the three Analogies – the principles of permanence of substance and of causality – and, due to the openness of the formulation, potentially to all transcendental principles (A149ff./B188ff.). Kant describes the examples as "really general laws of nature that exist completely *a priori*." The problem arises due to his conclusion: "There is then in fact a pure natural science" (IV:295). Here, the tasks of transcendental philosophy and pure natural science are not separated. In addition, in *Prolegomena*, Kant describes the table of principles as "Universal Principles of Natural Science" (*allgemeine[r] Grundsätze der Naturwissenschaft*; IV:303; cf. 306–307, 322), whereas in the first *Critique* – in both editions! – the same principles are listed under explicit *exclusion* of "general natural science" (*der allgemeinen Naturwissenschaft*; A172/B213). One might interpret talk of "universal principles" of natural science or of "general natural science" charitably and say that there is no conflict here. However, talk of "pure natural science" cannot be so interpreted. While in *Prolegomena* it is identified with transcendental philosophy, in *MFNS* it is reserved for the metaphysical principles of physics.[16]

There is much to recommend the view that, in *Prolegomena*, Kant had not yet fixed his set of relevant concepts and claims. Only by the time of *MFNS*, and the second edition of *CPR* that appeared one year after *MFNS*, did things settle down. This can be corroborated by pointing to a related shift in Kant's views concerning which concepts and principles can be considered "pure." In *Prolegomena*, he claims that a "propaedeutics of natural science," while containing the synthetic *a priori* principles mentioned above, is actually a mixed bag:

> ... there is also much in it that is not completely pure and independent of sources in experience, such as the concept of motion, of impenetrability (on which the empirical concept of matter is based), of inertia, among others, so that it cannot be called completely pure natural science ....
>
> (IV:295)

However, in *MFNS*, and in the B-edition of the first *Critique*, Kant affirms that these concepts and principles do belong to pure natural science. Thus,

> ... mathematical physicists could in no way avoid metaphysical principles, and, among them, also not those that make the concept of their proper object, namely, matter, *a priori* suitable for application to outer experience, such as the concept of motion, the filling of space, inertia, and so on.
>
> (IV:472; cf. CPR, B17–18)

Concepts such as those of motion or impenetrability, and principles such as that of "persistence of the same quantity of matter, inertia, and of equality of action and counter-effect" thus become an essential part of pure natural science or "*pura physica* (or rationalis)" (B21 n.). In this way, Kant distinguishes the contents of transcendental philosophy from those of metaphysical foundations of natural science.[17]

One might think that this is just a terminological improvement; no less, but also no more. But it matters in concrete ways. As indicated at the outset, in the Preface to *MFNS* Kant declares that the table of categories in the first *Critique* provides "the schema of the system of the metaphysics of corporeal nature" (IV:472–478). Only on that basis can he claim that he is using the categories to *formulate* metaphysical principles of natural science, and the transcendental principles as *premises* in the proofs of those principles. But for this to be possible, Kant must be conceptually clear about the difference between transcendental philosophy and metaphysical foundations of natural science. He must not conflate the two projects.

Consider again the Second Analogy. In *CPR*, it is stated as that all changes must have a cause. In *MFNS*, this principle becomes projected onto the concept of matter – the movable in space (IV:480) – thus delivering the Second Law of Mechanics as follows: "Every change in matter

Kant's Conception of Metaphysical Foundations of Natural Science

has an external cause" (IV:543). It is here that the Second Analogy is formulated in the "every-event-some-cause" version, leading to this specific formulation of the Second Law of Mechanics. Similarly, Kant projects the First Analogy of Experience, which asserts the permanence of substance throughout all changes (B224),[18] onto the concept of matter, resulting in the "First Law of Mechanics. In all changes of corporeal nature, the total quantity of matter remains the same, neither increased nor diminished" (IV:541). The same holds true for the Third Analogy and the Third Law of Mechanics (IV:544–545). In each case, Kant explicitly starts the proof of the respective mechanical law by citing the relevant transcendental Analogy.[19] The three laws of mechanics are precisely those which Kant, in the B-edition of the *Critique*, refers to as the "persistence of the same quantity of matter, inertia, and of equality of action and counter-effect" and which form an essential part of pure natural science (B21n). This all speaks both for a clear difference *and* a connection between transcendental and metaphysical principles.

That connection must not be overstated, however. There are several difficulties for the view that Kant simply projects each of the transcendental categories and principles of the first *Critique* upon the concept of matter. For instance, the Mechanics chapter does not merely consist of the three laws of mechanics, also called Propositions 2, 3, and 4, respectively. Before them, we find a Proposition 1, concerning the measurement of the quantity of matter by means of the quantity of motion at a given speed (IV:537). There is no transcendental principle that Kant would invoke here to formulate or prove that metaphysical principle. Also, the first part of *MFNS*, the Phoronomy – approximately, a kinematics – does not connect well to the first transcendental rule, the principle of the Axioms of Intuition (*CPR*, A162/B202). This principle appeals exclusively to the possibility of making judgments about "extensive" magnitudes, while in the Phoronomy chapter Kant writes of the constructability of velocities, which he calls "intensive" magnitudes (IV:493). In the *Critique*, intensive magnitudes are the topic of the second principle, the Anticipations of Perception (A166/B207; cf. Watkins 1998a; Pollok 2001, 28–30). Thus, there are sections of *MFNS* that do not have clear counterparts in *CPR* and vice versa. For the system of *MFNS*, Kant lets himself be guided or inspired by the system of transcendental categories, but also breaks away from it. So much for question (2) introduced earlier on.

For the record, let us note what relevance the difference and the – problematic – connection between transcendental and metaphysical inquiries has for question (4): how does the replacement of Newtonian mechanics in subsequent history of science affect the critical evaluation of Kant's transcendental philosophy? In his view, the explication and justification of the synthetic *a priori* presuppositions of Newtonian mechanics are the task of *MFNS* rather than the first *Critique*. Accordingly, if in the course of history, Newton's theory has been replaced by a different one, then one might need new metaphysical foundations for the new theory. For instance, post-Kantian science gave up the thesis of the permanence of material substance, replacing it with a principle of energy conservation. In this, interpreters such as Friedman (1992b, 2001b) are certainly right. However, a refutation of transcendental principles of cognition *überhaupt* can hardly be inferred from this. The transcendental principles are premises of some of *MFNS*'s proofs for the metaphysical principles, e.g. the laws of mechanics. However, if we find the conclusions of these proofs objectionable, mistakes might be found in our modes of reasoning or in the premises – of which are transcendental principles form only a part. Also, the problematic details of the connection between transcendental and metaphysical principles should give us reason for doubt too. In sum: yes, there are important connections between Kant's transcendental philosophy and his project of *MFNS*; but the differences matter just as much.

6.4 The "Idea" of Metaphysical Foundations of Natural Science

6.4.1 *Systems of Sciences in General*

We have considered but a small part of *MFNS*'s content and its relation to transcendental philosophy. Even if we went through all four chapters of the book in the same way, we would not gain a clear and completely general understanding of Kant's conception of metaphysical foundations of natural science. We could not infer from all the differences and relations a complete understanding of the boundaries of that science.

Remember that Kant claims that all sciences aim at systematicity. This means, minimally and ignoring for the moment other requirements, that each doctrine must consist of (1) a unified set of cognitions, which (2) is clearly distinguishable from other sets. Kant repeatedly stresses that we "do not enhance but distort sciences, if we allow them to trespass upon one another's territory" (*CPR*, Bviii). Otherwise, "none of [the sciences] can be thoroughly dealt with in a manner that suits its nature" (IV:265; cf. IV:472–473; VII:7). One might call (1) the "internal" and (2) the "external" systematicity of a science (Sturm 2009, ch. 3). With respect to *MFNS*, so far we have discussed only aspects of (1), but not of (2). Kant indeed often speaks about (2). Consider familiar examples from other areas of his oeuvre. Pure general logic ought not to be confused with psychology, since even though both deal with rules of thinking, they differ in their aims and in the epistemic status of their claims. Pure general logic details the necessary *a priori* laws of thought, whereas psychology provides the

Kant's Conception of Metaphysical Foundations of Natural Science

empirical, and hence continent and descriptive, laws of mental processes (A54/B78; IX:13–14). Metaphysics and mathematics, though both sciences of reason, use different methods and thus ought to be kept apart (A714/B742; IX:23). Ethics and anthropology have human action as their subject-matter, but only the former deals with *categorical* Oughts concerning action (IV:389, 411–412; VI:216–217; for more, see Sturm 2009, 169–70, 522–6). Kant thinks that sciences cannot make rational progress if we fail to separate them from one another. But while an "architectonic mind" brings such distinctions to light, such a mind also "methodically recognizes how all sciences are connected and how they mutually support one another" (VII:226; cf. A832/B860; Sturm 2009, ch. 3). Accordingly, while it could be profitable for both transcendental philosophy and the metaphysical foundations of natural science if we can distinguish between them in principled ways, we need not deny that Kant sees places or steps in his arguments where they connect with one another. But all this must be done in such a way that avoids confusions.

Our question, therefore, must be: what justifies the distinction between transcendental principles of cognition *überhaupt* and metaphysical foundations of natural science? For this, we need a general conceptual explanation. In Kant's terminology, we must identify the elements or features of the "idea" of that doctrine or discipline. One achieves a clear (enough) distinction between different but neighboring disciplines by defining or explicating the idea of each of them, which allows one to avoid distortions and trespassing among different sciences.

6.4.2 Defining Sciences: Subject-Matter, Method, and Aim

What elements make up such an idea? At the beginning of *Prolegomena*, Kant claims:

> If one wishes to present a body of cognition as *science*, then one must first be able to determine precisely the differentia it has in common with no other science, and which is therefore its *distinguishing feature*; otherwise the boundaries of all the sciences run together ...
> (IV:265)

The "distinguishing feature" of a science, he adds, may consist "in a difference of the *object*, or the *source of cognition*, or even of the *type of cognition*, or several if not all of these things together" (IV:265). So, two distinguishing features of the idea of any given special science are its "object" or subject-matter, or its proper *ontological* domain, and its "sources" or the "type" of cognition the science provides, that is, its *epistemological* characteristics.

As Kant makes clear on other occasions, ends or aims – i.e., *axiological*[20] features – can also play a crucial role for defining sciences. In the Preface to *MFNS*, he declares:

> For if it is permissible to draw the boundaries of a science, not simply according to the constitution of its object and its specific mode of cognition, but also according to the *end* that one has in mind for this science itself in uses elsewhere; and if one finds that metaphysics has busied so many heads until now, and will continue to do so, not in order thereby to extend natural cognitions (which takes place much more easily and surely through observation, experiment, and the application of mathematics to outer appearances), but rather so as to attain cognition of that which lies wholly beyond all boundaries of experience, of God, Freedom, and Immortality; then one gains in the advancement of this goal if one frees it from an offshoot that certainly springs from its root, but nonetheless only hinders its regular growth, and one plants this offshoot specially, yet without failing to appreciate the origin of [this offshoot] from it; and without omitting the mature plant from the system of general metaphysics.
>
> (IV:477; emphasis in 1st edition, though not the 2nd and 3rd editions; cf. Refl 4294; Refl 4459)

Thus, after repeating that the specification of the "object" as well as a suitable "source" and/or "type" of cognition are necessary parts of the definition of any science, Kant adds the role of its "end" (*Zweck*) too. Note how this passage also suggests that he has, up to then in the Preface, been occupied with determining the object and the type and sources of cognition of *MFNS*, and is only now turning to its end: natural science aims primarily at extending our "natural cognitions" (*Naturerkenntnisse*). Taken by itself, this claim appears uninformative or trivial; but we can see that Kant's point is to make clear that the end of a special science need not be one concerning its "uses elsewhere," or a practical one. Some philosophers, however, wish to put natural science to further use, namely – paradigmatically – to the ends of religion and morality. Such ends are not merely epistemic ones. Similarly, in the Architectonic of *CPR*, Kant declares that each doctrine that aims to be a science requires "a single supreme and inner end, which first makes possible the whole" (A834/B862). He distinguishes such an "inner end" – one that helps to define a given science – from so-called "arbitrary external ends" (A833–834/B861–862).

The three – ontological, epistemological, and axiological – differentiating features of the idea of any special science are flexibly interrelated at various levels, can be rationally discussed and improved, and ideally make possible a well-ordered and principled system of all sciences (Sturm 2009,

Kant's Conception of Metaphysical Foundations of Natural Science 163

ch. 3). In this vein, Kant presents his own systematic classification of metaphysics, of its parts and subparts, and relates it to various special sciences within the Architectonic of *CPR*. Not all details of that system remain stable in his development, as the discussed shifts from the first edition of *CPR* to *Prolegomena* to *MFNS* have made clear. That is as it should be: Kant believes that ideas of sciences, though rooted in reason, can be refined, reformed, or even replaced as research makes progress (A834/B862).

6.4.3 *Defining the Metaphysical Foundations of Natural Science*

Let us apply this account by returning to *MFNS*'s Preface. Whatever else this text does, it must explain the aim, the methods, and the subject-matter – using my terminology, the 'external' systematicity – of *MFNS*. Let us state each feature and compare it to parallel determinations of transcendental philosophy.

Having explicated the notions of nature, science, and natural science in the early pages of the Preface (IV 4:467–470), Kant argues that neither empirical psychology nor chemistry satisfy (at least not sufficiently[21]) the conditions for the status of natural science "properly so called" (IV:470–471). This leads him to the conclusion that only physics, or the "doctrine of body," deserves this honorific title. This, then, provides the determination of the subject-matter of *MFNS*: it deals only with cognition of corporeal nature. In contrast, the *CPR*, in its constructive first half, is an inquiry that identifies and justifies the synthetic *a priori* principles of an "object of cognition in general" (*Gegenstand der Erkenntnis überhaupt*). What is the point of this contrast?

To answer this question, we must clarify what Kant thinks about the "type" or "sources" of cognition presented in *MFNS*. In the first seven paragraphs (IV:467–470), Kant aims to define the concept of natural science in such a way that it becomes clear that physics is not a mere empirical science; it also needs *a priori* principles, and among these he is specifically concerned with metaphysical principles. The function of these principles of natural science can be stated by a comparison with the transcendental "logic of truth" of the first *Critique* (see Section 6.3.2). The transcendental inquiry aims to state and justify general synthetic *a priori* conditions for our judgments to be objective – to be determinately true or false of objects of experience, no matter what the specific ontological nature of these objects is (whether they be material or mental, what we mean by "material," "mental," and so on). The metaphysical principles of natural science, in turn, do not show that and why empirical judgments of physics are true, but the conditions under which such judgments can be determinately true or false (Brittan 1978, 28–35; Friedman 2001a, 73–75). Put differently, the distinctive task of *MFNS* is to explain the (real, not merely logical) possibility of empirical, and more specifically mathematical, natural science: such as measuring the distance between the Moon and the Earth (IV:482), studying rectilinear and curved motions (IV:483, 562) as well as the attraction between planets and their satellites (IV:515), determining the axial rotation of the Earth relative to the stars (IV:560), and perhaps even Kant's own nebular hypothesis concerning the formation of the universe (albeit he never gives that hypothesis a precise mathematical form). The concept of matter is explicated using the categories such that relevant judgments obtain a determinate meaning, so that they can thereafter be confirmed or rejected empirically and/or mathematically. One consequence of this is that the term 'Foundations' in the title of *MFNS* has to be understood moderately, and perhaps it is not even just to translate the German term *Anfangsgründe* by 'Foundations' in the first place: Kant presents principles from which research can *begin*, not a set of basic, intuitively known axioms from which one could already derive, by means of demonstration, all the knowledge contained in the relevant science. He does not follow an overly demanding form of rationalism, like that derived from the Aristotelian definition of *episteme*.[22] Instead, he is trying to show the conditions under which quantitative empirical research of matter in motion is possible.

However, a defense of the metaphysical principles as necessary conditions under which physical judgments can be determinately true or false may be too weak. After all, perhaps one can explain the possibility of empirical science, for instance, of Kepler's laws or of other features of the Solar system, by means of *different* principles? In this way, the classical objection against Kant's transcendental principles (see Section 6.3.1) resurfaces, once again, at the level of metaphysical foundations of natural science: they would lose their status as unchangeable truths. This objection leads us to more specific epistemological features of *MFNS*. The principles developed in its four main chapters are, in Kant's view, synthetic and *a priori*; and since they are *a priori*, they must be necessary in some sense.[23] That is what needs to be justified. They are synthetic because they involve predicate concepts that are not contained in the subject concepts at issue the respective principles. In other words, they are not analytic. The empirical concept of matter as the movable in space is determined completely through the four types of categories, thus creating synthetic judgments about physical objects (cf. also IV:495, 523, 551, 558). More importantly for countering the objection, these judgments are *a priori*, not only because they are presuppositions of existing cognition in natural science (such as the empirical laws of planetary motion, or the law of gravity), but because their apriority is justified by nonempirical arguments. The transcendental principle that "every alteration has a cause" becomes applied to material objects such that its metaphysical counterpart states that "every alteration

Kant's Conception of Metaphysical Foundations of Natural Science 165

in matter has an external cause" (IV:543). But this principle is not simply derived from the transcendental one by means of a substitution of the subject term. Instead, it requires additional premises pertaining directly to the metaphysical nature of matter. Accordingly, in *MFNS* Kant supports each metaphysical principle of physics by a non-empirical proof of its own. These proofs are what one needs to look at in order to assess the claims of apriority and necessity of the principles (Watkins 1998a; cf. de Bianchi 2022; Messina 2022; Stan 2022). Insofar as the proofs do not deliver, we must be prepared to revise the epistemic or modal status of these principles, if not the principles themselves. Conversely, the weight of the objection of revisability can be judged fairly only if one takes a closer look at Kant's proofs. A mere pointing to historical changes in philosophical frameworks will not do.

The analysis I have given of the epistemological or methodological specificities of Kant's idea of *MFNS* is not complete. Among other things, a complete analysis would have to include his considerations concerning the distinction between two types of "rational" cognition, namely metaphysical and mathematical cognition. It would have to explain Kant's claims (1) that these are distinct types of cognition (IV:469), (2) that any mathematical cognition of physical objects presupposes a metaphysical construction of the concept of these objects, and (3) that the widespread rejection of metaphysics by natural scientists should not be accepted (IV:472–479; cf., e.g. Plaass 1965, ch. 3; Pollok 2001, 68–70, 110–37; Friedman 2013, 29–34). Simply noting these additional features of Kant's conception of *MFNS* shows, once again, how this work does not aim to provide mere applications of the transcendental principles, but also pursues an agenda of its own.

Finally, we come to the end or ends of *MFNS*. As shown above (Section 6.4.2), Kant distinguishes between two types of ends: first, an end that is purely epistemic or internal to a given science; and second, an external end that "one has in mind for this science itself in uses elsewhere" (IV:477). With respect to *MFNS*, this difference must be understood as follows. On the one hand, the integration of cognitions in physics is not possible without an "inner end" of *MFNS*. That end is to state and justify, in a systematic way, the *a priori* concepts and principles essential for the metaphysical determination of matter. This "inner end" is intelligible, at least given certain Kantian assumptions, and supported by intelligible notions of the method by which it can be achieved. Aim and method fit together. On the other hand, Kant delivers a clear plea to keep traditional preoccupations of metaphysics – with which he still deals, if critically, in the Dialectic of *CPR* – out of the inquiries into the metaphysical foundations of natural science. The "inner end" stands for itself; it need not and should not be burdened with the "external end" to discover "that which lies wholly beyond all boundaries of experience, of God, Freedom, and Immortality."

Thus, for instance, natural science should no longer be burdened with purposes of physicotheology.[24] Kant also illustrates the relation between traditional, precritical metaphysics, and the critically informed metaphysical foundation of natural science by an organismic metaphor: "one gains in the advancement of this goal" – namely the traditional metaphysical interest in God, Freedom, and Immortality – "if one frees it from an offshoot that certainly springs from its root, but nonetheless only hinders its regular growth" (IV:477). He is fully aware of historical dependencies and contingencies in the development of disciplines, but he encourages the reader to promote a division of cognitive labor, whenever possible for the rational progress of our scientific understanding of reality and our place in it.

In sum, a close analysis of the conception of *MFNS* shows that Kant explicates it in three basic dimensions – determination of subject-matter (ontology), of type and source of knowledge (epistemology/methodology), and of ends (axiology). These form the coherent set of elements in his definition of pure natural science. Thereby, he conceptually distinguishes the metaphysical foundations of natural science from his transcendental philosophy. This is one way of making sense of two somewhat obscure claims of Kant's concerning his division of metaphysical disciplines: (1) the *CPR* is engaged in "criticizing" the very faculty of reason or in a "self-cognition" of reason, including a critical discussion of rationalistic metaphysics, which implies that transcendental philosophy is not itself part of the "system of pure reason" (A841/B869); and (2) that *MFNS*, in contrast, is a specific part of the system of pure reason: a metaphysics of the concept of matter.

Thus, *MFNS* depends on *CPR*, but not the other way around. We understand that now, in a more principled manner than through the discussion of the relation between the Analogies of Experience and their counterparts in *MFNS*, the Laws of Mechanics. While we should never overlook the connections between Kant's two projects, we should also not ignore that he understood each resulting doctrine as a unified whole of cognitions. If we ignore disciplinary differences, it may be to the disadvantage of each system – potentially a step back on its way to becoming a successful science.[25]

Notes

1 Many interpreters also take the Preface to contain Kant's official and final statement of the general concept of science, *Wissenschaft* (e.g. Walsh 1940; van den Berg 2014, ch. 2). Following Plaass (1965), I have expressed my doubts about such a reading: it does not pay attention to the fact that Kant distinguishes between *Wissenschaft* and *Naturwissenschaft* (cf. Sturm 2009, ch. 3, 2015; Sturm and de Bianchi 2015). The explication given in the *Metaphysical Foundations* was developed for this book's project, and the concepts of both science and natural science kept on changing afterwards too.

Kant's Conception of Metaphysical Foundations of Natural Science 167

2. In September 1785, Kant wrote to Christian Gottfried Schütz, the early promotor of his critical philosophy. Kant promised that *MFNS* would contain an appendix on metaphysical foundations of empirical psychology (X:406). However, no such Appendix exists. Instead, Kant presents criticisms concerning psychology and chemistry – arguments not found in his previous publications (nor in other source materials) and, in the case of chemistry, later retracted (Carrier 2001; McNulty 2016). We will later see an even more important example of Kant's changing views (Section 6.3.2).
3. The most extensive interpretations of the Preface are Plaass (1965), a still highly valuable book devoted to the Preface, and Pollok's historical-critical commentary on *MFNS*, which devotes more than 100 pages to the Preface alone (Pollok 2001, 45–179). Friedman (2013) contains extensive remarks on the Preface in the book's introduction (focusing on the broader historical background of the place of *MFNS* within Kant's critical system; Friedman 2013, 1–34) and its conclusion (discussing the "complementary" perspectives of *MFNS* and *CPR*; Friedman 2013, 563–608).
4. At A480/B508, Kant adds pure moral philosophy to the group of *Vernunftwissenschaften*.
5. Also note that, in *MFNS*, Kant speaks of "rational science" (*rationale Wissenschaft*), a term that is not identical in meaning with that of *Vernunftwissenschaften*. Rational science only requires that its cognitions be unified inferentially, as "grounds" and "consequences" (IV:468). There are no restrictions concerning the subject matter or the epistemic status of cognitions contained in such a science; only the form of their connection matters.
6. Kant sometimes uses the term "philosophy" in a broader sense, such that metaphysics is only a branch of it. He also uses the term "general metaphysics" as inclusive of critical or transcendental philosophy. I will ignore these complications here; see Baum (2015a, b).
7. Another question that I will leave aside here concerns whether *MFNS* is, in Kant's view, to be identified with the "metaphysics of nature," or to be seen as providing only a part of it. See X:406, Pollok (2001, 1–4), and Baum (2015c).
8. Debates about parts of Euclid's geometry, such as the parallel line postulate, had already begun in the eighteenth century, with works by Johann Heinrich Lambert, Abraham Gotthelf Kästner, and Johann Schultz, among others. For how far Kant took notice of these developments, see Heis (2020).
9. Besides these options, there are other interpretations (see Watkins 2005; Kannisto 2017).
10. Watkins (1998a) and Lyre (2006, 7–9) note mismatches between Newton's and Kant's laws of motions. Watkins (1997, 1998b) and Stan (2013) argue that Leibniz and Wolff played a role in Kant's understanding of the laws of mechanics. Friedman, who originally emphasized Newton's influence, has since accepted this point (Friedman 2013, xiv; cf. 26n37).
11. Kant has a moderate correspondence concept of truth, combined with a pluralism about criteria of truth. This account of truth shapes and guides his transcendental inquiry (Sturm 2018).
12. I thus follow Guyer and Wood's translation. For a fine textual analysis of the terminology – even though I do not follow it in all its details – see Willaschek and Watkins (2017).
13. For these and other options, see Allison (1994, 298–300) and Watkins (2005, 196–202). As Watkins makes clear in his reconstruction of Kant's argument for the Second Analogy (2005, 207–15), the epistemic nature of the Analogy

does not reduce to a merely psychological or phenomenological claim. Since Kant ultimately aims to show how cognition (*Erkenntnis*) is possible, and since cognition implies a claim to *truth*, cognizing a causal relation implies that there is something objective that makes such a claim true if it is true.

14 Einstein claimed that we must reject quantum mechanics because of its (alleged) indeterminism; therefore, relativity theory does not look like in violation of the moderate claims about causation that I ascribed to Kant in the first place. Quantum mechanics, in contrast, has indeed often been viewed as raising problems for the universal validity of Kant's principle of causality: among other things, because it is believed that the theory does not allow for an arbitrarily precise description of initial states of affairs and that it does not predict what will happen but only offers (objective) probabilities. But these issues are highly disputed. For balanced discussions, see Brittan (1978, 205–8), Brittan (1994), Mittelstaedt (1994), and Hoefer (2016).

15 A complete discussion of the interpretations that defend a close link between the Analogies of Experience and the laws of motion as Kant presents them in *MFNS* would have to distinguish between the following three claims (at least): (1) The Analogies of Experience require for their *meaning* instances of concrete application to spatial, i.e., material objects (what might be called the "semantic thesis"). (2) The sole function of the Analogies of Experience is to provide a *foundation* for Newton's laws of motion (the "grounding thesis"). This thesis might be understood as either claiming (2a) that Kant justifies the laws of motion on the basis of the Analogies alone and that this justification is their sole function, or, more plausibly, (2b) that the sole function of the Analogies is to provide necessary conditions, or presuppositions for the validity of these laws (Friedman 1992b, 2013; Lyre 2006). (3) The Analogies of Experience (and, especially, their proofs) are incomplete without metaphysical laws of motion (the Newtonian laws of motion, but, again, as Kant formulates them in *MFNS*) (the "completion thesis"). On this reading, the principles of *MFNS* are not merely required as presuppositions of empirical science; they also help to fill a gap in Kant's own transcendental philosophy (Westphal 1995). It would go beyond the scope of this chapter to discuss these interpretations.

16 When this problem is not noticed, scholars (e.g. Röd 1991, 118; Lyre 2006, 4) fail to distinguish between Kant's considered concepts of transcendental philosophy and of pure natural science. For more details, see Sturm and de Bianchi (2015).

17 There remain two passages in the second edition of *CPR* where Kant still seems confused about a related issue: at B3, he calls the transcendental principle of causality *a priori* but "not entirely pure," whereas barely one page later, he cites it as an example of a "pure *a priori*" judgment (B4–5). On this, see Cramer (1985).

18 I use the version of the First Analogy in the B-edition of *CPR*; in the A-edition (A182), the Analogy is formulated differently, but the point discussed here is not affected by this. In his own copy of the A-edition, Kant added comments on the proof of this Analogy, but did not change the proof as such.

19 Watkins (1998a, 572f., 575, 585) rightly notes problems here. For instance, a projection of the Second Analogy must involve more than a mere "substitution": if Kant only wanted to replace the notion of an object of cognition in general, one would only move from "all changes must have a cause" to "all changes of matter must have a cause." That the cause must be an external one has to do with a non-trivial assumption concerning matter: namely, its inertia

(cf. the addendum to the Second Law of Mechanics; IV:543). In the case of the Third Analogy, Kant weakens the connection between the transcendental and the metaphysical levels too.
20 This term is Laudan's (1984, 140): an axiology is a "set of cognitive goals or ideals."
21 I add this qualification since (a) Kant revises his views about chemistry a few years later (Carrier 2001; McNulty 2016) and (b) his criticism of empirical psychology does *not*, as many interpreters claim, result in an unrestricted impossibility claim; instead, the claim is a restricted one (see Nayak and Sotnak 1995; Sturm 2001, 2009, ch. 4).
22 I here broadly agree with Plaass (1965) and disagree with van den Berg (2014).
23 For instance, Kant rejects the view that the law of equality of action and reaction is empirical. He thereby distances himself from Newton, who "by no means dared to prove this law *a priori*, and therefore appealed rather to experience" (IV:549).
24 The attempt to deal with physicotheology had occupied the young Kant in many sections of his early masterpiece, the *Universal Natural History and Theory of the Heavens* (1755).
25 For comments and criticisms, I am highly grateful to Angela Breitenbach, Gabriele Gava, Rudolf Meer, Michael Bennett McNulty, Bernhard Thöle, Eric Watkins, to an anonymous referee, and to Max Dresow, who also provided linguistic assistance.

Bibliography

Adickes, Erich. 1924–1925. *Kant als Naturforscher*. 2 vols. Berlin: De Gruyter.
Allison, Henry. 1994. "Causality and Causal Law in Kant: A Critique of Michael Friedman." In *Idealism and Freedom*, edited by Henry Allison, 80–91. Cambridge: Cambridge University Press.
Baum, Manfred. 2015a. "Metaphysik." In *Kant-Lexikon*, edited by Günther Mohr, Jürgen Stolzenberg, and Marcus Willaschek, Vol. II, 1530–40. Berlin: De Gruyter.
Baum, Manfred. 2015b. "Metaphysik, allgemeine/spezielle." In *Kant-Lexikon*, edited by Günther Mohr, Jürgen Stolzenberg, and Marcus Willaschek, Vol. II, 1540–1. Berlin: De Gruyter.
Baum, Manfred 2015c. "Metaphysik der Natur." In *Kant-Lexikon*, edited by Günther Mohr, Jürgen Stolzenberg, and Marcus Willaschek, Vol. II, 1541–2. Berlin: De Gruyter.
Beck, Lewis White. 1978. "A Prussian Hume and a Scottish Kant." In *Essays on Hume and Kant*, edited by Lewis White Beck, 111–29. New Haven: Yale University Press.
Breitenbach, Angela. 2022. "Kant's Normative Conception of Natural Science." In *Kant's 'Metaphysical Foundations of Natural Science': A Critical Guide*, edited by Michael Bennett McNulty, 36–53. Cambridge: Cambridge University Press.
Breitenbach, Angela, and Michela Massimi, eds. 2017. *Kant and the Laws of Nature*. Cambridge/England: Cambridge University Press.
Brittan, Gordon. 1978. *Kant's Theory of Science*. Princeton: Princeton University Press.
Brittan, Gordon. 1994. "Kant and the Quantum Theory." In *Kant and Contemporary Epistemology*, edited by Paolo Parrini, 131–55. Dordrecht: Springer.

Buchdahl, Gerd. 1969. *Metaphysics and the Philosophy of Science*. Cambridge: Belknap Press.
Buchdahl, Gerd. 1986. "Kant's 'Special Metaphysics' and the *Metaphysical Foundations of Natural Science*." In *Kant's Philosophy of Physical Science*, edited by Robert Butts, 127–61. Dordrecht: Reidel.
Butts, Robert. 1986. "Introduction: Kant's Quest for a Method in Metaphysics." In *Kant's Philosophy of Physical Science*, edited by Robert Butts, 1–22. Dordrecht: D. Reidel.
Carrier, Martin. 2001. "Kant's Theory of Matter and His Views on Chemistry." In *Kant and the Sciences*, edited by Eric Watkins, 205–30. New York: Oxford University Press.
Cassirer, Ernst. 1921. *Zur Einsteinschen Relativitätstheorie*. Berlin: Bruno Cassirer (English translation: *Einstein's Theory of Relativity*. Chicago: Open Court, 1923).
Cramer, Konrad. 1985. *Nicht-reine synthetische Urteile a priori*. Heidelberg: Winter.
de Bianchi, Silva. 2022. "How Do We Transform Appearance into Experience? Kant's Metaphysical Foundations of Phenomenology." In *Kant's 'Metaphysical Foundations of Natural Science': A Critical Guide*, edited by Michael Bennett McNulty, 197–214. Cambridge: Cambridge University Press.
Falkenburg, Brigitte. 2000. *Kants Kosmologie*. Frankfurt: Klostermann.
Friedman, Michael. 1992a. *Kant and the Exact Sciences*. Cambridge: Cambridge University Press.
Friedman, Michael. 1992b. "Causal Laws and the Foundations of Natural Science." In *The Cambridge Companion to Kant*, edited by Paul Guyer, 161–99. Cambridge: Cambridge University Press.
Friedman, Michael. 2001a. *Dynamics of Reason*. Stanford: CSLI Publications.
Friedman, Michael. 2001b. "Matter and Motion in the *Metaphysical Foundations* and the First *Critique*: The Empirical Concept of Matter and the Categories." In *Kant and the Sciences*, edited by Eric Watkins, 53–69. New York: Oxford University Press.
Friedman, Michael. 2002. "Kant, Kuhn and the Rationality of Science." *Philosophy of Science* 69: 171–90.
Friedman, Michael. 2013. *Kant's Construction of Nature: A Reading of the 'Metaphysical Foundations of Natural Science'*. Cambridge: Cambridge University Press.
Gloy, Karen. 1976. *Die Kantische Theorie der Naturwissenschaft*. Berlin: DeGruyter.
Hacking, Ian. 1983. *Representing and Intervening*. Cambridge: Cambridge University Press.
Heis, Jeremy. 2020. "Kant on Parallel Lines: Definitions, Postulates, and Axioms." In *Kant's Philosophy of Mathematics, Vol. 1: The Critical Philosophy and Its Roots*, edited by Carl Posy and Ofra Rechter, 157–80. Cambridge: Cambridge University Press.
Hoefer, Carl. 2016. "Causal Determinism." *Stanford Encyclopedia of Philosophy*. https://plato.stanford.edu/entries/determinism-causal/
Kannisto, Toni. 2017. "Kant on the Necessity of Causal Relations." *Kant-Studien* 108: 495–516.
Kitcher, Philip. 1984. "Kant's Philosophy of Science." In *Self and Nature in Kant's Philosophy*, edited by Allen W. Wood, 185–215. Ithaca: Cornell University Press.
Körner, Stephan. 1955. *Kant*. Harmondsworth: Penguin.

Krüger, Lorenz. 2005. "Causality and Freedom." In Lorenz Krüger, *Why Does History Matter to Philosophy and the Sciences? Selected Essays*, edited by Thomas Sturm, Wolfgang Carl, and Lorraine Daston, 155–67. Berlin: De Gruyter.
Laudan, Larry. 1984. *Science and Values*. Berkeley: University of California Press.
Lefèvre, Wolfgang, ed. 2001. *Between Leibniz, Newton, and Kant: Philosophy and Science in the Eighteenth Century*. Dordrecht: Kluwer.
Lyre, Holger. 2006. "Kants *Metaphysische Anfangsgründe der Naturwissenschaft*: gestern und heute." *Deutsche Zeitschrift für Philosophie* 54: 1–16.
Massimi, Michela, ed. 2013. *Philosophy of Natural Science from Newton to Kant*. Special Issue, *Studies in History and Philosophy of Science*, 44.
McNulty, Michael Bennett. 2016. "Chemistry in Kant's Opus Postumum." *HOPOS* 6: 64–95.
Messina, James. 2022. "Space, Pure Intuition, and Laws in the Metaphysical Foundations." In *Kant's 'Metaphysical Foundations of Natural Science': A Critical Guide*, edited by Michael Bennett McNulty, 98–118. Cambridge: Cambridge University Press.
Mittelstaedt, Peter. 1994. "The Constitution of Objects in Kant's Philosophy and in Modern Physics." In *Kant and Contemporary Epistemology*, edited by Paolo Parrini, 115–29. Dordrecht: Springer.
Nayak, Abhaya C., and Eric Sotnak. 1995. "Kant on the Impossibility of the 'Soft Sciences'." *Philosophy and Phenomenological Research* 55: 133–51.
Plaass, Peter. 1965. *Kants Theorie der Naturwissenschaft*. Göttingen: Vandenhoek & Ruprecht. (in English: *Kant's Theory of Natural Science*, transl. by A. Miller and M. Miller. Dordrecht: Kluwer, 1994).
Pollok, Konstantin. 2001. *Kants 'Metaphysical Foundations of Natural Science': Ein Kritischer Kommentar*. Hamburg: Meiner.
Reichenbach, Hans. 1920. *Relativitätstheorie und Erkenntnis Apriori*. Berlin: Julius Springer. (=Reichenbach, Hans. 1979 *Gesammelte Werke*, edited by Andreas Kamlah and Maria Reichenbach. Vol. 3, 193–302. Braunschweig: Vieweg.)
Röd, Wolfgang. 1991. "Kants Reine Naturwissenschaft als kritische Metaphysik." *Dialectica* 45: 118–31.
Schönfeld, Martin. 2000. *The Philosophy of the Young Kant*. New York: Oxford University Press.
Stan, Marius. 2013. "Kant's Third Law of Mechanics: The Long Shadow of Leibniz." *Studies in History and Philosophy of Science* 44: 493–504.
Stan, Marius. 2022. "Phoronomy: Space, Construction, and Mathematizing." In *Kant's 'Metaphysical Foundations of Natural Science': A Critical Guide*, edited by Michael Bennett McNulty, 80–97. Cambridge: Cambridge University Press.
Strawson, Peter. 1966. *The Bounds of Sense*. London: Methuen.
Sturm, Thomas. 2001. "How Not to Investigate the Human Mind: Kant on the Impossibility of Empirical Psychology." In *Kant and the Sciences*, edited by Eric Watkins, 163–84. New York: Oxford University Press.
Sturm, Thomas. 2009. *Kant und die Wissenschaften vom Menschen*. Paderborn: Mentis.
Sturm, Thomas. 2012. "Kant über die dreifache Beziehung zwischen den Wissenschaften und der Philosophie." *Internationales Jahrbuch des Deutschen Idealismus* 8: 60–82.
Sturm, Thomas. 2015. "Wissenschaft." In *Kant-Lexikon*, edited by Mohr, Günther, Jürgen Stolzenberg, and Marcus Willaschek, Vol. III, 2670–5. Berlin: De Gruyter.

Sturm, Thomas. 2018. "Lambert and Kant on Truth." In *Kant and His German Contemporaries*, edited by C. Dyck and F. Wunderlich, 113–33. Cambridge: Cambridge University Press.

Sturm, Thomas. 2020. "Kant on the Ends of the Sciences." *Kant-Studien* 111: 1–28.

Sturm, Thomas, and Silvia de Bianchi. 2015. "Naturwissenschaft." In *Kant-Lexikon*, edited by Günther Mohr, Jürgen Stolzenberg, and Marcus Willaschek, Vol. II, 1643–50. Berlin: De Gruyter.

Thöle, Bernhard. 2004. "Immanuel Kant – Wie sind synthetische Urteile *a priori* möglich?" In *Klassiker der Philosophie heute*, edited by Ansgar Beckermann and Dominik Perler, 376–98. Stuttgart: Reclam.

van den Berg, Hein. 2014. *Kant on Proper Science*. Dordrecht: Springer.

Walsh, William H. 1940. "Kant's Conception of Scientific Knowledge." *Mind* 49: 445–50.

Watkins, Eric. 1997. "The Laws of Motion from Newton to Kant." *Perspectives on Science* 5: 311–48.

Watkins, Eric. 1998a. "Kant's Justification of the Laws of Mechanics." *Studies in History and Philosophy of Science* 29: 539–60.

Watkins, Eric. 1998b. "The Argumentative Structure of Kant's *Metaphysical Foundations of Natural Science*." *Journal of the History of Philosophy* 36: 567–93.

Watkins, Eric. 2005. *Kant and the Metaphysics of Causality*. Cambridge: Cambridge University Press.

Watkins, Eric, and Marius Stan. 2014. "Kant's Philosophy of Science." *Stanford Encyclopedia of Philosophy*. https://plato.stanford.edu/entries/kant-science/#PhyCriPerMetFouNatSci.

Westphal, Kenneth R. 1995. "Does Kant's *Metaphysical Foundations of Natural Science* Fill a Gap in the *Critique of Pure Reason*?" *Synthese* 103: 43–86.

Willaschek, M., and E. Watkins. 2017. "Kant on Cognition and Knowledge." *Synthese* 197: 3195–213.

7 Systematicity, the Life Sciences, and the Possibility of Laws Concerning Life

Hein van den Berg

7.1 Introduction

According to Kant, different natural sciences should be related to each other in such a way that they constitute a systematic unity. However, the life sciences have often been taken to threaten Kant's ideal of the systematic unity of different natural sciences. Whereas in physics we can provide mechanical explanations of natural phenomena, organisms resist mechanical explanation. Moreover, organisms require a special kind of teleological judgment, which is not employed in the exact sciences. For reasons such as these, authors such as Zammito (2003) and Guyer (2001) argue that Kant's views on organisms and the life sciences, as articulated in the third *Critique*, are difficult to square with his ideal of the systematic unity of natural science.

In this article, I will argue that there is a sense in which sciences such as physics, chemistry, and the life sciences constitute a unity. On the basis of an analysis of Wolff's and Kant's views on the hierarchy of the sciences, I argue that one sense in which different sciences constitute a unity is when more fundamental sciences provide statements which are used in less fundamental sciences to prove statements. For example, metaphysics is a more fundamental science than physics, i.e., physics presupposes results from metaphysics, and statements from metaphysics are used to prove statements in physics. In the same way, I argue, the life sciences, according to Kant, borrow statements from physics and chemistry in order to prove statements in the life sciences. I will express this state of affairs by saying that physics and chemistry *ground* the life sciences. Insofar as physics and chemistry ground the life sciences, these different sciences constitute a unity. Hence, Kant allowed for the ideal of a systematic unity among physics, chemistry, and the life sciences and, in the case of some features of organisms, took physics and chemistry to explain phenomena in the life sciences. However, the unity of physics, chemistry, and the life sciences is limited since Kant takes the purposeful unity of organisms to be

DOI: 10.4324/9781003166450-10

mechanically inexplicable. I further argue that although there is in some sense a unity between physics, chemistry, and the life sciences, the life sciences do not contain laws that are specific to these sciences. The reason that the life sciences of Kant's time do not, according to Kant, have laws is that biological regularities described in Kant's time (i) concern the purposeful unity of organisms which according to Kant is mechanically inexplicable, and (ii) these regularities could not be systematically related to the a priori foundations of natural science. Thus, whereas Kant allows for the idea that some features of organisms could be explained in terms of the laws of physics and regularities of chemistry, the scientific practice of his time did not allow him to fully articulate the ideal of a systematic unity of physics, chemistry, and the life sciences.

I proceed as follows. In the first section, I describe communalities between Wolff's and Kant's views on the hierarchy of the sciences. I show that Wolff and Kant both adopt the idea that some sciences borrow statements from preceding sciences in order to provide proofs. In the second section, I show that according to Kant we must reflect on organisms in mechanistic terms, which implies that we must provide mechanical explanations of organisms in so far as this is possible. This entails, as I will show on the basis of two case studies, that statements from physics and chemistry are used in the life sciences in order to provide proofs. In this sense, physics, chemistry, and the life sciences constitute a systematic unity. In the third and final section, I consider, in discussion with Breitenbach (2017), the question whether Kant allowed for the possibility of laws in the life sciences. I argue that although such laws may be in principle possible for Kant, he could not take the life sciences of his day to possess laws. The reason is that the regularities discussed in the life sciences of Kant's time concerned the mechanically inexplicable purposeful unity of organisms and could not be systematically related to the a priori principles of natural science.

7.2 Wolff and Kant on the Hierarchy of the Sciences

According to Kant, not only individual sciences should constitute systematic wholes. The relations among different sciences should also be constituted in such a way that these different sciences constitute a systematic unity. This is what Thomas Sturm calls "external systematicity": "Ideally, an 'architectonical mind' works towards reaching a complete system of special sciences, whereby we understand how metaphysics, mathematics, physics, chemistry, biology, medicine, geography, anthropology, history, law, and so on are different yet stand in well-ordered relations to one another" (Sturm 2020, 7). In the present section, I will not provide an exhaustive analysis of Kant's conception of external systematicity (see for one of the most extensive accounts, Sturm 2009). Rather, I will focus on one specific aspect of this view: the idea that more fundamental sciences provide concepts and propositions that are used by less fundamental sciences. This aspect of Kant's thought comes into sharp focus if we compare Kant's views on the hierarchy of the sciences with Christian Wolff's views on the hierarchy of the sciences.

An influential conception of the hierarchy of sciences was articulated by Christian Wolff, who dominated the philosophical landscape in the early eighteenth century. According to Wolff, sciences constitute a hierarchy, with more fundamental sciences providing concepts or propositions that are used in less fundamental sciences. As Wolff explains, for example, in the *Preliminary Discourse* (1963 [1728]), the science of ontology provides concepts and propositions that are used in demonstrations in sciences such as physics. As Wolff puts the point:

> Such general notions are the notions of essence, existence, attributes, modes, necessity, contingency, place, time, perfection, order, simplicity, composition, etc. These things are not explained properly in either psychology or physics because both of these sciences, as well as the other parts of philosophy, use these general notions and the principles derived from them. Hence, it is quite necessary that a special part of philosophy be designated to explain these notions and general principles, which are continually used in every science and art, and even in life itself, if it is to be rightly organized. Indeed, without ontology, philosophy cannot be developed according to the demonstrative method.
> (Wolff 1963 [1728], 40)

In line with this view on the hierarchy of sciences, Wolff argues that metaphysics must provide the foundations of physics if we are to give proper demonstrations in physics. The reason for this is that metaphysics provides grounds or reasons that explain phenomena discussed in physics. Hence, principles from metaphysics must lie at the basis of demonstrations in physics:

> If everything is to be demonstrated accurately in physics, then principles must be borrowed from metaphysics. Physics explains those things which are possible through bodies (#59). If these things are to be treated demonstratively, then the notions of body, matter, nature, motion, the elements, and other such general notions must be known. For such notions contain the reason of many things. Now these notions are explained in general cosmology and ontology (##73, 78). Therefore, if all things are to be demonstrated accurately in physics, principles must be borrowed from general cosmology and ontology.
> (Wolff 1963 [1728], 48)

We can illustrate Wolff's views by looking at his *German Physics*. In the first chapter of his German physics, Wolff explicates the essence and nature of bodies, which is a metaphysical topic. In the subsequent chapters, Wolff applies metaphysical propositions to results from experimental physics to provide demonstrations in physics. For example, Wolff proves in this way that bodies cannot be completely dense (*vollkommen dichte*), i.e., there are no bodies without any empty spaces (Wolff 2003 [1723], 67). Wolff starts by noting that observation and experiment teaches us that gold is the most dense body we know. However, gold has empty spaces between its parts. If the question is whether a completely dense body is possible, Wolff first cites the proposition of physics that a body is completely dense if it is continuously made up of matter, and the parts of the body are only different from each other qua location. From this it follows that all the parts of matter are similar to each other. Wolff then cites a proposition from metaphysics, according to which it is impossible that the smallest parts of a body are similar to each other. From this he concludes, within physics, that a matter cannot be completely dense (Wolff 2003 [1723], 70). This reduction shows how Wolff uses propositions from metaphysics and physics to demonstrate propositions in physics.

Although Kant's views on the hierarchy of sciences differ from those of Wolff (I will return to this point below), he shares the idea that sciences constitute a hierarchy and that concepts and propositions of more fundamental sciences can be used in less fundamental sciences. Evidence for such a reading of Kant comes, for example, from the *Critique of Judgment* and the *Jäsche Logic*. In the former, Kant argues that more fundamental sciences provide so-called auxiliary propositions (*lemmata*) that are used in less fundamental sciences:

> The principles of a science are either internal to it, and are then called indigenous (*principia domestica*), or they are based on principles that can find their place only outside of it, and are *foreign* principles (*peregrina*). Sciences that contain the latter base their doctrines on auxiliary propositions (*lemmata*), i.e., they borrow some concept, and along with it a basis for order, from another science
> (V:381. See for discussion of this quote in the context of Kant's views on teleology, van den Berg 2013.)

How should we precisely understand this quote? A "principle" is a technical term that denotes an a priori judgment from which other judgments are derived and which is itself not derived from other judgments. In the *Jäsche Logik*, Kant defines principles as follows:

> Immediately certain judgments a priori can be called principles, insofar as other judgments are proved from them, but they themselves

cannot be subordinated to any other. On this account they are also called *principles* (beginnings).

(IX:110)

Such principles can thus be either internal to a science or in a science we use principles that are foreign to this science. Foreign principles are called *lemmata*. In the *Jäsche Logik*, Kant defines lemmata as follows: "Propositions that are not indigenous to the science in which they are presupposed as proved, but rather are borrowed from other sciences, are called *lemmas (lemmata)*" (IX:113). This conception of lemmata was standard in Kant's time. In his *Neues Organon* (1764), for example, Lambert defines lemmata as statements which are not proven at the place in which they are used, but are borrowed from a preceding science (*vorgehende Wissenschaft*) (Lambert 1764, 99). Hence, according to Lambert and Kant lemmata are (i) presupposed as proved in a science and (ii) borrowed from another science. This characterization perfectly fits Wolff's views on the hierarchy of sciences. As we have seen with respect to the relationship between metaphysics and physics: Wolff (a) presupposed statements from metaphysics as proved in physics and (b) borrowed these statements to prove other statements in physics.

We can, following van den Berg (2013, 731), also give an example from Newton's *Principia* to illustrate Kant's views on the role of statements in a science that are borrowed from another science. In Book III of the *Principia*, Newton borrows several mathematical or kinematical statements, demonstrated in the first books of the *Principia*, to prove statements within natural philosophy. For example, Newton starts Book III with listing phenomena, among which phenomenon 1, which states that the satellites of Jupiter "by radii drawn to the center of Jupiter, describe areas proportional to the times, and their periodic times – the fixed stars being at rest – are as the 3/2 powers of their distances from the center" (Newton 1999 [1726], 797). This is, as Newton explains, an *a posteriori* statement based on astronomical observation. Hence, from Kant's point of view, this is a statement that is *internal* to natural philosophy. In proposition 2 of Book I, Newton had demonstrated the mathematical or kinematical hypothetical proposition that "every body that moves in some curved line described in a plane and, by a radius drawn to a point, either unmoving or moving uniformly forward with a rectilinear motion, describes areas around that point proportional to the times, is urged by a centripetal force tending toward that same point" (Newton 1999 [1726], 446). This is a mathematical or kinematical a priori statement and is thus from Kant's point of view *external* to natural philosophy. Newton applies this a priori statement to phenomenon 1 to derive proposition 1 of Book III, which states, among others, that the forces by which the satellites of Jupiter are drawn away from rectilinear motions are directed to the center of Jupiter

(Newton 1999 [1726], 802) (note that I have only treated part of proposition 1 of Book III and only part of its proof. My account nevertheless accurately describes Newton's procedure). Hence, Newton applies mathematical or kinematical a priori statements, principles *external* to natural philosophy, to a posteriori statements or phenomena, statements *internal* to natural philosophy, in order to derive statements of natural philosophy. This Newtonian example shows that Kant's views on lemmata capture an important aspect of scientific practice.

Up to this point we have pointed out similarities between Wolff's and Kant's views on the hierarchy of the sciences. It is important to note that there are also important differences. One of the most important differences is that for Kant, as Watkins argues (2019, Chapter 4), the principles of natural science are not derived from a more fundamental science but are established by *transcendental arguments* that show how experience of objects of outer sense is possible (see also Sturm 2022). Hence, Kant and Wolff have different views on how to establish the principles of (natural) science: the transcendental perspective of Kant is, not surprising, completely absent in Wolff. Notwithstanding this difference, and other differences which I will not elaborate here, Wolff's, Lambert's, and Kant's views on statements borrowed from preceding sciences are substantially the same.

7.3 Kant and Mechanical Explanations in the Life Sciences

As van den Berg (2014, Chapter 3) has argued, Kant construes mechanistic explanations of nature as ideal explanations that provide proper cognition of nature. Thus, for example, when commenting on the mechanistic maxim in the third *Critique*, a maxim we must follow in science, Kant states that it "indicates that I *should* always reflect on them *in accordance with the principle* of the mere mechanism of nature, and hence research the latter, so far as I can, because if it is not made the basis for research then there can be no proper cognition of nature" (V:387). Moreover, as Breitenbach (2017, 246) has stressed, Kant always emphasizes that we must reflect on organized beings in terms of mechanisms. According to Kant, we must reflect on organisms mechanistically if they are to count as natural beings, which entails, since we also conceptualize organisms teleologically as natural purposes, that we judge mechanisms as means for certain ends:

> [T]he mere teleological ground of such a being is equally inadequate for considering and judging it as a product of nature unless the mechanism of the latter is associated with the former, as if it were the tool of an intentionally acting cause to whose ends nature is subordinated, even in its mechanical laws.
>
> (V:422. Also quoted in Breitenbach 2017, 246)

However, the idea that we must reflect on organisms mechanistically and that mechanical explanations are proper explanations of nature is threatened by Kant's infamous claim that organisms are mechanically inexplicable. There cannot be, as Kant famously put it, a Newton who makes comprehensible even the generation of a blade of grass (V:400). How can we reconcile the view that we must reflect on organisms mechanistically while also doing justice to Kant's idea that organisms are mechanically inexplicable?

Some commentators have taken Kant's claim that organisms are mechanically inexplicable, coupled with his regulative conception of teleology, to imply that Kant could not view life sciences as genuine sciences. Thus, Zammito (2006, 755) states that "The third *Critique* essentially proposed the reduction of life science to a kind of pre-scientific descriptivism, doomed *never* to attain authentic scientificity, never to have its 'Newton of the blade of grass'." Similarly, Richards argues that "the *Kritik der Urteilskraft* delivered up a profound indictment of any biological discipline attempting to become a science" (Richards 2000, 26). Moreover, several authors have argued that the life sciences threaten Kant's ideal of the systematic unity of different (natural) sciences. Guyer (2001, 260), for example, argues that organisms threaten Kant's ideal of the unity of science insofar as "we have good reason to suppose that we can never succeed in bringing all of nature under a single principle attributing a single fundamental power to a single kind of substance." Similarly, Zammito argues that the life sciences are irreconcilable with Newtonian science and thus threaten Kant's ideal of the unity of science: "*any* science involving 'internal purposiveness' becomes irreconcilable with 'Newtonian science'. Indeed, this is the point toward which my whole exposition has been aiming, for it brings into glaring salience the problem of reconciling biology *at all* with Kant's prescriptions for science. Organisms rupture the 'top down'/'bottom up' integration of Kant's scientific system" (Zammito 2003, 102). In contrast to authors such as Richards and Zammito, Breitenbach (2017) argues that Kant allows for naturalistic explanations in the life sciences and allows for the possibility of biological laws, thus opening up the possibility that biology can become a science. Finally, van den Berg (2014) argues that although Richards and Zammito are correct that the life sciences do not constitute proper sciences for Kant, mechanical explanations of many features of organism are possible, since Kant's claim that organisms are mechanically inexplicable must be read as denying the possibility of mechanical explanations of specifically (i) the purposive complex unity of organisms and (ii) the fact that traits of organisms are adaptations but not of any feature of organisms *tout court* (see for an instructive account of biological method in Kant also Geiger 2022). In the following, I argue, drawing on van den Berg and Breitenbach, that Kant indeed allows for the mechanical explanation of many features of organisms and that this implies that physics, chemistry, and the life sciences constitute a systematic unity insofar as judgments from physics and chemistry are used in the life science to provide explanations, in line with Kant's views on the hierarchy of the sciences described in the previous section. However, as we shall see, the unity of physics, chemistry, and the life sciences is limited in scope, insofar as Kant argues that the purposive unity of organisms is mechanically inexplicable.

We have already seen that Kant takes mechanisms to be proper explanations of nature and that we must reflect on organisms in mechanistic terms if they are to be regarded as products of nature. Indeed, Kant prescribes to the life sciences the method of subordinating mechanism to teleology, investigating mechanisms in organisms but viewing these mechanisms as means toward certain ends. This already strongly suggests that Kant thinks that at least some features of organisms can be explained mechanically. Why would Kant insist that we investigate mechanisms in the life sciences if nothing can be mechanically explained? That mechanistic investigation plays a role in the life sciences is also strongly suggested by the scientific context in which Kant operated, in which organisms were in fact investigated mechanically. In the following, I analyze this scientific context.

Let us start by giving a physiological example: the investigation of the human eye. The human eye provides a prototypical example of a complex and purposeful organized organ. According to Kant, we cannot mechanically explain the purposeful unity of the human eye, i.e., we cannot mechanically explain how the different parts of the human eye (the cornea, iris, lens, etc.) came to be purposefully organized in the order and structure which they have, where every part is adapted to the other parts. However, this does not mean that mechanical explanations play no role in explaining the functioning of the human eye. In fact, mechanical explanations of the functioning of the human eye were accepted in Kant's time, which becomes clear if we consider Karsten's *Anleitung zur gemeinnützlichen Kenntniß der Natur, besonders für angehende Aerzte, Cameralisten und Oeconomen* (1783), a textbook on physics that Kant used for his lectures. Karten's treatment of the human eye is guided by the conviction that the eye functions like lens glasses and that therefore the laws of optics can be used to explain the process of vision. He first provides an anatomical description of the eye, describing the different layers of skin of the eye, the lens, the retina, and so forth (1783, 133–6). Then he argues that the light, which falls on the eye is refracted just as in the case of a glass lens. Through this process of refraction, which Karsten explains in great detail, we can explain that an image is formed on the retina, which accompanies the sensation of seeing (1783, 137). Karsten's explanation of the functioning of the human eye is a prototypical example of a mechanical explanation:

we explain the functioning of the eye in terms of the functioning of its parts (see for this account of mechanical explanation McLaughlin 1990; van den Berg 2014). In addition, and important for our present purposes, Karsten makes clear that we can use the laws of optics, a part of natural philosophy, to explain the functioning of the human eye. Hence, we can say that statements from optics are borrowed and applied to the physiological investigation of the human eye. In other words, optics grounds physiology, and the two sciences constitute a unity in this specific sense.

Let us secondly look at contemporary scientific accounts of growth via nutrition (here I draw on van den Berg 2014, 133–7). Kant describes the growth via nutrition of a tree as follows:

> This plant first prepares the matter that it adds to itself with a quality peculiar to its species, which could not be provided by the mechanism of nature outside of it, and develops itself further by means of material which, as far as its composition is concerned, is its own product. For although as far as the components that it receives from nature outside of itself are concerned, it must be regarded only as an educt, nevertheless in the separation and new composition of this raw material there is to be found an originality of the capacity for separation and formation in this sort of natural being that remains infinitely remote from all art.
>
> (V:371)

This quote makes a lot of sense when compared with the contemporary late eighteenth-century *chemical* investigation of the nutrition of plants. As described in Gehler's *Physikalisches Wörterbuch* (1798–1801), the scientist Senebier published his *Recherches sur l'influence de la lumière solaire pour métamorphoser l'air fixe en air pur par la vegetation* in 1783. In this work, as Gehler describes, Senebier argued that the growth of plants occurs in part by the decomposition of carbon dioxide gas into carbon. The carbon is retained in the plant and is used for the generation of parts of plants. Through this chemical process, oxygen is made, which is exuded as oxygen gas (Gehler 1798–1801, vol V, 683–4).

Senebier's theory explains Kant's claim that plants develop by means of material that qua compositions is its product. Senebier recognized that materials providing nutrients for plants are drawn from inorganic nature. These materials are given from without and, are thus, in Kant's terms, an educt. However, plants decompose inorganic compounds and thus generate (new) parts of plants. The chemical composition of these products is newly created. Similarly, Kant claimed that the composition of materials in plants is newly produced. In this sense, the plant is a product and not an educt. For our present purposes, it is important to note that Senebier borrowed statements from chemistry in order to prove statements concerning the nutrition of plants. Hence, chemistry grounds the scientific study of plants, and in this sense these sciences constitute a systematic unity.

To conclude this section, note that Kant's view that physics and chemistry ground the life sciences is limited in scope. The reason is that not all properties of organisms allow of mechanical explanation. As we have already seen in our discussion of the human eye, Kant argues that the purposeful unity of organisms defies mechanical explanation. Hence, physics and chemistry cannot explain the purposeful unity of organisms or organs. However, if we presuppose this purposeful unity as a given, we can mechanically investigate the causal processes that play a role in a purposefully organized organism or organ. Thus, for example, we can take the organization of the human eye as given and consequently investigate how the different parts of the eye mechanically function in securing the possibility of vision. In the next section, we will see that Kant's claim that we cannot properly explain the purposeful unity of organisms implies that he denied that the life sciences of his time possess genuine laws.

7.4 Kant and Laws Concerning Life

On the basis of two case studies we have seen that the life sciences borrow statements from optics, a branch of natural philosophy, and chemistry. Kant, I propose, recognized that sciences such as physics and chemistry ground the life sciences. The practice of providing mechanical explanations whenever possible can be understood as the practice of using statements from physics and chemistry to provide proofs in the life sciences.

Does this mean the life sciences constitute proper sciences or that there are laws regarding life according to Kant? Breitenbach (2017) argues that according to Kant the life sciences can become genuine sciences and that Kant allows for the possibility of what she calls biological laws. Breitenbach, stressing that Kant leaves room for naturalistic or (in my terms) mechanistic explanations of organisms, describes these biological laws as follows:

> Such laws would have to fulfill two desiderata. First, in order to be a genuine law of nature, they would have to be thoroughly naturalistic; that is, they would have to make use exclusively of concepts that determinately apply to natural phenomena. They would have to employ causal, nonteleological concepts. Second, in order to qualify as specifically biological laws they would have to employ some specifically biological concepts. Such concepts would have to be suitable naturalistic, too.
>
> (Breitenbach 2017, 247–8)

Breitenbach notes that Kant did not have knowledge of such biological laws, but that his philosophy allows for the possibility of such laws. I agree with the conclusion that Kant leaves room for the possibility of such laws. However, I think it is interesting to adopt a more historical perspective and to inquire why Kant did not think that the life sciences of his time could have genuine laws, a topic Breitenbach does not discuss. This will be my focus in what follows. I argue that Kant did not think that the life sciences of his time have genuine laws because (i) these sciences were often concerned with explaining the purposeful organization of organisms, a feature of organisms that is mechanically inexplicable according to Kant, and (ii) because the regularities that contemporary life scientists proposed only had inductive support and could not be systematically related to the a priori principles of natural science. In order to make this argument, we will first have to discuss Kant's conception of empirical laws.

There are different competing accounts of Kant's views on empirical laws, which have been aptly summarized by Kreines (2009), Messina (2017), and Breitenbach (2018) – who follows Messina – and McNulty (2015). In the following, I will follow Messina's and Breitenbach's systematization of different accounts of Kant's views on empirical laws.

According to the Best System interpretation, developed by Kitcher among others, the "particular laws of nature are those empirical generalizations that would figure in the best systematization of the empirical data at the ideal end of inquiry" (Breitenbach 2018, 111). Being part of a system of laws also confers necessity to a law, according to this account (Breitenbach 2018). According to a competing interpretation, called the Derivation Account and developed by Michael Friedman, generalizations are empirical laws if they can be derived from the a priori laws of nature (Breitenbach 2018). It is important to add that, according to the Derivation Account, empirical laws of nature are of course not derived *solely* from a priori principles. Additional empirical principles are required to. Thus, for example, according to Friedman's (1992) analysis of Kant's views on the Newtonian deduction of the law of gravitation (which is an *empirical* law), the law of gravitation is deduced on the basis of mathematical principles (a priori principles), metaphysical principles (a priori principles), and empirical generalizations captured by Newton's phenomena (see for Newton's procedure of applying mathematical a priori principles to empirical phenomena section I of this paper). Finally, according to the Necessitation Account, "the necessity of particular laws is grounded in the essential natures of things" (Breitenbach 2018, 112).

These different interpretations all have some level of textual support. In the following, I argue for the validity of the Derivation Account. Let us first consider the textual evidence for the Derivation Account. In the *Metaphysical Foundations*, Kant claims that proper natural science treats its object according to a priori principles (IV:468). A rational doctrine of nature is a proper science if it is based on a priori natural laws, which secures the apodictic certainty of our cognition, i.e., the a priori laws secure that we are in the possession of knowledge in the strict sense (*Wissen*) (IV: 468.). Moreover, Kant argues that laws involve necessity: laws involve the necessity of determinations of an object, and this requires that natural science is based on an a priori part (IV:468–469). (See for a thorough account of the necessity involved in laws, Watkins 2019, chapter 1). More specifically, laws are principles of the necessity of that which pertains to the existence of an object, and this requires, according to Kant, that laws are based on metaphysical principles (IV:468–470). All these remarks suggest that natural laws must be grounded by a priori principles and thus support the Derivation Account.

In the first *Critique*, Kant also provides arguments that explicitly support the Derivation Account. As McNulty stresses (2015, 3), Kant argues in the first *Critique* that the necessity of laws requires a priori grounds. As such, Kant's argument in the first *Critique* is similar to his argument in the *Metaphysical Foundations*. Kant says:

> Even laws of nature, if they are considered as principles of the empirical use of the understanding, at the same time carry with them an expression of necessity, thus at least the presumption of determination by grounds that are a priori and valid prior to all experience. But without exception all laws of nature stand under higher principles of the understanding, as they only apply the latter to particular cases of appearance. Thus these higher principles alone provide the concept, which contains the condition and as it were the exponents for a rule in general, while experience provides the case which stands under the rule.
>
> (A159)

Thus, the necessity of laws requires that they are subsumed under a priori grounds or principles of the understanding. According to the Derivation Account, it is the fact that laws of nature are derived from a priori principles, which are necessary statements, which secures the necessity of the laws of nature. This interpretation makes sense of Kant's claim that the necessity of laws requires them to have a priori grounds. The Best System interpretation, by contrast, has difficulty explaining why laws of nature can be necessary (McNulty 2015, 3; Breitenbach 2018, 112). As McNulty explains: "it is unclear how the mechanisms of the systematizers – the approximation of final science, increasing inferential density of a doctrine – could necessitate the judgments of a science. Verification by lower, entailed judgments could only inductively justify a judgment, and its entailment by higher principles would only necessitate the judgment in the case that these higher principles, themselves, carry necessity" (McNulty 2015, 3). Insofar as the Derivation Account is better able to explain the necessity of laws, it is to be preferred to the best system interpretation. Finally, proponents of the Necessitation Account such as Watkins (2005, 2021, 2019), who was one of the first to articulate this account, and Kreines (2009) claim that the Necessitation Account provides us with a *metaphysical* account of what laws are and how they are ontologically grounded by natures, whereas, for example, Friedman's Derivation Account provides us with an *epistemological* account of how we can have knowledge of laws. As two different perspectives on laws, the Necessitation Account and the Derivation Account, are fully compatible (see for a clear explanation of the Necessitation Account also Stang 2016). However, some adherents of the Necessitation Account, such as Kreines (2009), claim that we often cannot achieve knowledge of particular laws. As Breitenbach remarks on this point (2018, 113) "given Kant's extensive discussion of empirical laws in the context of his account of cognition, it would be somewhat surprising if our principled ignorance of particular laws were his last word on the matter." I fully agree: Kant's discussion of a priori laws of natural science in the *Metaphysical Foundations*, coupled with his views, described by Friedman (1992), of how the a priori laws ground empirical laws such as the law of gravitation strongly suggest that we have cognitive access to laws (not just the a priori laws discussed in the *Metaphysical Foundations*).

My exposition so far seems to leave the Derivation Account as a correct epistemological account of Kant's views on how we can have knowledge of empirical laws. However, authors such as McNulty (2015) and Breitenbach (2017) object to the Derivation Account because very few laws can be derived from a priori principles together with appropriate empirical principles. Hence, the implication of the Derivation Account is that according to Kant there are very few laws. In sciences such as physics there seem to be laws, such as the law of gravitation, but in sciences such as chemistry and the life sciences, the sciences treated by McNulty and Breitenbach, there seem to be no laws. I do not think this objection against the Derivation account is decisive, because I think Kant had good reasons not to attribute empirical laws in the strict sense to sciences such as chemistry and the life sciences. In my opinion, denying the status of laws to chemistry and the life sciences in Kant's time is simply a reflection of the scientific practice of these sciences in the late eighteenth century. In what follows, I will thus provide a historical argument for why chemical laws and biological laws are not properly laws according to Kant, explaining Kant's views on chemical and biological laws on the basis of the scientific context of his time.

Let us first focus on the laws of chemistry, which have been discussed by McNulty. McNulty claims that Kant regularly refers to chemical laws and notes that this is a problem for the Derivation Account (2015, 2). Indeed, Kant refers to chemical laws in the *Metaphysical Foundations*. A close look at these passages suggests, however, that Kant does not regard these laws as laws in a proper sense. Thus, after arguing that proper natural science must be based on a pure a priori part, Kant states:

> Hence, the most complete explanation of given appearances from chemical principles still always leaves behind a certain dissatisfaction, because one can adduce no *a priori* grounds for such principles, which, as contingent laws, have been learned merely from experience.
>
> (IV:469)

Hence, Kant explicitly states that chemical principles are, because they are based solely on experience, merely contingent. We have already seen that natural laws in the proper sense are necessary according to Kant. Hence, we can read the above passage as denying that chemistry has laws in the proper sense. Kant himself makes this inference fully explicit, when he argues:

> If, however, the grounds or principles themselves are still in the end merely empirical, as in chemistry, for example, and the laws from which the given facts are explained through reason are mere laws of experience, then they carry with them no consciousness of their *necessity* (they are not apodictically certain), and thus the whole of cognition does not deserve the name of a science in the strict sense; chemistry should therefore be called a systematic art rather than a science.
>
> (IV:468)

That chemistry does not have proper laws make sense if we consider the scientific practice of chemistry in Kant's time. In the *Metaphysical Foundations* (1786), Kant developed a dynamic theory of matter in which he explained fundamental properties of matter (such as the filling of a space or relative impenetrability) in terms of the interactions between the fundamental forces of matter (attraction and repulsion). However, Kant makes it clear that he cannot explain the specific variety of matter, including chemical phenomena, in terms of these fundamental forces. Thus, Kant claims: "But one should guard against going beyond that which makes possible the general concept of matter as such, and wishing to explain a priori its particular, or even specific, determination and variety" (IV:524). Hence, Kant strictly distinguishes between the a priori investigation of the *Metaphysical Foundations* and the specific empirical investigation into the specific variety of matter (which includes chemistry). As Michael Friedman

(1992) has argued, there is according to Kant a gap between the a priori investigation of the *Metaphysical Foundations* and the empirical research into the specific variety of matter, belonging to what we may call experimental physics. We cannot explain specific phenomena such as cohesion and chemical phenomena in terms of the actions of attraction and repulsion. As Friedman explains this gap:

> Whereas the *Metaphysical Foundations* deals with the universal forces of matter in general (the original forces of attraction and repulsion), it says nothing at all about any additional, more specific forces of matter-which, therefore, as far as the *Metaphysical Foundations* is concerned, are left solely to empirical physics. As far as the Metaphysical Foundations is concerned, any additional, more specific forces are thus left entirely without an a priori foundation.
> (Friedman 1992, 238)

This gap, which Kant remarks on in the *Metaphysical Foundations* and which Kant tried to remedy at the end of his life in his *Opus postumum*, was recognized in the scientific literature of Kant's time. This becomes clear if we focus on the phenomenon of cohesion. As van den Berg (2014, 182) notes: when writing on attraction, Gehler, in his *Physikalisches Wörterbuch*, states that although we conceive of cohesion as a form of attraction, we only have proper knowledge of Newton's law of universal attraction or gravitation. Laws concerning other types of attraction, such as attraction in contact (cohesion), have not been established with certainty. We only know that cohesion operates according to different laws than Newtonian attraction. According to Gehler, then, we do not have knowledge of the cause of cohesion and we are unable to properly explain it (Gehler 1787–1796, I, 171–2). Accordingly, we do not have knowledge of proper laws governing cohesion. Kant identified similar problems as Gehler with respect to cohesion and can thus also be attributed the view that we do not have knowledge of proper laws governing cohesion. The situation is the same, I submit, with respect to chemistry: since there is a gap between the *Metaphysical Foundations* and chemistry we cannot (yet) properly explain chemical phenomena in accordance with the principles of physics, and accordingly we do not have knowledge of proper chemical laws.

Let us now turn our attention to the life sciences. The question we are confronted with is whether we have knowledge of laws concerning organic phenomena. In my view, Kant denies we have knowledge of laws in the life sciences in the late eighteenth century. The reason is that the life sciences in Kant's time were fundamentally concerned with the purposive unity and functioning of organisms, phenomena which are mechanically inexplicable according to Kant. This becomes clear if we consider Blumenbach's famous *Über den Bildungstrieb und das Zeugungsgeschäfte* (1781), a famous work in which Blumenbach argued for epigenesis. In this work, Blumenbach postulated the teleological vital force of the *Bildungstrieb* to account for the generation of organisms, the maintenance of organisms and regeneration of organic parts, and the nutrition and growth of organisms (1781, 12). Importantly, the features of organisms that Blumenbach wished to explain with the *Bildungstrieb* all concerned the purposive organization and maintenance of organisms. For example, the process of embryogenesis, guided by the *Bildungstrieb*, concerned the coming to be of organic and purposive structures. According to Kant, the coming to be of organic structures is mechanically inexplicable, and hence we do not have knowledge of laws concerning this phenomenon. In his account of nutrition, Blumenbach mentions that nutrition serves the self-maintenance of organisms. Organic bodies are continually subject to decay, and would, Blumenbach writes citing Bernoulli, be completely destroyed after three years if the process of nutrition did not serve to balance the organic body (Blumenbach 1781, 70–71). Nutrition serves to balance the decay of parts and thus enables the continued maintenance of the organic body. To account for this self-maintenance, Blumenbach postulated the *Bildungstrieb*. Hence, what Blumenbach stresses in his account of nutrition is not the chemical processes involved in the taking up of nutrition, as for example discussed by Senebier, but the purposive harmony and self-organization of organic bodies. To account for this harmony and self-organization Blumenbach invoked the teleological posit of the *Bildungstrieb*. According to Kant, such purposive harmony and self-organization are mechanically inexplicable, and hence we cannot properly explain such phenomena nor do we have knowledge of laws governing such phenomena. Finally, the process of regeneration again concerned the purposive self-maintenance of organisms and is again a phenomenon that according to Kant is mechanically inexplicable. To conclude: the study of Blumenbach's seminal work shows that the life sciences in Kant's time were fundamentally concerned with the purposive features of organisms, features which are mechanically inexplicable according to Kant and for which we cannot articulate genuine laws.

Finally, we may inquire into the relation between the a priori principles of natural science and the regularities discussed in the life sciences. It follows from the fact that there is a gap between the *Metaphysical Foundations* and chemistry, as discussed above, that there is also a gap between the *Metaphysical Foundations* and the life sciences. The reason for this is that the life sciences, as we have seen, often borrow statements from chemistry in order to explain organic phenomena. For this reason, the regularities of the life sciences will not be systematically related to the a priori principles of natural science and accordingly there are no proper laws in the life sciences. This is the case even for these phenomena, such as nutrition, where chemical statements allow us to provide partial explanations in the life sciences.

We arrive at the same conclusion if we look at some of the regularities discussed in the life sciences of Kant's time. In the second edition of *Über den Bildungstrieb* (1789), at the end of the book, Blumenbach listed some of what he calls laws (*Gesetze*) concerning the *Bildungstrieb*. These include laws such as (i) the strength of the *Bildungstrieb* is inversely proportional to the age of the organism, and (ii) when the *Bildungstrieb* takes a counternatural course, there arise *Misgeburten* (1789, 93–107). For these laws, Blumenbach listed only empirical and inductive evidence. There is no attempt to relate these regularities to regularities in chemistry or physics, or, in Kant's terms, to achieve a systematic unity of physics, chemistry, and the life sciences. Accordingly, these laws also lack a priori grounding and are not laws in the proper sense. This is not a reason to fault *Blumenbach*. Blumenbach simply did not have at his disposal regularities from physics or chemistry with which the mainly embryological phenomena with which he was concerned could be explained. Hence, although we have seen examples from physiology (the human eye) and nutrition where physics and chemistry could aid with giving explanations, the situation is completely different for embryology, where there is a disunity of the sciences. With respect to several disciplines in the life sciences, Zammito and Guyer are thus correct that organisms pose a threat to the systematic unity of science. For this reason, we again cannot speak of the existence of proper laws in the life sciences of Kant's time.

7.5 Conclusion

On the basis of an analysis of Wolff's and Kant's views on the hierarchy of sciences, we have shown that some sciences, such as metaphysics, provide statements that are used to provide proofs in other sciences, such as physics. I have argued that, according to Kant, this situation also exists between physics, chemistry, and the life sciences: statements from physics and chemistry are used to provide explanations in the life sciences. In this sense, physics, chemistry, and the life sciences constitute a systematic unity. Hence, Kant recognized the ideal of a systematic unity between physics, chemistry and the life sciences, and for some features of organisms, the ideal could be worked out. However, I have also argued that the scientific practice in the life sciences of Kant prevented him from fully working out the systematic unity between physics, chemistry, and the life sciences. Many important features of organisms, in particular the purposive structure and self-maintenance of organisms, which were studied for example in the embryological works of Blumenbach, resisted mechanical explanation, and accordingly statements from physics and chemistry could not be used to provide explanations of these features. Moreover, there existed a gap between the *Metaphysical Foundations*, articulating the a priori principles of natural science, and chemistry and the life sciences. Accordingly, the regularities studied in the life sciences could not be systematically related to the a priori principles of natural science, and the life sciences of Kant's time did not contain proper natural laws. Hence, Kant's philosophy of the life sciences reflected the scientific practice of his day. He articulated an ideal of the systematic unity of different natural sciences, but could not fully work out this ideal due to the fact that several phenomena in the life sciences were treated in isolation from physics and chemistry. It was the task taken up by Kant's successors, such as Schelling, to fully articulate and give content to the ideal of a true system of the natural sciences.

Acknowledgments

Hein van den Berg is supported by The Netherlands Organisation for Scientific Research under grant number 277-20-007. Hein wishes to thank the editors, Jeroen Smid and Boris Demarest and members of the group Concepts in Motion for support with this paper.

Bibliography

Blumenbach, J.F. 1781. *Über den Bildungstrieb und das Zeugungsgeschäfte*. Göttingen: Johann Christian Dieterich.
Blumenbach, J.F. 1789. *Über den Bildungstrieb*. Göttingen: Johann Christian Dieterich.
Breitenbach, A. 2017. "Laws in Biology and the Unity of Nature." In *Kant and the Laws of Nature*, edited by Michela Massimi and Angela Breitenbach, 237–55. Cambridge: Cambridge University Press.
Breitenbach, A. 2018. "Laws and Ideal Unity." In *Laws of Nature*, edited by Walter Ott and Lydia Patton, 108–22. Oxford: Oxford University Press.
Friedman, M. 1992. *Kant and the Exact Sciences*. Cambridge: Harvard University Press.
Gehler, J.T.S. 1787–1796. *Physikalisches Wörterbuch oder Versuch einer Erklärung der vornehmsten Begriffe und Kunstwörter der Naturlehre mit kurzen Nachrichten von der Geschichte der Erfindungen und Beschreibungen der Werkzeuge begleitet in alphabetischer Ordnung*. Leipzig: Schwickert.
Gehler, J.T.S. 1798–1801. *Physikalisches Wörterbuch oder Versuch einer Erklärung der vornehmsten Begriffe und Kunstwörter der Naturlehre mit kurzen Nachrichten von der Geschichte der Erfindungen und Beschreibungen der Werkzeuge begleitet in alphabetischer Ordnung*. Neue. Auflage. Leipzig: Schwickert.
Geiger, I. 2022. *Kant and the Claims of the Empirical World: A Transcendental Reading of the Critique of the Power of Judgment*. Cambridge: Cambridge University Press.
Guyer, P. 2001. "Organisms and the Unity of Science." In *Kant and the Sciences*, edited by Eric Watkins, 259–81. Oxford: Oxford University Press.

Karsten, W.J.G. 1783. *Anleitung zur gemeinnützlichen Kenntniß der Natur, besonders für angehende Aertze, Cameralisten und Oeconomen.* Halle: Renger.
Kreines, J. 2009. "Kant on the Laws of Nature: Laws, Necessitation, and the Limitation of Our Knowledge." *European Journal of Philosophy* 17: 527–58.
Lambert, J.H. 1764. *Neues Organon oder Gedanken des Wahren und dessen Unterscheidung vom Irrthum und Schein.* Band 1. Leipzig: Wendler.
McLaughlin, P. 1990. *Kant's Critique of Teleology in Biological Explanation: Antinomy and Teleology.* Lampeter: Edwin Mellen Press.
McNulty, M.B. 2015. "Rehabilitating the Regulative Use of Reason: Kant on Empirical and Chemical Laws." *Studies in History and Philosophy of Science Part A* 54: 1–10.
Messina, J. 2017. "Kant's Necessitation Account of Laws and the Nature of Natures." In *Kant and the Laws of Nature*, edited by Michela Massimi and Angela Breitenbach, 131–49. Cambridge: Cambridge University Press.
Newton, I. 1999 [1726]. *The Principia. Mathematical Principles of Natural Philosophy.* 3rd ed. Translated and edited by I. Bernard Cohen and Anne Whitman. Berkeley: University of California Press.
Richards, R.J. 2000. "Kant and Blumenbach on the Bildungstrieb: A Historical Misunderstanding." *Studies in History and Philosophy of Biological and Biomedical Sciences* 31: 11–32.
Stang, N. 2016. *Kant's Modal Metaphysics.* Oxford: Oxford University Press.
Sturm, T. 2009. *Kant und die Wissenschaft vom Menschen.* Paderborn: Mentis.
Sturm, T. 2020. "Kant on the Ends of the Sciences." *Kant-Studien* 111(1): 1–28.
Sturm, T. 2022. "Kant's Conception of the *Metaphysical Foundations of Natural Science*: Subject Matter, Method, and Aim." In *Cambridge Guide to the Metaphysical Foundations of Natural Science*, edited by Michael Bennet McNulty, 13–35. Cambridge: Cambridge University Press.
Van den Berg, H. 2013. "The Wolffian Roots of Kant's Teleology." *Studies in History and Philosophy of Biological and Biomedical Sciences* 44(4B): 178–205.
Van den Berg, H. 2014. *Kant on Proper Science: Biology in the Critical Philosophy and the Opus postumum.* Dordrecht: Springer.
Watkins, E. 2005. *Kant and the Metaphysics of Causality.* New York: Cambridge University Press.
Watkins, E. 2019. *Kant on Laws.* Cambridge: Cambridge University Press.
Watkins, E. 2021. "Replies to the Comments of Fabian Burt, Marius Stan and Marcus Willascheck." *Studi Kantiani* 34: 159–73.
Wolff, C. 1963 [1728]. *Preliminary Discourse on Philosophy in General.* Translated by Richard Blackwell. Indianapolis: Bobbs-Merril.
Wolff, C. 2003 [1723]. *Vernünftige Gedancken von den Würckungen der Natur.* Hildesheim: Olms.
Zammito, J. 2003. "'This Inscrutable Principle of an Original Organization': Epigenesis and 'Looseness of Fit' in Kant's Philosophy of Science." *Studies in History and Philosophy of Science* 34: 73–109.
Zammito, J. 2006. "Teleology Then and Now: The Question of Kant's Relevance for Contemporary Controversies over Function in Biology." *Studies in History and Philosophy of Biological and Biomedical Sciences* 37: 748–70.

8 Kant's Aethereal Hammer
When Everything Looks Like a Nail

Michael Bennett McNulty

Many early modern philosophers were quick to emphasize their rejection of the then-traditional Scholasticism. Indeed, several prominent figures consciously characterize their approaches to philosophy as innovative and as overthrowing the given paradigm. An aspect of this break involves a reconceptualization of natural explanation. In particular, the adherents of the new, mechanical philosophy looked askew upon the essences and substantial forms utilized by their Scholastic predecessors. Seventeenth-century luminaries such as Galileo Galilei (1957, 273–9), Descartes (AT, III:506, IXB:321–323), Boyle (BW, V:298–304), and Locke (N, 407–20, 438–71) all maligned both the explanatory value of the Aristotelian framework and our epistemic access to such exotic explanatory entities.

According to a classic criticism made by detractors of the Scholastic approach, their explanations founder due to the conflation of explanans and explanandum. That is, the attempt to explain a phenomenon via postulated substantial forms simply assumes what is to be explained. To state that a causal relation is explained by an essence or substantial form is, in the eyes of many a philosopher of the period, to stop short in explanation. An event cannot be explained simply by a substance's essential possession of a power to cause that event. Molière famously lampooned such explanations in his play, *The Imaginary Invalid* (premiered in 1673), by depicting the "Bachelor" as explaining opium's power to put one asleep by appeal to a "dormitive virtue," the nature of which is to make one drowsy (1876, 319). Mechanical philosophers of the early modern period, in particular, sought to replace Scholastic explanations of natural phenomena with ones in terms of the shape, size, and impact of constituent pieces of matter. Along these lines, opium causes one to sleep not because of its dormitive virtue, but rather due to the complex interplay between the matter that makes up the opium and that of one's body. In the eyes of the 17th-century mechanical philosophers, with their innovative new approach to natural philosophical explanation, natural philosophy was thereby wrested from the shackles of Scholasticism and could then blossom independent of the traditional dogmata of Aristotelianism.

A less commonly told story in the history of philosophy regards the later reintroduction and precipitate proliferation of essentialist explanations in subsequent natural philosophy. Throughout the 18th century, natural philosophers postulated a wide variety of subtle, imponderable fluids, or *aethers*, whose flows, aggregations, periodic disturbances, or expression of short-range forces explain particular classes of natural phenomena.[1] In their overview of aether concepts from pre-Socratic Ionians to the turn of the 20th century, historians of science Cantor and Hodge offer the following schema for aethers.

> [A]n ether was a spatially and temporally extended entity exerting but not merely identifiable with certain forces and supposed to fit most of the following descriptions: It may be present in spaces empty of ordinary solids, fluids, and gases; it is not perceivable as such ordinary materials are; it transmits actions or effects including or like those of magnetism, electricity, heat, and nervous impulses; it can penetrate and pass through ordinary solid, fluid, and gaseous materials; changes in its distribution or its state can cause observable changes in ordinary bodies.
> (Cantor and Hodge 1981, 2)

Aethers thus were conceived of as exceedingly subtle substances that are postulated, especially in order to account for a particular range of natural phenomena.[2] But, nonetheless, in this schematic characterization of aether theories, there is ample room for variation. An aether may be thought of as material or immaterial, as continuous or particulate, as mechanical or not, as genuinely existing or as a mere instrument, and so forth.

Although the scholastic terminology of essences and substantial forms was naturally eschewed, there is a structural similarity between aether-based and Scholastic explanations.[3] Aethers were postulated as reified powers to explain particular classes of natural events, such as gravitation, optical phenomena, cohesion, elasticity, thermal phenomena, electricity, magnetism, physiology, and perception. But with the aethers, explanation commonly halted. For aether theorists, it's simply the *nature* or *essence* of, say, caloric to cause thermal phenomena or of the luminiferous aether to cause optical phenomena. One may understandably discern little conceptual daylight between the luminiferous aether's power to communicate light and opium's dormitive virtue.

Despite their detractors, aethers proved popular in Enlightenment physics and proliferated throughout the 18th century. Historian J.L. Heilbron wistfully reflects on this movement in early modern natural philosophy.

> By 1750 repulsion had been reified in air, aether, fire, and electricity. In the next few decades, physicists accepted a second electrical fluid and other force-carrying imponderables like phlogiston, caloric, and the agents of magnetism. Special carriers of attraction likewise multiplied. The number of fundamental fluids became an embarrassment.
> (Heilbron 1982, 62)

Heilbron views this process as rolling back the earlier advances made by banishing Scholastic essences from natural philosophy, summing up his sentiment by ruing that "physics ended the century richer in essences than it had begun" (Heilbron 1982, 62).[4]

For his part, Immanuel Kant was a keen supporter of this aethereal tradition in early modern physics. In his own philosophical work, he enthusiastically reified natural powers as subtle, imponderable fluids. Kant, of course, famously ruminates on the notion of the aether in the drafts of his *Opus postumum* (culminating in the notorious "aether proofs" of 1799s *Übergang* 1–14 drafts), but he also demonstrates a comfort with subtle fluids earlier in and throughout his philosophical career. The matters of elasticity, fire, and light were at the heart of Kant's master's thesis, *Succinct Exposition of Some Meditations on Fire*. The reification of chemical powers in elements is referenced in the appendix to the Transcendental Dialectic of the *Critique of Pure Reason*, allusions to subtle-fluid-based explanations of heat and cohesion are found in the *Metaphysical Foundations of Natural Science*, and reference to the aethereal explanation of optical phenomena occurs in the *Critique of the Power of Judgment*. Beyond the published works, throughout student notes from Kant's lectures on physics (especially those from the 1770s) along with his handwritten *Reflections* on physics and chemistry (correspondingly, those from the 1770s), Kant liberally postulates and theorizes about aethers and subtle fluids.[5] Such references, when woven together, paint a complex, if not disorienting, image of Kant as philosopher of nature. Kant is standardly known for his aprioristic grounding of physics in the fundamental forces of attraction and repulsion and, more remotely, in the categories (see Friedman 1992; Watkins 1998). However, when faced with the task of explaining particular empirical phenomena, he hastens not to the fundamental forces, but to subtle fluids.

Yet, despite his enthusiastic endorsement of aether theorizing, Kant keenly recognized its downsides and sought both to mitigate them and to reign in the excesses of the tradition. In particular, Kant appreciated that natural philosophy does not end with the postulation of fundamental materials corresponding to observed powers. Rather, philosophers must proceed subsequent to such postulation of subtle fluids to *unify* and to *systematize* them. Collectively, Kant's writings on natural philosophy reveal an ultimate goal of reducing the entire catalogue of subtle fluids under a *single*, unifying aether. Although Kant's conception of the aether (especially as described in the *Opus postumum*) is notorious, by situating it within the *tradition* of postulating subtle fluids, we can recognize that throughout his life Kant believed his own theory of the aether to redress problems facing that tradition.

This is the story I intend to detail in the course of this chapter. In addition to situating Kant within the history of science and thereby restituting his theory of the aether, I argue that consideration of the aether in Kant's natural philosophy serves as a case study par excellence for his conception of natural scientific systematicity. In comparison with the sketchy examples of scientific systematicity handled in the appendix to the Transcendental Dialectic, Kant's treatment of the aether serves as a fleshed out, detailed, and live illustration in his theory of the systematicity of nature, one that offers new morals for our understanding of Kant's philosophy of science.

In Section 8.1, I give a brief overview of the aether-tradition in 18th century, especially with eye toward figures in Kant's purview. In Section 8.2, I describe Kant's theory of the aether and the associated unificationist stratagem regularly executed by Kant. Finally, in Section 8.3, I extract the upshots of this account for our understanding of Kant's theory of systematicity and argue that Kant's treatment of subtle-fluid hypotheses reveals the project of reducing or combining causal-explanatory grounds to be among the ideals for scientific systematicity.

8.1 Aethers in 18th-Century Physics

As mentioned above, during the 18th century, aethers were especially popular tools for comprehending nature. In this section, I catalogue some of the primary uses of aether theories in Kant's day, although I make no claim to comprehensiveness – see, instead, historical studies such as Schofield (1970), Cantor and Hodge (1981), Heilbron (1982), and Hankins (1985). My aims in this section are rather twofold: first, to demonstrate the popularity and pervasiveness of aether theories in the broad context of 18th-century physics, and second, to highlight sources proximate to Kant, which, owing to their relative unimportance for broad trends in the history of science, evade the consideration of the aforementioned historians.

I highlight first reflections on theories of subtle fluids by the following major figures: Isaac Newton, Stephen Hales, Hermann Boerhaave, and

Leonard Euler. This is partially due to their influence in 18th-century natural philosophy at large but also in virtue of their especial impact on Kant. In Kant's natural-philosophical works, such natural philosophers are each explicitly noted as influences. For this reason, some scholarship already situates aspects of Kant's theory of the aether with respect to them, including superb work by Massimi (2011), in the wake of which my project exists.[6] I further fill out this picture by discussing theorizing about aethers from sources more proximate to Kant. In particular, I present the views expressed in the textbooks that Kant used in his lectures on physics from 1770 to 1785, so as to better exhibit the immediate context for his unificatory account of the aether.

8.1.1 The Broad Context

Any account of aether theories in the 17th and 18th centuries must grapple with Isaac Newton's place in the story. Although Newton was a famous skeptic of medium theories of light and a partisan of the competitive, emission theory, he repeatedly notes the utility of aether theories. His conjectures about aethers, especially as expressed in the 2nd edition of the *Opticks* (1718), the General Scholium added in the 2nd edition of the *Principia* (1713), and in a posthumously published letter to Boyle (written in 1678/9),[7] were well known and influential in the 18th century. In such writings, Newton mulls over with the explanatory power of a subtle, universally dispersed material. In the letter to Boyle, he writes "that there is diffused through all places an æthereal substance, capable of contraction and dilatation, strongly elastic, and, in a word, much like air in all respects, but far more subtle" (Boyle 1744, 70). For Newton, aether is present in all space (not only void space, but additionally the interstices within bodies), is fundamentally expansive, and repulses all matter.[8] However, aether is less dense within bodies, and hence there expresses less expansive force. For that reason, on the surface of bodies, there will be a differential (greater) expression of expansive force. In the letter to Boyle, Newton suggests that this is the ground of the cohesion of bodies: the aether external to a body, which is denser than the aether within, expresses a net compressive force on the body.[9]

But, additionally, Newton conjectures in the *Opticks* and the letter to Boyle that this conception of the aether explains some optical phenomena, like refraction, via its expression of a short-range, repulsive force. Newton claims that the density of the aether does not vary discontinuously. Rather, there is a spatially extended aether-density gradient at the surfaces of bodies. This means that when a light ray approaches the superficies of a body (say, glass), as the particles of light enter the diminishing density gradient, they will be progressively less impeded by the repulsive force of the

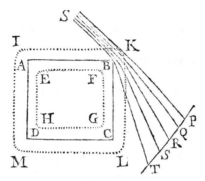

Figure 8.1 Refraction of light passing through an aether-gradient (Boyle 1744, 70)

aether and accelerate (Figure 8.1). The converse happens when the light ray emerges from the denser medium back into the air. As the particles pass through the intensifying density gradient, they decelerate due to the increased expansive force.

In his *Vegetable Staticks* (1727), Stephen Hales conceives of air as a self-repulsive matter and the fundamental principle of elasticity. Hales' meticulous analyses of plant physiology led him to believe that plants are nourished not only through water imbibed through the roots, but, additionally, through air taken in the leaves and stem. Air can thus become "fixed" in bodies. So fixed, air continues to express the expansivity characteristic of its aerial form. That is, when joined with the gross matter, air exerts an expansive force on all the parts of matter. For Hales, this expression of repulsive force on all the parts of a body resists said parts collapsing into one another under the influence of the attractive, gravitational force. In this sense, air is the leavening agent for gross matter. According to Hales, air is ultimately the fundamental bearer of repulsion, its flows, aggregations, and combinations explain a wide variety of repulsive phenomena. Through the subsequent decades, Hales' account proved critical to the recognition of the causal power of air as well as for the development of further, essentialist, subtle-fluid theories of repulsion and elasticity.

Herman Boerhaave's material fire constituted one such subtle-fluid theory of repulsion, one that was especially influential on 18th-century theories of heat, Kant's included. For Boerhaave, fire is an omnipresent, expansive, subtle fluid that grounds phenomena of dilution, fluidity, and fermentation (Heilbron 1982, 61–62). As Boerhaave puts it, "by Fire I shall always mean that being, however otherwise unknown, which is endued with this property, that it penetrates all Bodies, both solid and fluid, and by this very action, extends them into larger spaces" (Boerhaave 1735, 106). Boerhaave's fire is always in motion, and, indeed, its motion is the ground of its causal efficacy. The "agitation" of this fluid causes observed temperature and burning flames.[10] He famously used the example of striking steel with a flint, whereupon sparks fly off, to demonstrate that motion is intimately connected with observed fire (Powers 2012, 78).[11] Although Boerhaave was committed to the mechanical approach to explanation, he declines to give anything more than a tentative hypothesis as to the mechanical efficacy of the elementary fire, suggesting that fire consists of minute spherical particles that are without weight.[12]

Leonard Euler was well known in his day for his medium theory of light, according to which light is a wave in a luminiferous aether. In his famous "Nova Theoria Lucis et Colorum" (1746), Euler crafts the classic dichotomy between medium and emission theories of light, simplifying the dialectical position for his own advocacy of the medium theory. Euler defends this theory from the classic criticisms that it faced in the time period – like the problem of the rectilinear propagation of light – and presented his own powerful objections to the emission tradition.[13] Throughout his work on light, Euler's rests his theory upon a strong parallel between light and sound (an analogy that Kant himself echoes in his own comments on light).

Euler's aether is a subtle fluid that is compressed to a high degree by mundane matter and, for that reason, is highly elastic (Hakfoort 1995, 95). Light propagates as a wave in this medium. The analogy between light and sound also becomes relevant in the treatment of colors according to Euler's wave theory. Euler conceives of a derivative analogy between the seven colors diffracted from white light and the seven tones making up a heptatonic scale. According to this account, colors are different frequencies of vibrations in the aether, just as tones are particular frequencies of vibration in air. By conceiving of light as persistent waves in the aether – opposed to Hooke or Huygens, who thought of light as independent *pulses* within it – Euler thus proposes the first major wave theory of light. Furthermore, by offering an account of diffraction far superior to that available for emission theorists, Euler's wave theory truly inaugurates the wave-particle debates that dominated optics in subsequent decades.

For Euler, the aether is also the cause of the cohesion of bodies, just as Newton suggested in the letter to Boyle. According to Euler the compression of a body by the highly elastic aether causes a net compressive force on the surface of the body. In the "Nova Theoria," Euler even calculates the density and elasticity of the aether, depending on the analogy between the propagation of sound in air and that of light in aether.[14]

8.1.2 The Proximate Context

Kant himself was well exposed to many of the foregoing innovations. He well knew of the relevant work of Newton, Hales, Boerhaave, and Euler from his early days, as is demonstrated by his substantial engagement with their aether theories in *On Fire*.[15] He also had other exposure to major Dutch Newtonians – van Musschenbroek, 'sGravesande, and Niewentyt – who, like Boerhaave, were important figures in developing and disseminating the theory of the material fire.[16]

But for a more proximate source, take Johann Gottfried Teske. Teske was a professor at the University of Königsberg in Kant's time as a student. Among other topics, Teske was especially interested in theories of fire, penning his inaugural dissertation on the topic. In this dissertation, Teske (1729) defends a conception of fire similar to Boerhaave's according to which there is a matter of heat, which is subtle and everywhere present, and whose rapid motions cause heat or fire. Kant's master's thesis, *Succinct Exposition of Some Meditations on Fire* (1755), written under the mentorship of Teske, is similarly concerned with the topic of the matter of fire, as I discuss below.

Johann Peter Eberhard, whose *Erste Gründe der Naturlehre* (1774) Kant used in association with various of his physics lectures, regularly utilizes aethereal explanations of natural phenomena.[17] He claims that fire is a subtle fluid: "Elementary fire is an elastic fluid being, that is dispersed through the entire universe, and whose parts never contact each other" (1774, 358; see also 366, 392–4). Elementary fire, according to Eberhard, must also be distinguished from the burnable – the fuel, as it were, for fire (i.e. phlogiston) (1774, 359) – and from light, which is also "a fluid elastic being, which, like the elementary fire" (1774, 366). Eberhard contended that the elementary fire is an elastic, fluid being dispersed everywhere in the universe and that its parts seek to distance themselves from one another. It is subtle and penetrates the interstitial space of bodies.

Although he accepts the existence of the elementary fire as a subtle fluid, Eberhard does not use it to explain many other phenomena. He considers the possibility that the elementary fire explains the elasticity of fixed bodies (1774, 622–3), but believes it unlikely that all the various manifestations of elasticity have a *single* cause (1774, 616–7). Eberhard also contends that, among the various accounts of cohesion, the best has it that cohesion is due to an inner force of matter and neither the assistance of gravity nor some external matter, such as an aether or the elementary fire (1774, 114–6). That said, with the respect to coeval theories of light, Eberhard contrasts Newton's and Euler's accounts, before ultimately opting for Euler's wave theory, again exhibiting a comfort with subtle fluids (1774, 428–31).

Johann Christian Polykarp Erxleben also catalogues many of the aethereal explanations used during the day in his *Anfangsgründe der Naturlehre* (1772), which Kant used in later physics lectures. However, Erxleben is substantially more skeptical of aethereal explanations of phenomena than

Eberhard (or than Kant later is). For example, Erxleben rejects the aethereal explanation of elasticity out of hand, writing that

> However, because, on the one hand, we have entirely no reason to regard the aether as the cause of elasticity and, on the other hand, cannot say why the aether is elastic, we would perhaps do better simply to confess straight away that the cause of elasticity is altogether unknown to us.
>
> (Erxleben 1772, 116)

With respect to the explanation of cohesion via the impact of the aether, Erxleben is similarly skeptical. He instead suggests that corporeal particles, themselves, possess inner forces whereby they cohere to one another (1772, 33–34). Erxleben appears tentative with respect to a material theory of heat, rhetorically asking whether there is a "a distinct *matter of fire*, an *elementary fire*, a most fine, subtle being, which is homogeneously distributed through the interstitial space of all bodies, and in whose vibrations heat consists" (1772, 359). Although he appears to accept the theory, Erxleben also brings up problems facing the material theory of heat – for example, noting that that if heat is material, then a body must gain weight during heating (1772, 360).[18] Notably, he claims that the assumption that the matter of fire is identical to the matter of light is "still not as entirely certain as some researchers of nature believe" (1772, 359). Kant, of course, consistently supports this thesis.

Erxleben offers a detailed consideration of the different optical theories, accepting a material theory of light (1772, 243), defending Euler's wave theory and discussing his calculations of the density and elasticity of the aether (1772, 247–8). He additionally discusses theories of electricity, accepting an "electric matter," which is "fine, fluid, and elastic matter, which penetrates the interstitial spaces of all bodies" (1772, 422–3). He also presents a one-fluid theory of electricity, claiming that a body is electric when it takes in this electric matter, and he praises Franklin's conception of electricity (1772, 423ff). Finally, there is also a "fine, fluid matter" of magnetism, which "can penetrate all dense bodies" (1772, 443).[19]

What are the upshots of these considerations? First, Kant had acquaintance with the tradition of postulating aethers from many sources. Second, Kant was not at all a particular outlier for positing an aether theory or for his enthusiastic employment of the aether. Third, Kant was exposed to views skeptical of the utility of aethereal explanations, like those of, especially, Erxleben. Nonetheless, as we will see, he both accepted the aethereal approach to explanation and sought to improve upon it.

8.2 The "Provocative Efficacy" of the Aether

As I mentioned above, Kant consistently propounded a program of unifying natural phenomena under the explanatory umbrella of the aether. Given his prolific use of the aether to explain an abundance of natural phenomena, Edwards and Schönfeld, two of the leading scholars that spotlight Kant's theory of the aether, characterize his aether as possessing a "provocative efficacy" (2006, 115). The aether played this central explanatory role not only in the *Opus postumum* but also in various works produced throughout Kant's career and before the notorious aether proofs.

Kant's program for unifying disparate natural phenomena via a single aether was first presented in his aforementioned master's thesis, *On Fire*. This work circles around three subtle fluids: the matter of elasticity, the matter of fire, and the luminiferous aether. The influences on Kant's theory are fairly clear: Kant mentions Newton's and Euler's theories of the aether (I:377–8), Hales' *Vegetable Staticks* and its treatment of elastic air (I:381), and Boerhaave (I:378). Massimi (2011) connects Kant's concerns in *On Fire* with the preceding context of subtle fluid theories. She characterizes *On Fire* as an attempt to synthesize Hales' elastic air, Boerhaave's matter of fire, and Newton's aether and briefly suggests that Kant is in the tradition of subtle fluid theorists called "materialists" by Schofield (1970).[20]

In section I of *On Fire*, Kant discusses states of aggregation, cohesion, and elasticity (I:371–5). At this point, Kant countenances a particulate conception of corporeal constitution. That is, bodies consist not of a continuum of matter, as he would claim in the *Metaphysical Foundations of Natural Science*, but of an aggregate of "minute spherical particles" (I:371). However, Kant notes, were it the case that bodies *only* consisted of such particles, then fluids would be impossible. Under the influence of the force of gravity, any aggregate of tiny spheres would collapse into a one-particle thick film. But, liquids, being cohesive, form drops on a flat surface that contain layers of particles. Kant maintains that such cohesion is only possible by an interstitial, elastic matter among the spherical parts of the liquid.[21] Kant goes on to claim that this elastic matter is also present in solids and explains their elasticity.

In section II, Kant proceeds to the main point of the dissertation: that the elastic matter discussed in section I is identical to the matter of fire, which, in turn, is identical to the matter of light (I:376–84). The matter of fire, for Kant, is the cause of heat through its vibrations. Although he says little more about the matter of heat in this context, it is unequivocally a subtle substance, which insinuates into bodies and communicates its vibrations. In proposition VII, Kant offers two proofs of the identity of the elastic matter with the matter of fire (I:376–7). First, he notes that increasing heat increases the volume of a body and conversely that decreasing heat decreases the volume. From this Kant infers that heat is the same as the elastic matter; the direct effect of adding more heat to a body is therefore its expansion. Second, Kant conceives of the formation of bubbles as demonstrating his point. As more heat is added to a liquid on the brink of boiling, it creates more strenuous undulations, which eventually break loose some of the matter of fire, which then enters the forming bubbles. Such bubbles expand, because what is present in them – the matter of fire – is elastic.

In proposition VIII, Kant supports his claim that the aether, or the matter of light, is identical to the matter of fire (or, equivalently, that of elasticity). He argues, for instance, that the bodies that have the most capacity to refract light are also those that have the most capacity to absorb heat; oils constitute a paradigmatic case of this.[22] Kant infers – based, it seems, on considerations of parsimony – that there must be a single attractive force expressed by such bodies that attracts fire and light, which shows that "the matter of heat and the matter of light agree as closely as possible or, rather, that they are not different" (I:377). Kant also argues that consideration of the constitution of glass reveals the identity of the matter of fire and that of light. The heating of sand involved in the production of glass is understood by Kant as fusing the sand with the matter of fire. The resulting transparency is then due to the identity of the matter of fire and light; the matter of fire added to the sand makes possible the easy transport of what is the same matter of light.

Thus Kant was a member of the tradition in natural philosophy of postulating aethers and yet sought to improve upon it. The chief conclusion of this work was that the disparate aethers – the elastic matter, the matter of heat, and the luminiferous aether – are actually identical. I mean to emphasize that this project of Kant's was *not* a one-off. That is, Kant not only accepted this particular point throughout his career – that these three aethers are identical – rather, additionally, he attempted the same explanatory stratagem time and time again in his natural philosophy. That is, he repeatedly set forth the goal of unifying subtle fluids and even continues to expand the explanatory breadth of the aether.

The aether can also be found hanging in the margins of Kant's well-known, critical works bearing on his philosophy of nature. For instance, various explicit mentions of the aether as well as several implicit allusions are found in Kant's *Metaphysical Foundations of Natural Science*.[23] He references his earlier developed aethereal explanation of elasticity in the General Remark to the Dynamics: "Expansive elasticity, however, can be either original or derivative. Thus air has a derivative elasticity in virtue of the matter of heat, which is most intimately united with it, and whose own elasticity is perhaps original" (IV:529–30). In the Dynamics, Kant approvingly mentions Euler's medium theory of light (IV:520n), suggesting that it squares well with his theory of matter as a dynamic plenum. Later in the Remark, Kant discusses his theory of chemical dissolution and mentions the matter of fire as chemically dissolving into bodies and refers to a subtle, magnetic fluid (IV:532).[24]

But the most interesting and compelling evidence for the lasting, foundational role of the aether in empirical physics (before the drafting of the *Opus postumum*) appears in a series of writings from the 1770s: student notes from Kant's physics lectures along with a series of handwritten reflections from Kant's *Nachlass*.[25] First, Kant continues to attribute thermal phenomena to the aether. Fundamentally, according to the student notetaker for the *Berlin Physics*, Kant espouses a kinetic theory of heat, according to which heat consists in the vibrations of substance (XXIX:83–84).[26] However, the aether *qua* matter of heat is *originally* undulating. When it is combines with a matter, the aether communicates its vibrations to that matter and hence heats it. Second, Kant also contends that the aether is the medium for luminal phenomena, coming to the defense of Euler and analogizing the role of the aether vis-à-vis light to that of the air to sound (XXIX:84–85; see also the *Danzig Physics*, XXIX:146, 150). Furthermore, Kant maintains that the aether is the *"fluidum originarium"* – the originally fluid material (*Berlin Physics*, XXIX:86; see also *Danzig Physics*, XXIX:150). That is, substances only have the state of fluidity *via* the aether.[27] This is an especially striking thesis, as Kant is continuously committed to the idea that matter is principally fluid and only achieves other states through transformation. He makes just this point in the lectures, asserting that solidity is only possible through removing aether from a fluid body (*Berlin Physics*, XXIX:86). Thereby is the aether a causal prerequisite for *any* aggregative state of matter.

In particular, Kant attributes the *cohesion* of fluids to the aether. The student notetaker ascribes to Kant the view that "In fluid bodies, the parts are not separated, rather only shifted through foreign force. No *coacervation* of totally fine parts can bring forth a fluid" (*Berlin Physics*, XXIX:86). The point here is that a fluid is not just a collection of small particles. Rather, the parts of a fluid must form a continuous, cohesive medium. Only the aether can bring such a cohesive unity to the parts. This is the same point that Kant makes in section I of *On Fire*, as mentioned above. Elsewhere in the lectures, he claims that the cause of cohesion is the compressive force of the aether through the gravitational force (XXIX:82). That is, the reason that a body resists displacement of its parts is that those parts are being pushed together by the compressive force of the external aether. Kant reiterates this conception of the aether's role in explaining cohesion in the aforementioned reflections from the 1770s, writing in Reflection 44 that "The aether is compressed by the attraction of all matter of the universe and is the womb of all bodies and the ground of all cohesion" (XIV:295) and that "The cause of cohesion is not internal and proper to matter in general as matter (is superficial force). (We have a universal attraction and a general medium, which is expansive and is the cause of all corporeal forms and cohesion thorough that press[ure].)" (XIV:315–6).[28]

Kant clearly thinks of electricity and magnetism as communicated by subtle fluids (as can be gleaned from the above-mentioned passage from the General Remark to the *Metaphysical Foundations of Natural Science*). Sometimes he also suggests that these subtle fluids are identical to the aether: "Everything that the heat holds together, that also holds together electricity. The spring is also elastic and holds warmth together. Heat and electricity appear to be based on the same ground, they are different from each other only in the different motions" (*Berlin Physics*, XXIX:87). Yet the notes also echo Erxleben's skepticism with respect to electricity and magnetism. "Electricity is a subtle matter that penetrates all bodies, and whereby one can derive many properties if one will become nearer acquainted with it. It is for us very little known, and it is for us still even such a mystery as the magnetic force" (*Berlin Physics*, XXIX:91). However, a suggestion at the end of Kant's 1763 essay, *Attempt to Introduce the Concept of Negative Magnitudes in Philosophy*, offers a more resounding endorsement of the identity of the different subtle fluids. Based on analogies among the phenomena and similar causal accounts, Kant states that, "In general, the force of magnetism, electricity, and heat seem to occur in virtue of the self-same mediating matter" (II:187; see also his *Physical Monadology*, I:486; Adickes 1924–1925, II:77–101).

Given that Kant attributes such a wide variety of natural phenomena to the aether, it ought to be no surprise that the student notetaker for the *Berlin Physics* attributes to Kant the view that "the supreme cause of all derivative forces is the aether" (XXIX:82). Although Kant's general conceptualization of "derivative forces" is slippery, any forces beyond the fundamental repulsive force – through which matter fills its space, that is, resists the penetration of other bodies – and the fundamental attractive (gravitational) force count as derivative.[29] Kant explicitly calls cohesion and elasticity derivative forces, though any other physical phenomenon presumably rests on such derivative forces. The explanatory extent of the aether is, for Kant, truly massive.[30]

8.3 Morals and Systematicity

As I mentioned at the outset, Kant's treatment of the aether constitutes a concrete example of his conception of natural-scientific systematization par excellence. The aether comprises *the most detailed*, comprehensive illustration of scientific systematicity, especially in the case of physics.

While Kant offers brief, suggestive, but ultimately nondescript examples of systematization in the appendix to the Transcendental Dialectic, the case of the aether is fleshed out, detailed, and supplements Kant's abstract characterizations of systematicity. In this section, I tease out morals from this case study for our understanding of Kant's overarching conception of systematicity.

First, I maintain that the case study of the aether reveals *causal* and *explanatory* unification to be a core component of systematization in the practice of natural science. The representation of systematization that one gleans from the appendix to the Dialectic of the *Critique of Pure Reason* (A642–668/B670–696), or, fittingly, from the *Jäsche Logic* (IX:139–140), is one that is primarily logical in essence. That is, systematizing a doctrine or body of cognitions of the understanding is a process of establishing or revealing logical connections among the concepts and judgments – relations of conceptual containment and of inference. The classic image of a system for Kant – a fully articulated, continuous Porphyrian tree – fits this logical conception of systematicity.[31]

Our case study demonstrates, however, that in actual scientific inquiry – on the ground, as it were – Kant also aims at the unification of the causal, explanatory structure of natural science. I contend that there are, in this context, at least two ideals of systematicity. The systematicity of what Kant calls natural description – the classification of concepts of natural things – involves logically interconnecting concepts. So, for example, in natural description, the primary mode of systematization involves categorizing concepts as genera and species, as in classical Linnaean taxonomies.[32] But what is distinctive about natural science, in comparison with such mere doctrines of nature such as natural description, is that it involves *causal relations* and *laws* (*Metaphysical Foundations of Natural Science*, IV:467–468). The distinctive aspect of systematicity in natural science hence lies in the interconnection of causes and explanations; an aspect of this project involves the combination and reduction of causes. All of this we see in the aether case study: the aether is posed as the ultimate cause and explanation of a huge variety of phenomena.[33]

Furthermore, Kant clearly states that among the essential goals of natural philosophy is the project of causal unification or reduction, reinforcing my contention that causal/explanatory systematization is an essential part of natural science. For instance, Kant makes this point in the General Remark to the Dynamics of the *Metaphysical Foundations of Natural Science*. "[A]ll natural philosophy consists, rather, in the reduction of given, apparently different forces to a smaller number of forces and powers that explain the actions of the former, although this reduction proceeds only up to fundamental forces, beyond which our reason cannot go" (IV:534). The aether is the keystone to this process of explanatory reduction, which

Kant contends is *the* project of natural philosophy. Through the aether's unique characteristics – particularly, its subtlety, its omnipresence, its extreme manifestation of the repulsive force – it grounds and explains physical phenomena. Thus, to reiterate, my first moral from the consideration of the aether is that it highlights the importance of systematizing the *causes* and *explanatory* bases of natural philosophy. This is an essential element of the systematization of natural science, I claim, for Kant.[34]

Second, this case study reveals the extent to which Kant's quest for systematicity, at least as embodied in the aether, is substantially connected with and responsive to coeval science. Given the substantial proliferation of aethers in 18th-century physics, a sensible response, while still abiding broadly by the approach, is to pare things down a bit and eliminate some of the imponderable, insensible posits. Indeed, as I have mentioned, a number of physicists of Kant's day either sought to restrict the number of subtle fluids or tacitly assumed the identity of some of the class. Even Newton, in his more speculative moments, attributed *various* distinct phenomena to the aether, such as cohesive and optical phenomena. For Boerhaave and van Musschenbroek, fire was the "*chief* agent of change," and "heat, light and electricity were all forms of fire" (Hankins 1985, 52; my emphasis). Nollet famously argued that the matter of fire and that of electricity were identical (Home 1979). As Adickes observes, both Silberschlag and Crusius had similarly comprehensive and unifying theories of the aether as well (1922, 338–40, 344–5; see also 1924–1925, II:78n). Kant was thus not unique in theorizing a unificatory role of the aether in 18th-century physics, but he did keenly recognize this trend in subtle fluid theorizing and envisaged its natural end point: the coalescence of *all* the subtle fluids in a *single*, all-pervasive aether. Hence the study of the role of the aether, for Kant, reveals his ideal of a single causal-explanatory basis to be no mere baroque, hyper-rationalistic albatross. Rather, the singular aether is both the natural end point of the subtle fluid theorizing and a reasonable postulate, given the state of 18th-century physics. Heilbron's lamentations about the frequent postulation of aethers, mentioned above, can at least partially answered in this manner: Kant, for his part, meant to improve on the tradition by reducing some of these postulates.

Third, consideration of the case of the aether (perhaps ironically) reveals the way in which Kant's systematization of science depends crucially upon actual empirical evidence and well-grounded reasonings. In the absence of evidence or a reason to combine various natural phenomena under a subtle fluid, the mere postulation of a hypothetical, all-explaining material provides no genuine consolidation of the various phenomena. Kant himself recognizes this, stating in 1788s *On the Use of Teleological Principles in Philosophy* that the simple "coining [of] the *common title* of various basic powers" provides no actual unity (VIII:181n). Simply postulating that there is some unifier for a system does not make good on the systematizing or unifying role that is meant to be filled. It's one thing to say that there is a common genus, let's call it **X**, for humans and clams, but determining that humans and clams are both Nephrozoa, a clade of Bilaterians, a subkingdom of Animalia, sharing a common ancestor around 560 million years ago, that is another thing altogether (OneZoom 2020). Similarly, one can simply postulate that electricity and magnetism are manifestations of one and the same force, but the genuine unificatory work is done by Maxwell's Laws, the set of rules that govern electromagnetism.

Kant's theory of the aether is commonly viewed as an embarrassment. In particular, the idea that the existence of the aether, as a material condition of the possibility of experience, can be deduced a priori is both prima facie absurd (after all, the aether was a scientific *hypothesis* and one that we now know to be *false*) and stands in tension with Kant's approach in the first *Critique*, according to which what can be known a priori of objects of experience are their *formal* conditions of possibility. Nevertheless, that the theory of the aether and its unifications transparently take in empirical information for (dis)confirmation goes some way to excuse the excesses of the aether hypothesis. Kant does not simply postulate it as the unknown unifier of all natural phenomena. Rather, he attempts – to greater and lesser success in different cases – to provide rational arguments based on actual empirical evidence for unifying particular phenomena in the same material. *On Fire*, for instance, is replete with arguments for unifying the matter of heat, the elastic matter, and the luminiferous aether in an individual substance. Although some of these arguments constitute a nadir in Kantian argumentation, it is noteworthy that the arguments he provides aspire to be grounded on empirical evidence. The aether's unificatory role is not based simply on a need for systematic unity; rather the systematic unifying is a happy consequence of experimental and observational evidence Kant adduces.

The case here parallels Georg Stahl's phlogiston experiments, which Kant praises in the second edition preface to the *Critique of Pure Reason* and which I have discussed in at length elsewhere (McNulty 2015). In Stahl's experiment, he calcined metallic lead to produce lead calx, or what we now call lead oxide. Then, he burned the calx in the presence of charcoal, which, according to Stahl's theory, was nearly entirely made up of phlogiston, the principle of combustion. Upon roasting the calx with the charcoal, he found that the metal had revivified with its characteristic properties. This showed that what was lost from the lead during calcination is exactly what is released from organic material in combustion. Thereby these phenomena are unified under a single causal-explanatory ground: phlogiston. Kant extols this experiment as one of those illustrating the secure path of science, whereby we interrogate natural by use of reason's principles. But the case also bears substantial resemblance to the case of the aether under consideration. Proposed unifications in chemistry or physics must be brought as principles to nature, held up against it, and tested with empirical evidence.[35]

Fourth, and finally, in the aether, we see a fascinating coalescence of top-down and bottom-up reasoning in physics, or perhaps better, determinative and regulative judgment, even in the period before the outset of the *Opus postumum*'s transition project, which purportedly aims at bridging this gap (Friedman 1992). In his philosophy of science, Kant is regularly seen as the arch-foundationalist, basing all natural science on top-down determinative judgment proceeding from the categories. That is, empirical physics is based on rational physics, which consists of synthetic a priori principles about matter, which are in turn grounded on the categories, or, specifically, their application to the concept of matter (see Watkins 1998). Consideration of Kant's aether theory upsets this picture. When faced with the explanation of empirical phenomena in physics, he combines bottom-up and top-down reasoning. That is, Kant reasons about observations of and experiments upon natural phenomena in order to ground his unifications of them. In this sense, bottom-up reasoning is critical to Kant's own theorizing about physics. But, nonetheless, Kant consistently conceives of the activity of the aether as *mediated* by the fundamental forces of matter, the primary objects of rational physics. Kant conceives of the aether as an exceptionally (and originally) expansive material, where its manifold actions on matter are effected via its enormous elasticity (see IV:534). This is to conceive of the aether as effective via the fundamental forces (see McNulty 2022). It is precisely for this reason, I suggest, that the aether made such a natural option for bridging the gap between the metaphysical foundations of natural science and empirical physics in OP (or, relatedly, between determinative and reflective judgment).

Thus, recognizing Kant as in a tradition of aether theorizing makes his theory of the aether more palatable and helps us to recognize its substantive use in empirical physics. But additionally, the aether theory serves as a concrete, fleshed-out case study in Kant's conception of natural scientific systematization, highlighting an ideal, for Kant, of reducing observed causal powers to the activity of a single, subtle fluid.[36]

Notes

1 Schofield (1970) tells this story well, contrasting "materialism," as the tradition of postulating subtle fluids as bearers of particular causal powers, with the aforementioned "mechanism." I largely avoid Schofield's terminology of materialism in this chapter, first, due to Cantor and Hodge's (1981, 35–37) objections to Schofield's carving up of 18th-century natural philosophy into these two camps, and second, due to its potential to confuse in the philosophical context, wherein materialism bears a distinct, ontological denotation.

2 I use "aether" and "subtle fluids" interchangeably to refer to such substances. However, note that, as Cantor and Hodge observe, not all aethers were conceived of as *continuous* fluids. Some were rather thought of as discrete or particulate (such as Newton's).
3 Hankins (1985, 53) notes an important contrast: the aethers or subtle fluids were purported to be quantifiable (though not ponderable). Indeed, Hankins depicts the major benefit of aether theories being that they facilitated the quantification of phenomena, especially in a manner superior to the competitive mechanical philosophy.
4 Hankins (1985, 50–53), in contrast, nicely lays out the attractions of the aethereal program. Hankins contends that subtle fluids assisted scientists in conceptualizing physical phenomena, best explained the conservation of the associated phenomena (heat, electricity), and made possible quantification.
5 Edwards (2000, 112–66) provides a likeminded account to my own, according to which Kant's conception of the aether remains a throughline for the various phases of his philosophy of nature.
6 Massimi (2011) relates Kant's early natural philosophy in the "materialist tradition," connecting it with work on subtle fluids from Newton, Hales, and Boerhaave. To this work I bear a debt of outlook and substance. In this chapter, I expand Massimi's perspective, especially by grappling with Kant's critical theory of the aether, recognizing a greater domain for aethereal explanations, and extracting implications for our understanding of Kant's theory of systematicity in the sciences.
7 This letter, published originally in Boyle's collected words (1744), and Newton's views on the aether became more widely known with the publication of Bryan Robinson's *Sir Isaac Newton's Account of the Aether* (1745) (a follow-up to his *A Dissertation on the Aether of Sir Isaac Newton* (1743)), which offered a full-throated endorsement of the aether theory and reproduced Newton's letter to Boyle. Robinson's work was influential; as Heilbron puts it, "British natural philosophers took it as evidence that Newton had always believed in, and had virtually demonstrated, the existence of an active, springy, non-material aether" (1982, 61).
8 Cantor and Hodge (1981, 19–24) observe that Newton's aether is non-mechanical. That is, it does not express effects on gross matter via motion. Rather, the aether is *active* and expresses forces that are immediately influential on matter.
9 Newton conjectures that the pressure of the aether on bodies explains a famous observation made in Boyle's air pump: the coherence of marble tiles in the evacuated chamber of the air pump (1744, 70). This experimental finding was curious in that day. Boyle conjectured that two cohering marble tiles would fall away from one another in the vacuum of the air pump (no longer being pressed together by air pressure). He found that the tiles remained cohered to one another in the evacuated chamber. Newton's postulation of the variably dense aether explains why.
10 Earlier, kinetic theories of heat were posed by Francis Bacon (2000, 130–5) and Galileo Galilei (1957, 273–9).
11 This connection is even deeper. Boerhaave claims that the elementary fire is the ultimate principle of motion throughout the universe (Love 1974, 549).
12 For more detailed and historical treatments of Boerhaave's theory of fire, see, for example, Love (1974), Powers (2012, 63–91, 115–40), and Boantza (2017).

13 See also Euler (1768, 61–78).
14 Finally, although they will play a less crucial role in our consideration of Kant's views on aethers, it is worthwhile to also note other aethers postulated in the time period. William Gilbert was the first to postulate a magnetic fluid. Descartes also countenanced a magnetic matter, though his consists of tiny screws, which enter the pores of magnetic objects. Based on whether the pores are right- or left-handed, the screws will either repel or attract the body (Cantor and Hodge 1981, 17–18). For Euler, magnetic objects have canals that only allow the penetration of magnetic matter from one direction, which causes motion, opposed to nonmagnetic object, which allow passage of magnetic matter in all directions (1768, 243–6). Gilbert also distinguished magnetism from electricity, which is also due to a subtle fluid (Heilbron 1982, 160–3). Du Fay postulated two electric fluids – vitreous and resinous – where Benjamin Franklin's major innovation was to pare the two down to one (Hankins 1985, 61–67). Finally, aethers even find their place in physiology, especially with the work of David Hartley. Hartley postulates a "medullary" fluid in the nervous system, vibrations in which cause perceptions in the brain (Laudan 1981, 159–64).
15 Kant also owned copies of the ultimate edition of Newton's *Principia* (1714) and of the second Latin edition of his *Opticks* (1719), both of which include aether conjectures (Warda 1922, 35).
16 Kant owned the German translation of van Musschenbroek (Warda 1922, 35). There were, of course, other, intermediate influences on Kant's aether theory. For example, Adickes (1924–1925, II:15–16) notes that Christian Wolff made use of the aether in natural explanations, holding that "the impact of the aether or other fluids is a cause for the cohesion of solid bodies," while fluidity depends upon another foreign matter that hinders the separation of parts.
17 The first edition was published in 1753. For the present chapter, I consulted the fourth edition.
18 This is a well-known and notorious problem for material theories of heat and combustion, particularly the phlogistic theory. For a detailed account of the problem of weight-gain during calcination, the widespread knowledge of the problem even before Lavoisier's famed work on combustion, and a survey of responses to it, see Guerlac (1961, 111–45).
19 To the question of why magnetic force can behave differently with respect to different poles of magnetic objects, Erxleben considers both the Cartesian and Eulerian accounts. For Descartes, magnetic matter consists of tiny screws, which enter the pores of magnetic objects. Based on whether the pores are right- or left-handed, the screws will either repel or attract the body. Erxleben calls Euler's theory "less contrived," as it does not appeal to contrived shapes of particles. For Euler, the pores of magnetic objects only allow the penetration of magnetic matter from one direction, which causes motion, opposed to nonmagnetic object, which allow passage of magnetic matter in all directions (1772, 444; see also Adickes 1922, 343–4).
20 Adickes (1924–1925, II:2; 1922) also situates Kant's theory of heat at the juncture between substantial theories of heat (à la Wolff, Boerhaave, van Musschenbroek, Crusius, Lavoisier, and Laplace) and mechanical theories of heat, according to which heat consists in the vibration or motion of matter (à la Bacon, Descartes, Locke, Boyle, Newton, and Leibniz).
21 Kant goes on to argue that, due to the elastic matter's omnipresence within the body and its elasticity, this matter communicates the force of the weight of the liquid in *all* directions, which entails Pascal's law of hydrostatics.

22 For a reading of this puzzling argument, see Adickes (1924–1925, II:41).
23 For more on the role of the aether in the *Metaphysical Foundations*, see Edwards (2000, 132–44).
24 Additionally, the aether plays an important role in a well-known revision to the third edition of the *Critique of the Power of Judgment* regarding Euler's theory of light. In the first and second editions of the book, Kant expresses that "I very much doubt" Euler's theory, whereas in the third edition, he then reports that of Euler's theory, "I have very little doubt" (V:224). See Förster (2000, 24–47) and the editorial endnote on the passage in the Cambridge Edition translation of the *Critique of the Power of Judgment* (Kant 2000, 370n).
25 The most comprehensive treatment of the 1770s reflections on the aether, one that complements that on offer in this chapter, is provided by Edwards (2000, 123–32).
26 In an editorial note, Lehmann surmises that Kant used Erxleben (1772) in association with the *Berlin Physics* lectures (XXIX:656). Thus, the departures from Erxleben's text – like the full-throated endorsement of aethereal explanations – must be due to Kant.
27 Kant reiterates this idea in a later handwritten note. "All bodies necessarily have had fluidity, in order to become solid. The original fluidity is that of the universal vehicle of all things: the aether" (Reflection 98, XIV:616).
28 See also Reflection 46 (XIV:418–427), Reflection 48 (XIV:435), and Reflection 50 (XIV:443–444).
29 For more detailed treatments of the distinction between fundamental and derivative forces, see Howard (2021) and McNulty (2022).
30 Although my concerns are especially located with Kant's theory of the aether and systematicity before his renewed interest in the late 1790s during the drafting of the *Opus postumum*, there are many descriptions of the unificatory role of the aether throughout those drafts. For a short list of some such references, consider the following. Kant explains cohesion via the pressure of the aether (XXI:374, 453), which cohesion, in turn, makes possible differences in densities (XXI:374) and goes on to claim that the original fluidity of the aether explains the derivative forces of dissolution and expansion by heat (XXI:374; see also XXI:378). Additionally, he maintains that vibrations of the aether cause solidity and, more generally, that solidity must have the same cause as cohesion (XXI:374). Kant continues to hold that heat and light are both modifications of one selfsame material, the aether (XXII:214), while also asserting that magnetism and electricity attract via an intermediary matter (XXII:215). There are *many more* such descriptions in the *Opus postumum*. For more on the role of the aether in Kant's late philosophy, see Friedman (1992, 213–341); Edwards (2000, 145–92); Förster (2000, 75–116); Hall (2014, 71–122); and Thorndike (2018, 92–111).
31 Indeed, the maxims of systematicity discussed in the appendix to the Transcendental Dialectic – those of genera, specification, and continuity – are explicitly dubbed "logical" principles (A658/B686).
32 For more on the logical nature of natural description, and Kant's preference for natural history as an approach to taxonomy that gets at the *real* species, see *Of the Different Races of Human Beings* (II:429); *Danzig Physics* (XXIX:100); Sloan (2006); and Sandford (2018).
33 Guyer (1990) also highlights this point, recognizing a regulative ideal of achieving an explanatory minimum.
34 I also add that the study of the aether reinforces the Kantian approach to reasoning about nature, especially as characterized by Robert Butts (1994).

According to Butts, Kant does not make use of standard induction, reasoning from particulars to the universal. Rather, Kant postulates generic, causal forms that unify the relevant particulars. Kant's reasoning vis-à-vis natural science is essentially unificatory, not inductive or generalizing. The aether constitutes a paradigm instantiation of this strategy.

35 Relatedly, there is a similarity between the empirically informed unification project via the aether and the project of unifying the chemical elements, conceived of as principles or bearers of properties (see Carrier 1990; McNulty 2015).

36 I thank the audience at the conference on Kant and the Systematicity of the Sciences hosted at Goethe Universität in 2019. I particularly benefited from the feedback of and discussion with Jennifer Mensch and Michael Olson. Additionally, I am grateful to Daniel Warren and Andrew Janiak for our conversations of the history of the aether and Kant's theory of empirical physics.

Bibliography

Adickes, Erich. 1922. "Zur Lehre von der Wärme von Fr. Bacon vis Kant." *Kant-Studien* 27(1–2): 328–68.

Adickes, Erich. 1924–1925. *Kant als Naturforscher*. Vols. I and II. Berlin: De Gruyter.

Bacon, Francis. 2000. *The New Organon*, edited by Lisa Jardine and Michael Silverthorne. Cambridge: Cambridge University Press.

Boantza, Victor. 2017. "Elements, Instruments, and Menstruums: Boerhaave's Imponderable Fire Between Chemical Masterpiece and Physical Axiom." In *The Romance of Science: Essays in Honour of Trevor H. Levere*, edited by Jed Buchwald and Larry Stewart, 9–46. Cham: Springer.

Boerhaave, Herman. 1735. *Elements of Chemistry*. Translated by Timothy Dallowe. Vol. 1. London: J. and J. Pemderton, J. Clarke, A. Millar, and J. Gray.

Boyle, Robert. 1744. *The Works of the Honourable Robert Boyle*, edited by Thomas Birch. Vol. 1. London: A. Millar.

Boyle, Robert. 1999–2000. *The Works of Robert Boyle*. 14 vols. Edited by Hunter, Michael, and Edward Davis. London: Pickering & Chatto. (BW)

Butts, Robert. 1994. "Induction as Unification: Kant, Whewell, and Recent Developments." In *Kant and Contemporary Epistemology*, edited by Paolo Parrini, 273–89. Dordrecht: Kluwer.

Cantor, G.N., and M.J.S. Hodge. 1981. "Major Themes in the Development of Ether Theories from the Ancients to 1900." In *Conceptions of Ether: Studies in the History of Ether Theories 1740–1900*, edited by G.N. Cantor and M.J.S. Hodge, 1–60. Cambridge: Cambridge University Press.

Carrier, Martin. 1990. "Kants Theorie der Materie und ihre Wirkung auf die zeitgenössische Chemie." *Kant-Studien* 81(2): 170–210.

Descartes, René. 1897–1909. *Oeuvres de descartes*. 11 vols. Edited by Charles Adam and Paul Tannery. Paris: Léopold Cerf. (AT)

Eberhard, Johann Peter. 1774. *Erste Gründe der Natur*. 4th ed. Halle: Renger.

Edwards, Jeffrey. 2000. *Substance, Force, and the Possibility of Knowledge: On Kant's Philosophy of Material Nature*. Berkeley: University of California Press.

Edwards, Jeffrey, and Martin Schönfeld. 2006. "Kant's Material Dynamics and the Field View of Physical Reality." *Journal of Chinese Philosophy* 33(1): 109–23.

Erxleben, Johann Christian Polycarp. 1772. *Anfangsgründe der Naturlehre*. Göttingen: Dieterich.

Euler, Leonard. 1746. "Nova theoria lucis et colorum." *Opuscula varii argumenti* 1: 169–244.
Euler, Leonard. 1768. *Lettres a une princesse d'Allemagne sur divers sujets de Physique & de Philosophie*. Vol. 1. Saint Petersburg: Imperial Academy of Sciences.
Friedman, Michael. 1992. *Kant and the Exact Sciences*. Cambridge: Harvard University Press.
Förster, Eckart. 2000. *Kant's Final Synthesis*. Cambridge: Harvard University Press.
Galilei, Galileo. 1957. "The Assayer." In *The Discoveries and Opinions of Galileo*, edited by Stillman Drake, 231–80. New York: Doubleday Anchor.
Guerlac, Henry. 1961. *Lavoisier – The Crucial Year: The Background and Origin of His First Experiments on Combustion in 1772*. Ithaca: Cornell.
Guyer, Paul. 1990. "Reason and Reflective Judgment: Kant on the Significance of Systematicity." *Noûs* 24(1): 17–43.
Hakfoort, Casper. 1995. *Opticks in the Age of Newton*. Cambridge: Cambridge University Press.
Hales, Stephen. 1727. *Vegetable Staticks*. London: W. and J. Innys and T. Woodward.
Hall, Bryan. 2013. *The Post-Critical Kant: Understanding the Critical Philosophy through the Opus postumum*. New York: Routledge.
Hankins, Thomas. 1985. *Science and the Enlightenment*. Cambridge: Cambridge University Press.
Heilbron, John Lewis. 1982. *Elements of Early Modern Physics*. Berkeley: University of California Press.
Home, Roderick Weir. 1979. "Nollet and Boerhaave: A Note on Eighteenth-Century Ideas about Electricity and Fire." *Annals of Science* 36: 171–6.
Howard, Stephen. 2021. "Kant on the Fundamental Forces of Matter: Why Attraction and Repulsion?" *Kantian Review* 26(3): 413–33.
Kant, Immanuel. 1900–. *Gesammelte Schriften*. Edited by Berlin-Brandenburgischen Akademie der Wissenschaften (previously, Akademie der Wissenschaften der DDR, Akademie der Wissenschaften zu Berlin, and Königlich Preußischen Akademie der Wissenschaften). 29 vols. Berlin: De Gruyter (previously, Reimer).
Kant, Immanuel. 2000. *Critique of the Power of Judgment*. Edited by Paul Guyer. Translated by Paul Guyer and Eric Matthews. Cambridge: Cambridge University Press.
Laudan, Larry. 1981. "The Medium and Its Message: A Study of Some Philosophical Controversies about Ether." In *Conceptions of Ether: Studies in the History of Ether Theories 1740–1900*, edited by G.N. Cantor and M.J.S. Hodge, 157–85. Cambridge: Cambridge University Press.
Locke, John. 1975. *An Essay Concerning Human Understanding*. Edited by Peter H. Nidditch. Oxford: Clarendon. (N)
Love, Rosaleen. 1974. "Herman Boerhaave and the Instrument-Element Concept of Fire." *Annals of Science* 31: 547–59.
Massimi, Michela. 2011. "Kant's Dynamical Theory of Matter in 1755, and Its Debt to Speculative Newtonian Experimentalism." *Studies in History and Philosophy of Science* 42(4): 525–43.
McNulty, Michael Bennett. 2015. "Rehabilitating the Regulative Use of Reason: Kant on Empirical and Chemical Laws." *Studies in History and Philosophy of Science* 54: 1–10.
McNulty, Michael Bennett. 2022. "Beyond the Metaphysical Foundations of Natural Science: Kant's Empirical Physics and the General Remark to the Dynamics."

In *Kant's Metaphysical Foundations of Natural Science: A Critical Guide*, edited by Michael Bennett McNulty, 178–96. Cambridge: Cambridge University Press.
Molière. 1876. *The Dramatic Works of Molière*. Translated by Henri van Laun. Vol. 6. Edinburgh: William Paterson.
Newton, Isaac. 1713. *Philosophiæ Naturalis Principia Mathematica. Editio Secunda Auctior et Emendatior*. Cambridge: Cornelius Crownfield.
Newton, Isaac. 1714. *Philosophiæ Naturalis Principia Mathematica. Editio Ultima Auctior et Emendatior*. Amsterdam: Sumptibus Societatis.
Newton, Isaac. 1718. *Opticks: Or, A Treatise of the Reflections, Refractions, Inflections and Colours of Light. The Second Edition, with Additions*. London: W. and J. Innys.
Newton, Isaac. 1719. *Optice, sive de reflexionibus, refractionibus, inflexionibus et coloribus lucis, libri tres. Editio Secunda, auctior*. London: G. and J. Innys.
OneZoom Core Team. 2020. *OneZoom Tree of Life Explorer*. Version 3.4.1. URL = http://www.onezoom.org
Powers, John C. 2012. *Inventing Chemistry: Herman Boerhaave and the Reform of the Chemical Arts*. Chicago: University of Chicago Press.
Robinson, Bryan. 1743. *A Dissertation on the Aether of Sir Isaac Newton*. Dublin: G. Ewing and W. Smith.
Robinson, Bryan. 1745. *Sir Isaac Newton's Account of the Aether*. Dublin: G. and A. Ewing and W. Smith.
Sandford, Stella. 2018. "Kant, Race, and Natural History." *Philosophy and Social Criticism* 44(9): 950–77.
Schofield, Robert. 1970. *Mechanism and Materialism: British Natural Philosophy in an Age of Reason*. Princeton: Princeton University Press.
Sloan, Phillip. 2006. "Kant on the History of Nature: The Ambiguous Heritage of the Critical Philosophy for Natural History." *Studies in History and Philosophy of Biological and Biomedical Sciences* 37(4): 627–48.
Teske, Johann Gottfried. 1729. *Dissertatio de igne ex chalybis silicisque collisione nascente*. Königsberg: Reusner.
Thorndike, Oliver. 2018. *Kant's Transition Project and Late Philosophy: Connecting the* Opus postumum *and Metaphysics of Morals*. London: Bloomsbury.
Warda, Arthur. 1922. *Immanuel Kants Bücher*. Berlin: Breslauer.
Watkins, Eric. 1998. "The Argumentative Structure of Kant's *Metaphysical Foundations of Natural Science*." *Journal of the History of Philosophy* 36(4): 567–93.

9 Systematicity and the Definition of a Science

Physics in Kant's *Opus postumum*

Stephen Howard

This chapter seeks to clarify the role of definitions in Kant's theory of the systematicity of the sciences. In the key discussion of the systematicity of science that opens the Architectonic chapter of the *Critique of Pure Reason*, after stating that "Nobody attempts to establish a science without considering there to be an idea that would underlie it," Kant adds that "the schema, indeed even the definition, which the scientist gives at the very beginning of their science, very seldom corresponds to the idea" (A834/B862).[1] In this passage – to which we return below – Kant indicates a relation between a science's definition and the schema of its idea. The idea of a science is a regulative concept, projected by reason, of the form of the whole of the science; it provides the scientist with a conception of the ordered domain of cognitions under investigation.[2] The schema is needed for the "execution [*Ausführung*]" of the guiding idea in scientific practice (A833/B861). Although the schema and the definition of a science do not precisely correspond to its idea, Kant's wording suggests that definitions, like schemata, play a role in organizing cognitions into the unified system that constitutes a science. Additionally, Kant claims in the *Prolegomena* that in order to present a set of cognitions as a science, "one must first be able to precisely determine the distinguishing features [*das Unterscheidende*], which it has in common with no other science and which are therefore unique [*eigentümlich*] to it" (IV:265). These unique distinguishing features sound very much like the characteristic marks (*Merkmale*) that appear in the philosophical definitions employed by Kant's German metaphysician predecessors.

Recent literature has pointed to the role of definitions in Kant's account of the systematicity of the sciences, but without consensus. Two opposed interpretations have been put forward. In Section 9.1, I set out these contrasting positions. Briefly put, Thomas Sturm considers definitions to be key to Kant's account of the systematicity of science, while Katharina Kraus considers them to play a minor and dispensable role. I shall present further evidence that seems to support Kraus' position: passages in which

216   Kant and the Systematicity of the Sciences

Kant denies that definitions can be legitimately used in philosophy. These passages apparently suggest that Kant would reject the view, ascribed to him by Sturm, that the definition of a science is a necessary condition of its systematic structure. However, a different picture emerges if we turn to Kant's late reflections on the systematicity of empirical physics in the *Opus postumum*. Although generally ignored in the literature to date on our topic, a series of drafts from 1799 to 1800 are remarkable because they show Kant repeatedly attempting to define physics. Section 9.2 provides a brief overview of these definitions. I take the 1799–1800 drafts to provide an extended example of how Kant applies his theory of definition "in practice," so to speak. Finally, Section 9.3 returns to Kant's account of the significance of definitions for the systematicity of science. I argue that Kant's application of his theory of definition in his late drafts validates the general direction of Sturm's interpretation, while also encouraging us to amend some of the details. Moreover, I suggest that Kant's definitional attempts in the *Opus postumum* can usefully illuminate some nuances of the canonical statements about systematicity and definition in the Architectonic and Discipline chapters of the *Critique*.

9.1 Sturm and Kraus on Definitions and the Systematicity of Science

The significance of definitions for Kant's account of the systematicity of science has been innovatively foregrounded by Thomas Sturm. Sturm points out that Kant distinguishes between the idea and the definition of a science (Sturm 2009, 162–3, 167–8). While noting that the two main published passages from which we can reconstruct this distinction (A834/B862 and IV:265, quoted above) are not unambiguous, Sturm argues that the idea and definition are "two thoroughly different *functions* of our reflection on the sciences" (Sturm 2009, 167–8). These two functions align with what Sturm contends are two distinct levels in Kant's conception of the systematicity of the sciences. On the one hand, "inner systematicity" concerns "how the integration and structuring of cognitions within a science is to be achieved." On the other hand, "outer systematicity" concerns "what distinguishes a science from another (and how the sciences as a whole are classified and ordered)" (Sturm 2009, 168, see also 135–6). Inner systematicity is the unity and completeness of a single science, while outer systematicity differentiates a science from others and locates it within the wider system of sciences.

In Sturm's view, the idea of a science furnishes its inner systematicity: it provides the *focus imaginarius* that guides the scientist to unify disparate cognitions into a whole (A832/B860; Sturm 2009, 138–9, 168). By contrast, the definition of a science provides outer systematicity: it distinguishes it from others by identifying its defining characteristic marks (*definierende Merkmale*) (Sturm 2009, 168). On Sturm's reading, three kinds of characteristic marks constitute the definition of a science: the *object, method* (or type of cognition: *Erkenntnißart*), and *aim*. Kant points to the first two of these elements in the *Prolegomena* (IV:265) and all three in the *Metaphysical Foundations* (IV:477; Sturm 2009, 163).

In addition, Sturm argues that outer systematicity precedes inner systematicity, and so the definition is more fundamental than the idea:

> To be able to specify the systematic connection of cognitions in a science presupposes that one possesses the 'idea' of the science. But this in turn presupposes that one possesses a definition of this science, even if it is a preliminary definition that will be improved in the course of the research. That is to say that without a definition, the demarcation of the science from others remains unclear. Likewise, a science is for Kant only completely systematized when its cognitions are understood and explained as a 'whole.'
> (Sturm 2009, 136; see also 162)

The inner systematicity of an individual science, which is furnished by its idea, presupposes that the science possesses the outer systematicity that is provided by its definition. It is thus a science's definition, which identifies its object, method, and aim, distinguishing it from others and locating it within the encyclopaedia of the sciences, that is for Sturm the ultimate basis of its systematicity.

Contrary to Sturm's interpretation, Katharina Kraus has suggested that the definition of a science is insignificant to Kant's theory; what is instead important is the idea of reason that guides the scientist. Kraus points to the key passage from the Architectonic, already partially quoted above:

> Nobody attempts to establish a science without considering there to be an idea that would underlie it. But in the elaboration of the science, the schema, indeed even the definition, which the scientist gives at the very beginning of their science, very seldom corresponds to the idea; for this [idea] lies, like a seed, in reason, in which all parts lie hidden, still deeply enveloped [*noch sehr eingewickelt*] and barely recognizable through microscopic observation.
> (A834/B862)[3]

As Kraus reads this passage, the definition is "artificially" given at the outset of the science and "has to be corrected along the way" until we attain the more adequate idea of the science (Kraus 2018, 79n). For Kraus, only the idea (explicated by its schema) is needed to provide a science with its domain, intrinsic structure, intrinsic purpose, and methodology (Kraus 2018, 79). Kraus implicitly reverses Sturm's order of precedence by discussing outer systematicity after inner systematicity, and she claims that both types of systematicity are provided by the idea, not the definition, of a science (Kraus 2018, 80, 82).

Kraus argues against Sturm only in passing, but we can find further evidence in support of her claims in Kant's discussions of the methodological difference between mathematics and philosophy. Kant consistently rejects philosophers' attempts to imitate Euclidean geometry by beginning with definitions. In writings published almost two decades apart, the 1764 *Inquiry* and the chapter on the Discipline of Pure Reason in the *Critique*, Kant presents very similar arguments.[4] In the *Inquiry*, he states,

> In mathematics I begin with the definition [*Erklärung*] of my object, for example of a triangle, or a circle, or whatever. In metaphysics I may never begin with a definition. Far from being the first thing I know about the object, the definition [*Definition*] is nearly always the last thing I come to know.
>
> (II:283)

Definition and *Erklärung* are used interchangeably here.[5] Kant claims that in philosophy one discovers the definition at the end of an investigation, whereas mathematicians begin by defining their concepts (see also A730–731/B758–759). This is due to the differing functions of definition in mathematics and philosophy, grounded on the differences between the two disciplines.

To grasp these differences, let us briefly summarize Kant's account in the *Inquiry* and the Discipline chapter. Mathematicians define their concepts by arbitrarily combining concepts (II:276, 280; A729/B757).[6] In this process, the defined object is made possible and brought into being (II:280, 281; A729/B757). This is because, as the Doctrine of Method in the *Critique* makes clear, mathematics constructs its concepts by exhibiting them *a priori* in intuition (A713/B742; see A729–730/B757–758). Because mathematical definitions produce their objects, they "can never err" (A731/B759). The situation is the opposite in philosophy on every point. Philosophical concepts are given, not constructed (II:283; A729–730/B757–758). Definitions in philosophy cannot therefore produce an object but can only try to clarify obscure given concepts (II:283; A730/B758). In philosophy, concepts can be given either *a priori* or empirically, and in neither case can the concept be defined, strictly speaking: in the former case, because we cannot be sure that the definition is exhaustive; in the latter case, because we cannot determine the boundaries of the definition with certainty, i.e., which marks belong to the definition (A727–728/

B755–756).⁷ Whereas mathematical definitions can never err, philosophical definitions are a common source of error (II:292; A732/B760). One should not therefore begin with definitions in philosophy. In the *Inquiry* this is even the "first and the most important rule" for attaining metaphysical certainty. Rather than "venture" (*wagen*) the definition from the beginning of an investigation, one should only "concede" (*einräumen*) it at the end, on the basis of "the most certain judgements" (II:285; see II:289; A730/B758).⁸ Philosophical definitions can, at best, be posited at the end of an investigation, if they must be used at all.

In these passages from 1764 and 1781, Kant criticizes those who, like Wolff, employ a broadly Euclidean method in philosophy. Such a method, as Wolff puts it, "starts from the definitions, proceeds to the axioms, and from here to the theorems and the problems."⁹ In logic lecture notes, Kant describes Wolff's "mathematical" method for philosophy in precisely this way and states that it is misguided (XXIV:783, quoted in Gava 2018, 274). Gava proposes that "it seems plausible to conclude that, when Kant complains that Wolff was wrong in applying the mathematical method to philosophy, he simply meant that Wolff should not have started with definitions" (Gava 2018, 277). Due to differences between Wolff's and Kant's conceptions of mathematical and philosophical method, Gava suggests that the two thinkers' disagreement on the status of definitions concerns only the proper order of philosophical argument. Gava claims that this "superficial dissimilarity" masks a deeper continuity, in that both affirm that philosophy should begin with "analyses of given concepts" (Gava 2018, 293, cf. 302–3). Nevertheless, on Gava's reading, Kant gives up the Wolffian "mathematical" method according to which philosophy should begin with definitions.

There thus seems to be no doubt that Kant dismisses the use of definitions as starting points in philosophy. How, then, should we understand his claim in the Architectonic chapter that the scientist gives "the definition ... at the very beginning of their science?" This act of defining would certainly be a philosophical task, according to Kant's basic distinction between philosophical and mathematical cognition in the Doctrine of Method. Only mathematical definitions create concepts; the scientist who seeks to define a science is concerned with a *given* concept, namely their burgeoning science, and so this act of definition is a philosophical undertaking. For an initial explanation of the apparent contradiction in Kant's claims here, we can look at the continuation of the passage to which Kraus refers. After noting that the schema and the definition of a science rarely correspond to the idea, because the latter is hidden in reason like a seed, Kant adds,

> For the founder [*Urheber*] and often even their most recent successors circle blindly around an idea [*um eine Idee herumirren*], which

they have not made distinct to themselves and which thus cannot determine the proper content, the articulation (systematic unity) and boundaries of the science.

(A834/B862)[10]

The founder of a science and their followers alike lack a sufficiently clear idea of the science, and so they cannot determine its content, unity, and boundaries.[11] The founder's initial definition is thus a mere "description [*Beschreibung*]" (A834/B862). Kant goes on to note with regret ("*Es ist schlimm…*") that we only glimpse the idea in a clearer light "after a long time of rhapsodically collecting many relevant cognitions, under the direction of an idea lying hidden within us" (A834–835/B862–863). Employing a vivid image, he suggests that "systems seem to be formed like maggots through a *generatio aequivoca* from the mere confluence of gathered-together concepts" (A835/B863). That is, the systematicity of a science – and thus of the sciences *in toto* – is not determined in advance of the empirical investigation, but it emerges in a way analogous with "equivocal" or "spontaneous" generation in biology (this theory sought to explain how organized beings apparently emerge from disorganized matter, for example worms from rotting compost).[12] Kant seems to claim here that even though the idea of a science ultimately guides its systematic development – which means we can think of a guiding schema of the idea hidden "in merely self-unfolding reason [*in der sich bloß auswickelnden Vernunft*]" – the emergence of systematicity looks *to us* like the spontaneous generation of worms from compost. Science only gradually appears as systematic after the rhapsodic accumulation of much empirical data, and the idea of the science only becomes evident at a late stage in scientific investigation.

On this reading of these passages from the opening of the Architectonic, Kant distinguishes between the true idea of a science and its erroneous definition. The definition of a science would be a tentative description that should be replaced by the idea once we have sufficient empirical data to gain a clearer conception of it. The definition, which is given "right at the outset," would be erroneous precisely because it is stated upfront, in line with Kant's criticisms of the Wolffian philosophical method. The implication seems to be, in line with Kraus' view, that definitions are initial, inaccurate descriptions hazarded by the founder of a science, and the philosopher who seeks to secure the systematicity of the science should attempt instead to ascertain the idea that underpins the mass of empirical data accumulated in a well-developed science.

The evidence here seems to be firmly on the side of Kraus' interpretation, against that of Sturm. However, there is a highly relevant text that has been ignored in this context: Kant's attempt to make a "transition" from the metaphysical foundations of natural science to empirical physics in the drafts known as the *Opus postumum*. Surprisingly, given the conclusions we have arrived at so far, the late drafts show Kant making copious attempts to *define* the science, physics, at the heart of his late project. The next section will provide a sketch of these attempts, before we return, in the final section, to the light shed by the *Opus postumum* on Kant's theory of the significance of definitions for the systematicity of science.

9.2 Kant's Definitions of Physics in the *Opus postumum*

Although the drafts that constitute Kant's *Opus postumum* are notoriously messy and difficult to interpret, I propose that careful study reveals their primary concern to be with the systematicity of physics.[13] The "transition from the metaphysical foundations of natural science to physics" – which is how Kant consistently formulates the task of his project – is the attempt to provide *empirical* physics with systematic form. Kant struggled with the "transition problem" for over a decade, leaving at his death over 500 pages of drafts, notes, and revisions. Very briefly, the difficulty with which he wrestles is this: to what extent can the results of an empirical science, in this case physics, be determined *a priori*? As the context of the present volume will make clear, Kant's transition problem is closely linked to the central role of systematicity in his conception of a science. If empirical physics is to be a science, the manifold cognitions belonging to it should be unified *a priori* by an idea. This means that the idea of physics should to some extent determine, in advance of the theories, observations, and experiments of working scientists, the domain of the cognitions as well as the relations between these cognitions (see A832/B860). An interpretative question remains open, I would contend, regarding how strong a claim we take Kant to be making in the Architectonic about the *a priori* determination of (empirical) science. On my reading, the *Opus postumum* shows that Kant himself had not ascertained how strong his claim was. The drafts thus show Kant experimenting with a wide range of concepts and philosophical structures through which physics might be determined *a priori* to differing degrees.[14] In the terms he often uses in the drafts, this is the question of how far the specific results of empirical physics can be *anticipated*.[15]

In one of the phases of the drafts, written between August 1799 and April 1800 and known as "fascicles X/XI," Kant tackles the transition problem in a way that is highly relevant to our present concerns: he repeatedly attempts to define "physics." Such definitions did appear in earlier fascicles (e.g. XXI:307, 407; XXII:240), but Kant greatly intensifies his attempts in fascicles X/XI. I estimate that he proposes more than 100 different definitions of physics. These definitions appear under the headers of "Definition" or "What is Physics?," or they begin, "Physics is ...." In fascicles X/XI, Kant tends to designate each folio – which is a large page folded to make a booklet of four sides – with a letter: the drafts of the period run from "A" to "Z" and then "AA" and "BB." Unusually for the 1936–1938 Academy edition, these drafts of 1799–1800 are generally printed chronologically (XXII:295–539), so the reader can relatively easily follow Kant's attempts to define physics.[16]

It seems that at this point in his work, Kant believed that he could make progress with the problem of the transition to physics by clarifying its "arrival point," physics. Although he offers a great number of definitions in folios "A"–"K," it is first in folio "L" that he writes the title, "What is physics?" (XXII:363). In "N," this becomes the first part of a twofold structure: "1) What is physics? ... 2) How is physics possible?" (XXII:377–378; cf. XXII:380–381). The question of how to define physics here precedes the "transcendental" question of its conditions of possibility. Subsequent folios expand this formula into a fourfold task: what is physics; how is it possible; what is the transition; how is it possible? (XXII:461, 467, 496). By the time of folios "T," "U," "X," and "Y," Kant is repeatedly writing the headers "What is physics?," "Definition," and "Physics," followed by attempts at definitions.

No clear linear development of Kant's definitions of physics can be identified. Straightforward and complicated definitions appear side by side on a single page, and similar formulations recur in earlier and later drafts of the period. Nevertheless, we can say in general that Kant presents increasingly surprising formulations as the drafts of fascicles X/XI progress – surprising because they deviate from Kant's previous writings or from our everyday conception of physics. We can here begin with some more routine definitions from earlier folios in fascicles X/XI. Each begins "Physics is ...":

> the systematic doctrine of the empirical investigation of nature
> ("B," XXII:313)

> the *doctrinal system* (*systema doctrinale*) of the sum total (*complexus*) of the moving forces of matter as object of experience
> ("B," XXII:317)

> cognition of the objects of the senses in experience
> ("C," XXII:318)

> the doctrinal system of the moving forces of matter insofar as they are objectively contained in a natural system of the same
> ("C," XXII:319)

the science of the moving forces of matter insofar as they can be acquired through experience

("D," XXII:328)

the science of the principles of natural investigation insofar as it has to do with objects of the senses

("D," XXII:329)

empirical cognition in a system

("F," XXII:339)

the doctrinal system of empirical cognition insofar as its deduction is not possible *a priori*

("I," XXII:356)

doctrine of experience (through observation and experiment) of the moving forces of matter

("K," XXII:359)

To analyze these definitions, let us take up Sturm's reconstruction of Kant's theory, according to which a definition should specify a science's object, method, and aim. In the cases quoted, the definitions all indicate the *method* (or type of cognition) of physics: a systematic doctrine, doctrinal system, (empirical) cognition, science, or observational and experimental doctrine of experience. Moreover, the *object* of this method is always stated: the empirical investigation of nature, the sum total of the moving forces of matter, objects of the senses, the moving forces of matter, the principles of natural investigation, empirical cognition. These definitions regularly further specify the object of physics with "insofar as," usually limiting it to experience or objects of the senses, but also to that which is objectively contained in a natural system or that which is incapable of an *a priori* deduction. Kant often explores the scope of the objects of physics in the *Opus postumum*. One version of the "transcendental" question about conditions of possibility runs, "how is the estimation of the extent of the objects belonging to physics possible[?]" (XXII:318; see also XXII:397–398). We shall see that Kant will explore dramatically widening the scope of the objects of physics.

Two of the characteristics identified by Sturm, the method and object, thus consistently appear in Kant's definitions, but the third, the aim of the science, does not. Even if we consider the text following the one-sentence definitions that commonly expands on them, we rarely find reference to an aim of physics in the earlier folios of 1799–1800. As the drafts progress, however, Kant more often posits an aim, using the formulation "for the sake of [*zum Behuf*]." Sometimes this aim aligns with our everyday conception of physics, for example: "for the sake of explaining a phenomenon" (XXII:366). But Kant much more frequently depicts the aim of physics as "for the sake of experience" or "for the sake of the possibility of experience." A typical formulation runs: "It is not *from* experience but *for* it that we connect the objects of senses in physics" (XXII:327). This is a remarkably broad aim for physics: one would instead expect transcendental philosophy or critique to be the science concerned with the possibility of experience.

The definitions quoted above are generally unsurprising because they present physics as concerned with the objects of outer sense, in line with Kant's conception of physics in the *Critique* and the *Metaphysical Foundations* (A846/B874, IV:467, 470). The regular references to systematicity cohere with what we know from the *Critique* about Kant's conception of a science. The final definition quoted, from "K," is perhaps the most traditional: physics observes and experiments upon the moving forces of matter. In the only extended study of Kant's conception of physics in the *Opus postumum*, Hansgeorg Hoppe proceeds as if this is Kant's consistent conception of physics throughout the drafts; he ignores the great variety of other conceptions.[17] In any case, even the straightforward definitions quoted above are notable for their diversity. Kant does not find and stick to a single definition that satisfies him; he is clearly experimenting with different formulations. We shall return to why this may be.

We can now turn to some of the more surprising formulations. Although they go in directions that seem to conflict with Kant's previously established views, I will suggest that the roots of these departures can be traced to his earlier writings. Three developments seem to be particularly important. First, Kant introduces a new emphasis on the *totality* of the moving forces that constitute the object of physics. This is already evident in folio "B": "Physics is the doctrinal system of the totality [*All*] of moving forces of matter as outer sense objects insofar as it is an object of experience" (XXII:306); "Physics is therefore a systematic presentation of an absolute *whole* [*Ganze*] of the moving forces of matter in appearance (i.e., empirical representations) as a thoroughgoingly determined concept as something existing (of a thoroughgoingly determined thing) ..." (XXII:309). We might expect to find such a stress on totality in Kant's conception of cosmology rather than of physics.[18] But the notion of physics encompassing *all* moving forces results naturally from the combination of two of Kant's views: that physics is the science of the moving forces of matter, and that physics, as a science, is a system and so should be complete. Alternatively and more straightforwardly, we could say that the position results from Kant's conception of nature as "the sum total [*Inbegriff*] of objects of experience" (Bxi; cf. B163; IV:467), or from his conception of "world" and "nature" as two perspectives, mathematical and dynamical, on the same *whole* of all appearances (A418/B446).

Nevertheless, Kant seems to be aware that these definitions are in tension with an everyday conception of physics. In folio "B," after writing that the "outer sense-object" is the "sum total (*complexus*) of matter as moving forces in space," he uses the overtly cosmological language of the "world-whole [*Weltganze*]" and "the totality [*All*] (το παν)" to depict the composition sought in physics.[19] A further definition even suggests that, because its object is a totality, "physics" is a mere regulative idea:

> Physics, therefore, as the doctrinal system of the whole of the forces affecting the senses ... is only an idea of a science that is never fully attainable but is grasped in perpetual progress, for which we do indeed have principles for researching the elementary cognitions but never for encapsulating them in a completed system.
> (XXII:309–310)

In these definitions, Kant foregrounds an issue that has plagued his transition project: how can *all* the specific varieties of physical phenomena be accounted for in (the transition to) physics? In the earliest drafts of the *Opus postumum*, Kant believed he could exhaustively classify all physical forces and properties under the four headers of the categories in what he called an "elementary system." This attempt was unsuccessful.[20] By the time of fascicles X/XI, he begins to distinguish the elementary or natural system of physics, which is the "objective" or "material" part and which proceeds indefinitely, from the doctrinal system, which is "subjective," "formal," and capable of being completed.[21] This seems to be an attempt to deal with the inclusion of the notion of totality in the very definition of physics.

The second development in Kant's definitions of physics is related to the first. According to many of the straightforward definitions quoted above, physics is an empirical science. At the same time, physics, on Kant's view, should be systematic. In fascicles X/XI, Kant repeatedly insists that the notion of an "empirical system" is a contradiction in terms.[22] The reason is clear from Kant's conception of a system: the extent and interrelations of the manifold of a science are determined *a priori*, not empirically, by an idea of reason (A832/B860). How, then, can physics be both empirical and systematic? Regardless of whether we might think that Kant should have an answer to this question on the basis of his critical writings, the *Opus postumum* shows that he considered it a significant problem to be grappled with – in part through his definitions of physics. A particularly radical response explored by Kant is to reject the very notion of empirical science. Under the header "What is physics?," he writes,

> It is not an empirical science (for that would be a self-contradiction, because every cognition, insofar as it should be scientific, must be grounded on formal principles of the connection of the manifold of its representations).
>
> ("R," XXII:407)

In folio "D," Kant bluntly restates this claim that "empirical science" is a contradictory notion: "an empirical science is in general a concept that stands in contradiction with itself" (XXII:329). This radical attempt to jettison the very notion of empirical science rarely appears in the drafts, for good reason: few would agree that "empirical science" is a contradiction in terms. But Kant is here again attempting to resolve the transition problem – how to systematize an empirical science – through new definitions of the science.

The third development may be the most surprising for the reader familiar with Kant's critical-period views. Kant experiments with definitions of physics like the following:

> Physics is an aggregate of all empirical representations, inner and outer, in a system for the sake of experience.
>
> ("N," XXII:375)

> What is physics? It is the scientific doctrine of cognition of objects of the senses (of outer as well as inner) in experience.
>
> ("R," XXII:407)

> Physics (study of nature) is a sum total of both outer and inner representations of the senses in a system, i.e., from both outer as well as inner empirical intuitions as well as inner perceptions of the subject, i.e., sensations (called feelings when they contain pleasure or displeasure).
>
> ("X," XXII:500)

Kant suggests – and in the third definition seems to insist – that the object of physics is representations in both outer *and* inner sense. This clearly undermines the basis of his standard earlier distinction between physics and psychology. Kant now tries conflating physics with what in the *Critique* he called physiology, the umbrella term for physics and psychology (A381, A845/B873).[23] Although strange, this is again not an incomprehensible development in Kant's attempts to define physics. It results from the demand for the completeness of any system and Kant's distinction between the objective and the subjective parts of physics. Due to the impossibility of exhaustively classifying what he calls the objective or material part of physics, Kant's definitional attempts increasingly focus on the subjective or formal part. He regularly cites the scholastic motto, *forma dat esse rei* ("form gives being to a thing") and insists that in physics we only completely know what we insert into it (*hineinlegen*) ourselves. This point is familiar from Kant's critical writings: it appears in the famous discussion of Galilei, Torricelli, and Stahl in the B Preface to the *Critique* (Bxviii).[24] As fascicles X/XI progress, Kant gives increasing attention to perception, which he consistently defines (as in his earlier writings) as "empirical representation with consciousness." Kant's increasingly broad definitions of physics allow him to incorporate into his reflections the subjective part of physics, the consciousness of the perceiving subject, and even some speculative proposals about the action and interaction between the forces of matter and the mental "moving forces" of the subject.[25] Once more, Kant's radical redefinitions of physics seem to be attempts to solve the problem of systematizing an empirical science.

9.3 Kant's Theory of Definitions and the Systematicity of Science in Light of the *Opus postumum*

The disagreement in the literature discussed in Section 9.1 emerges, I believe, from genuine obscurities in the role of definitions in Kant's theory of the systematicity of science. Passages highlighted by Sturm suggest that Kant considers definitions vitally important for furnishing the sciences with systematic unity and order. But Kant's cautionary comments about the use of definitions in philosophy seem to support Kraus' view that definitions, in contrast to ideas, are not significant for Kant's theory of the systematicity of science.[26] The textual evidence in the *Critique* seems to favor Kraus' reading. The balance is shifted, however, by the 1799–1800 drafts from the *Opus postumum*.

These drafts, I have sought to show, reveal how Kant applies his theory of definition "in practice," when attempting to secure the systematicity of empirical physics in the transition project. In my view, Kant's definitional attempts in the *Opus postumum* validate the broad direction of Sturm's interpretation. Kant is worried that empirical physics may be an unprincipled aggregate or "*farrago*." A key tool in his systematizing attempts – apparently the primary tool in 1799–1800 – is the positing of the definition of physics. Because the drafts provide extensive evidence of Kant applying his theory of definition, they allow us to now flesh out Sturm's account and, I will argue, to adjust it in some ways. They also shed new light on Kant's comments in the *Critique* about the use of definitions in philosophy, revealing a more nuanced position that does not conflict with his enthusiastic attempts to define physics in the late drafts. We can consider three points.

First, on Sturm's reconstruction, a definition of a science should, in principle, specify its object, method, and aim. The *Opus postumum* shows that this is correct, but that the first two elements are much more important than the third. Kant always specifies the object of physics: this is often the moving forces of matter, although, as we have seen, his view shifts dramatically as he explores different approaches to the transition problem. He also always indicates how physics proceeds, whether as systematic doctrine, empirical science, observational and experimental practice, and so on. Yet while the object and method are integral to Kant's definitions, the aim seems to be more dispensable. Kant relatively rarely states the aim of physics, and when he does, it is surprisingly general, even vague: for example, physics aims at (is "for the sake of") the possibility of experience. Such an aim does not do much to determine the outer systematicity of physics, as the aim is shared by critique and transcendental philosophy. This is in line with a point Sturm has made elsewhere: Kant takes certain practical or epistemic aims to be general and shared between disciplines.[27] Kant's practice in the *Opus postumum* suggests that aims are of secondary importance compared to object and method for defining an individual science.[28]

Second, a bold interpretative claim made by Sturm is that the definition and the idea of a science serve the purpose of securing its outer and inner systematicity, respectively, and that the latter presupposes the former. In my view, the *Opus postumum* reveals Kant to hold somewhat looser views: he seems to collapse the distinction Sturm makes between outer and inner systematicity. The definitions we have examined seem intended to at once demarcate physics, by identifying features that distinguish it from other sciences, *and* provide it with the intrinsic unity and completeness with which Kant is consistently concerned in the transition project. We may even wonder whether a strict distinction is possible in principle: the determination of the inner systematicity of physics can contribute to its outer systematicity and vice versa. I take Kant to suggest such a conflation of these two tasks when he states in the *Critique* that an accurate conception of the idea of a science can "determine the proper content, the articulation (systematic unity) and boundaries of the science" (A834/B862). The idea, which Sturm aligns with inner systematicity, here helps determine the boundaries of a science, which Sturm considers to be an aspect of the outer systematicity that should be furnished by the definition. Moreover, Kant's more radical definitions in the *Opus postumum* threaten rather than secure the outer systematicity of physics, because they blur what seem to be well-established boundaries: between, in the examples we have considered, physics and cosmology, or physics and psychology.[29] Kant's definitions seem to aim more at securing what Sturm calls inner systematicity, by ensuring that physics is not a disordered aggregate or *farrago*.

The third point follows from the previous one. What is the relationship between the definition of a science and its idea? Here, we can return to Kant's critical comments about definitions in the Discipline chapter of the *Critique* and whether these are contradicted by his practice of defining physics in the *Opus postumum*. In the light of the attempts in the *Opus postumum*, a footnote to the discussion in the *Critique* seems particularly important:

> Philosophy is swarming with flawed definitions, especially those that indeed actually contain elements of the definition but are not yet complete. If one could not get started with a concept until one had defined it, then all philosophizing would be in a bad way. But since, as far as the elements (of the analysis) reach, a good and secure use can always be made of them, even deficient definitions, i.e., propositions that are actually not yet definitions but are true anyway and thus approximations to them, can be very usefully used. In mathematics definitions belong *ad esse* [to the being], in philosophy *ad melius esse* [to the improvement of the being]. It is great to attain them, but often very difficult. Jurists are still searching for a definition of their concept of right.
>
> (A731/B759n)

This note provides a more nuanced account of the use of definitions in philosophy than the discussion in the main text, and it is closer to the way that Kant uses definitions in the *Opus postumum*. Many definitions in philosophy, Kant states in the note, are flawed (*fehlerhaft*) or deficient (*mangelhaft*), either because they are only partial, containing some but not all the elements of the definition, or because they do not have the form of a definition, although they may nevertheless make some true statements about the object. A feature of Kant's thinking about definitions is here evident: he implicitly distinguishes between the true and the flawed definition. In the main text of this section of the *Critique*, Kant implies that the definition "strictly speaking [*genau zu reden*]," which would be the true definition, can be distinguished from the definition in a loose sense, which would include flawed definitions (A728/B756). Importantly, Kant shows in the footnote that flawed definitions can be useful. Because they contain some elements of the true definition or approximate it, flawed definitions can help the philosopher to improve their concept (and they so belong *ad melius esse*).

This helps make sense of what Kant is doing when seemingly obsessively redefining physics in the *Opus postumum*. Different (even if subtly different) definitions provide different angles of attack on the transition problem, namely, how to systematize empirical physics. Kant is conceivably exploring which formulations are most fruitful for his task. In this way, one can use definitions in philosophy without falling into the errors diagnosed in the *Inquiry* and the *Critique*. Kant's definitions do not inaugurate the science of physics; he has before him the long history of physics and so, as is always the case with philosophical definitions, he has a given concept that he can try to define. The definitions are tentative and iterative: they should progressively converge upon the true definition.

This true definition, which Kant is seeking to approximate in his late drafts, would be, in my view, itself an approximation of the idea of the science. This is at least a possible reading of Kant's claim that "the definition, which the scientist gives at the very beginning of their science, very seldom corresponds to the idea; for this [idea] lies, like a seed, in reason, in which all parts lie hidden ..." (A834/B862). The definition seldom corresponds to the idea because it is usually flawed, but it can be improved to the point of approximating the idea. I thus agree with Sturm (2009, 162–3, 167–8) that the idea and the definition of a science should be distinguished. But I do not consider them to have fundamentally different functions. Rather, I take Kant's attempts to define physics in the *Opus postumum* to suggest that he considers definitions to be iterative attempts to converge on the (never fully attained) idea of a science and to provide the science with both outer and inner systematicity.[30]

Notes

1 My translations of Kant are usually guided by the Cambridge edition when the passages appear therein, although with frequent modifications (regarding those made to the quoted lines from A834/B862, see note 3 below). All translations of the *Opus postumum* are mine.

2 A832/B860. For discussion, see Sturm (2009, 138–9, 142–6) and Gava (2014, 382–4).

3 For the first sentence of this passage, Guyer and Wood offer the more concise phrase, "Nobody attempts to establish a science without grounding it on an idea." However, Kant's wording and choice of tense seems intended to show how the *hypothetically posited* idea guides the individual scientist: "Niemand versucht es, eine Wissenschaft zu Stande zu bringen, ohne daß ihm eine Idee zum Grunde liege." In the second sentence, Guyer and Wood skip "er" (he), which refers to the scientist who attempts to establish a science.

4 As is well known, in the *Inquiry* Kant claims that mathematics proceeds through synthesis and philosophy through analysis; by the time of the *Critique*, he has significantly modified the latter claim about philosophy. This shift does not affect the continuities in the claims about the use of definitions that I discuss here.

5 As Kant later writes in the *Critique*: "The German language has for the expressions *Exposition, Explikation, Deklaration* and *Definition* nothing more than the one word: *Erklärung*" (A730/B758). Wolff uses *Erklärung* for his foundational propositions; the term is for him synonymous with *definition* in Latin (see Wolff's *Kurzer Unterricht* §1 and *German Logic* §36, quoted by Gómez Tutor 2004, 126, 141).
6 "Arbitrary" (*willkürlich*) here does not mean "random" but indicates that the mathematician freely chooses the concepts for a purpose. See Beck (1956, 186–7) and Sutherland (2010, 189n); Sutherland uses "elective" in place of "arbitrary."
7 Kant uses the phrase "strictly speaking [*genau zu reden*]" when denying that *a priori* concepts can be defined (A728/B756). In the final section, we shall return to Kant's implied distinction between strict and loose senses of definition.
8 Kant makes an exception in the *Inquiry* for "nominal definitions," although he adds that even nominal definitions can be confidently stated only in "a few cases" (II:285). The nominal/real distinction broadly follows Wolff's traditional distinction between definitions of words and of things, respectively (see Gómez Tutor 2004, 142–6; Gava 2018, 281; but cf. Beck 1956, 181). For Kant, nominal definitions arbitrarily situate objects in genus-species relations according to their specific differences. Real definitions permit the cognition of objects according to their inner determinations (see IX:143; XVI:608–611; Nunez 2016, 117). In what follows I leave aside Kant's real/nominal distinction, as well as his distinction between analytic and synthetic definitions. The philosophical definitions that are relevant to our discussion are *real* and *analytic*. They are analytic because they analyze *given* concepts; this does not prevent them from being synthetic judgements (see Beck 1956, 182).
9 *Kurzer Unterricht von der mathematischen Lehrart* §1, quoted in Gava (2018, 280). On Wolff's use of this method, see Gava (2018, 279–84), and, with detailed attention to the development of Wolff's views, Gómez Tutor (2004, 120–60). Classic studies of the historical methodological debate and Kant's intervention include Tonelli (1956) and Wolff-Metternich (1995).
10 I modify the Guyer/Wood translation. "*Herumirren*" recalls the "groping around [*Herumtappen*] … among mere concepts" for which Kant reproaches metaphysics in the B Preface (Bxv). Kant regularly uses *Herumtappen* in the fascicles of the *Opus postumum* that we will examine below, when criticizing unsystematic empirical physics. For a brief discussion, see Howard (2023, 38).
11 Kant here combines what Sturm calls inner and outer systematicity: we shall return to this.
12 The biological theory of *generatio aequivoca* is discussed in the third *Critique* (V:419).
13 I argue at length for the interpretation sketched in this paragraph in Howard (2023).
14 The most prominent of Kant's attempts, or at least those most often discussed in the literature, are: (1) an "elementary system" of all specific physical properties and moving forces of matter, structured according to the four headers of the classes of the pure categories of the understanding; (2) the determination of the "intermediary concept" of "ether" or "caloric," conceived of as a material transcendental condition whose possibility or actuality Kant seeks to prove; (3) the account of an act known in the literature as "self-positing" through which the I or subject posits itself as an object in time and space;

(4) the elaboration of a "system of ideas," God, world, and the human-in-the-world; and (5) a rethinking of the nature of transcendental philosophy. For an introduction to these topics, see Förster (2000).

15 See Howard (2023, 41–48).
16 Folios "AA" and "BB" are however placed at the beginning of fascicle XI, XXII:425–52, rather than where they should be, at the end (after XXII:539). Additionally, the drafts at XXII:409–421 are from a later date.
17 See Hoppe (1969, 87–88), for more passages that support his reading. I provide an extended critique of Hoppe's claims about Kant's conception of physics in the *Opus postumum* in Howard (2023, section 5).
18 See: "the sum total of all appearances (the world) is the object of *cosmology*" (A334/B391; cf. A419/447).
19 XXII:308. The Greek for "the all," το παν, appears in the definition of the world in Baumgarten's cosmology chapter (Baumgarten [1739] 2013, §354).
20 See Förster (2000, 13–19) and Howard (2023, 34–36).
21 See Howard (2023, 36–39).
22 For example, XXII:310, 328, 336, 345, 381, 384, 391, 395, 398, 448. Further passages are cited by Hoppe (1969, 77) and Gloy (1976, 185–6).
23 For further discussion, see Howard (2023, 39–41).
24 See Hoppe (1969, 115) and Howard (2023, 46–47).
25 See Howard (2023, 41–45).
26 Nunez (2016, 118, 137n) identifies a further, related obscurity in Kant's account, which my discussion below will address: within a couple of pages of the *Critique* Kant seems to make contradictory claims about whether definitions can be legitimately used in philosophy.
27 For illuminating discussion of Kant's account of the ends of sciences, see Sturm (2020). Sturm suggests that the general, practical aims of sciences "cut across disciplinary distinctions" (Sturm 2020, 21). Epistemic aims are more specific on Sturm's account, but they do not always serve to distinguish different sciences. For example, while the epistemic aim of formulating quantitative laws distinguishes physics from psychology (Sturm 2020, 23), this particular aim may not differentiate physics and chemistry. For, even if Friedman is wrong to suggest that in the *Opus postumum* Kant considers chemistry a proper, mathematically grounded science (see McNulty 2016, 82), Kant does not rule out that, in the future, we may discover a concept in chemistry that can be mathematically constructed (IV:470–471). The formulation of quantitative laws may therefore be the ultimate aim of both physics and chemistry.
28 I do not here claim to say anything that Sturm would disagree with: he points to the secondary character of aims when writing that "ends or aims – i.e., axiological features – *also* play a crucial role in the sciences" (Sturm 2020, 9, my emphasis; see also his further discussion of external and essential ends, Sturm 2020, 10–13). My point can nevertheless guard against a hasty reading that might take Sturm to be maintaining that definitions should *always* specify the object, method, *and* aim of a science.
29 One could also consider how Kant blurs the boundaries between different empirical sciences in the transition project when he proposes that chemistry and biology belong to physics (see Friedman 1992, 216–7, 290–316; Guyer 2001, 272–80; van den Berg 2014, 183–5, 211–9, 239–44).
30 For comments on and discussion of earlier versions of this paper, I would like to thank Thomas Sturm, Katharina Kraus, and Mark Textor.

Bibliography

Baumgarten, Alexander. [1739] 2013. *Metaphysics*, edited and translated by Courtney D. Fugate and John Hymers. London: Bloomsbury.
Beck, Lewis White. 1956. "Kant's Theory of Definition." *The Philosophical Review* 65(2): 179–91.
Förster, Eckart. 2000. *Kant's Final Synthesis: An Essay on the Opus postumum*. Cambridge: Harvard University Press.
Friedman, Michael. 1992. *Kant and the Exact Sciences*. Cambridge: Harvard University Press.
Gava, Gabriele. 2014. "Kant's Definition of Science in the Architectonic of Pure Reason and the Essential Ends of Reason." *Kant-Studien* 105(3): 372–93.
Gava, Gabriele. 2018. "Kant, Wolff and the Method of Philosophy." *Oxford Studies in Early Modern Philosophy* 8: 271–303.
Gloy, Karen. 1976. *Die Kantische Theorie der Naturwissenschaft. Eine Strukturanalyse ihrer Möglichkeit, ihres Umfangs und ihrer Grenzen*. Berlin: de Gruyter.
Gómez Tutor, Juan Ignacio. 2004. *Die wissenschaftliche Methode bei Christian Wolff*. Hildesheim: Olms.
Guyer, Paul. 2001. "Organisms and the Unity of Science." In *Kant and the Sciences*, edited by Eric Watkins, 259–81. Oxford: Oxford University Press.
Hoppe, Hansgeorg. 1969. *Kants Theorie der Physik. Eine Untersuchung über das Opus postumum von Kant*. Frankfurt am Main: Klostermann.
Howard, Stephen. 2023. *Kant's Late Philosophy of Nature: The Opus postumum*. Cambridge: Cambridge University Press.
Kraus, Katharina. 2018. "The Soul as the 'Guiding Idea' of Psychology: Kant on Scientific Psychology, Systematicity, and the Idea of the Soul." *Studies in History and Philosophy of Science* 71: 77–88.
McNulty, Michael Bennett. 2016. "Chemistry in Kant's *Opus postumum*." *HOPOS: The Journal of the International Society for the History of Philosophy of Science* 6: 64–95.
Nunez, Tyke. 2016. "Definitions of Kant's Categories." In *Kant: Studies on Mathematics in the Critical Philosophy*, edited by Emily Carson and Lisa Shabel, 113–39. London: Routledge.
Sturm, Thomas. 2009. *Kant und die Wissenschaften vom Menschen*. Paderborn: Mentis.
Sturm, Thomas. 2020. "Kant on the Ends of the Sciences." *Kant-Studien* 111(1): 1–28.
Sutherland, Daniel. 2010. "Philosophy, Geometry, and Logic in Leibniz, Wolff, and the Early Kant." In *Discourse on a New Method: Reinvigorating the Marriage of History and Philosophy of Science*, edited by Mary Domski and Michael Dickson, 155–92. Chicago: Open Court.
Tonelli, Giorgio. 1956. "Der Streit über die mathematische Methode in der Philosophie in der ersten Hälfte des 18. Jahrhunderts und die Entstehung von Kants Schrift über die 'Deutlichkeit'." *Archiv für Philosophie* 9: 37–66.
van den Berg, Hein. 2014. *Kant on Proper Science: Biology in the Critical Philosophy and the Opus postumum*. Dordrecht: Springer.
Wolff-Metternich, Brigitta-Sophie von. 1995. *Die Überwindung des mathematischen Erkenntnisideals: Kants Grenzbestimmung von Mathematik und Philosophie*. Berlin: de Gruyter.

10 Systematicity in Kant's Philosophy of History

Andree Hahmann

Can history for Kant be an object of systematic science? To answer this question, we need to analyse three closely related issues: First, we must clarify how a system is to be understood in historical science. Second, we need to examine the role of teleology within this system. Third, we have to see how history understood in this way fits into the systematics of Kant's critical philosophy.

I will discuss each of these issues in the following three sections of this chapter. The first section will relate Kant's account of history to methodological considerations of the emerging historical sciences of his time.[1] We will see that a crucial question that historians have addressed is how exactly a scientific system can be generated from an aggregate of historical facts. Key to answering this question is the assumption of an end point of historical development.[2] Against this background, the second section examines the teleological perspective on history in more detail. A look at Kant's pre-critical lectures on anthropology will help us to better understand how the different ends of history found in the various texts might relate to each other.[3] In the last section we will bring together the results of the previous sections and consider how Kant's philosophy of history depends on crucial assumptions of his critical system.[4] In this way, we can see more clearly whether and how history can be considered a systematic science for Kant.

10.1 System of History

That historical research must proceed systematically is an idea that has dominated the discussion in the developing science of history in Germany since the middle of the 18th century. From the beginning, this idea was linked to the popular demand that history should be presented pragmatically and have a demonstrable benefit for human action.[5] Among the influential authors who advocated this idea and specified it in their works were the two Göttingen historians Johann Christoph Gatterer (1727–1799) and August Ludwig von Schlözer (1735–1809).[6] For Gatterer, the pragmatic in history is "what in the actual sciences is called the systematic" (Gatterer 1767b, 12). In the following, we will focus on this aspect of the pragmatic, that is, on the question of what exactly constitutes the systematic in historical science.[7] In this context, Gatterer emphasises the special position of so-called universal history and refers to the comprehensive plan that underlies it.[8] Gatterer believes that one must take a comprehensive approach to history in order to adequately grasp historical causes and effects. Accordingly, a historian must present a "system of events" or "the whole system of causes and effects, means and intentions" that depict a historical process (1767a, 79–80). According to Gatterer, this can only be done through a "presentation (*Vorstellung*) of the general connection of things in the world (*nexus rerum universalis*)" (Gatterer 1767a, 85). The thought behind this is that no historical event can be adequately grasped on its own, since all historical events are correlated. Gatterer believes that this is what connects universal history with philosophy. However, this ultimately results in the necessity for a historian to become a philosopher in order to coherently present the idea of the general coherence of things in the world.[9]

Schlözer, a former student and colleague from Göttingen, takes up these thoughts of Gatterer and develops them further in his universal-historical approach. Schlözer begins his presentation by explaining how he conceives of universal history, that is, a "systematic ... world history" (1772, preface, p. 1). In his view, this implies that the "great events of the world are considered in relation to each other" (1772 §1, p. 2) and not separately or in isolation. Against this background, Schlözer first distinguishes universal history from the so-called special histories, which are dedicated to a specific region or human association. In contrast to the special histories, "universal history encompasses all parts of the world and all ages, and it gathers together all peoples in all countries. Its object is the world and the human race" (1772 §1, p. 3). The special histories, however, provide the material for the universal historian, who selects and compiles it "purposively" (1772 §7, p. 13). In addition to the comprehensive claim of universal history, the purposive selection represents the second main characteristic feature of a universal history. These two features distinguish universal history from world history. Accordingly, although world history and universal history refer to the same subject matter (all people, in all ages on the entire earth, which distinguishes them from special histories), universal history, unlike world history, treats events systematically.

What does systematic mean in this context? Schlözer makes it clear that a systematic approach presupposes that events are related to an overarching end. The end thus provides the criterion for the organisation of events:

> World history may be conceived from a twofold point of view: either as an aggregate of all special histories, the collection of which, even if complete (the mere juxtaposition of which), forms a whole; or as a *system* in which the world and mankind are the unity, and out of all the parts of the aggregate some in relation to this object, are specially selected, and purposively arranged.[10]
>
> (1772 §8, p. 14)

This makes it clear that Schlözer knows very well that it is the historian himself who first gives history a systematic form. The historian does this by selecting and connecting those events that are relevant to history from a universal perspective. Most important for this task is the end that is presupposed for history.[11] It follows that even if one presented all the special histories of the world and reported on every possible event, one would still not have a universal history (1772 §9, p. 18). What is missing, according to Schlözer, is the "general view that includes the whole" (1772 §10, p. 18). For only this comprehensive and purposive view can link the unconnected events into a system by selecting from the available material with a view to the development of this whole. Schlözer explicitly distinguishes at this point between the real connection, on the one hand, and the temporal connection of events, on the other (1772 §18, p. 46). Real connection means that events are related to each other like cause and effect, that is, they are rationally grounded in each other (1772 §19, p. 46). A temporal connection, on the other hand, includes those events that take place at the same time but are not necessarily grounded in each other (1772 §20, p. 48). The universal historian establishes the real connection by putting together what was not necessarily recognised as a connection before. The historian thus makes the historical lines of development visible to the observer. Schlözer even claims that in this way, that is, by synthesising the individual events according to rational laws, the universal historian gives history intelligibility by giving it, so to speak, a rational meaning (1772 §10, p. 22).[12] Seen from this perspective, history presents itself to Schlözer as a sign of divine providence. Accordingly, the harmony that becomes visible in the purposeful course of events (which the historian brings about through rational selection and constructive arrangement) gives rise to praise of providence, which must have ordered the world in accordance with reason, an order that reveals itself only to the inquiring eye of the historian (1772 §15, p. 37). The reference to providence, however, must not obscure the fact that the teleological assumption is first and foremost a scientific-methodological device aimed at creating a system out of unconnected events.[13] It is not a religious faith but the pursuit of scientific systematics that is responsible for the teleological view of history.

10.2 The Vocation of the Human Being as End of History

There are important points of contact between the views of the developing historical science and Kant's philosophy of history.[14] However, as Sturm has shown in his work, Kant believes that the study of history must be theoretically supported by an anthropology. Sturm also notes a further problem for historical science. Historiography cannot further theoretically substantiate the final end it must ascribe to history on the basis of its scientific-systematic assumptions.[15] It is true that we find some thoughts in Schlözer's work that point to a cosmopolitan perspective that the historian must adopt. However, Schlözer does not give a reason for this, which is probably due to the limited empirical framework of historical science. In what follows, we will see that in his lectures on anthropology, Kant not only provides the material foundations of historical science but also the outline of an overarching teleology that is essential for its systematic order.[16] Both themes are thought together by Kant in what he calls the "vocation of the human being."[17] We will see in the next section how Kant incorporates this result into his critical systematics and how it takes the form of an a priori guideline for the science of history.

Kant first appeared as an author interested in history in 1784 with a supposed occasional writing.[18] However, his first engagement with history goes back much longer than his published writings.[19] What is often overlooked by scholars is that the origins of Kant's philosophy of history can be found in his lectures on anthropology.[20] The connection between history and anthropology is evident in the lecture notes of the 1770s and Kant's drafts and reflections from that period. In fact, we will see in this section that Kant's philosophy of history, if we consider only the content of its main statements, was already fully developed by 1775. This is particularly telling and could also help us when it comes to a specific problem that preoccupies Kant scholarship, namely, the question of exactly what end one must assume in history. Several possible ends are controversially discussed among scholars. Allison, for example, assumes that it is a political end, "namely, the greatest possible freedom of each under law that is compatible with the freedom of all."[21] Kleingeld, on the other hand, emphasises that Kant ultimately aims at moral perfection as the end of history.[22] Consequently, history would aim at the complete moralisation of the human species which moves the philosophy of history close to the postulates of the *Critique of Practical Reason*,[23] thus making it a kind of extended philosophy of religion.[24] But Kant scholars also discuss a third possibility. If one takes seriously the initial conditions of historical development elaborated by Kant in the *Idea*, then it becomes clear that human history is guided by a kind of natural teleology. The historical process thus begins with natural teleological conditions and is also driven by an end inherent in human nature. This is a teleological presupposition that similarly extends to all living things, namely the idea that all predispositions ("Anlagen") should develop fully by nature (*Idea*, VIII:18; 22; 30; *Anthropology*, VII:329).[25] Consequently, the end of history would consist in the fact that the natural predispositions of the human species, here above all its rational predisposition, which is contained in it like a germ, must fully develop in the course of history.[26]

If we turn to the transcripts of Kant's lectures with this question, the first thing that strikes us is that in 1775 Kant distinguishes between two kinds of natures in human beings. In addition to what he understands by human nature in the narrower sense, there is also an animal nature that pursues a different end, namely the preservation of the human animal species. The latter connects humans with all other living beings (*Friedländer*, XXV/2.1:677; see also *Anthropology* VII:325). From this animal nature, Kant extracts human being's "incompatible"[27] (*Friedländer*, XXV/2.1:677) characteristics, that is, his cunning behaviour, which is dangerous for his fellow human beings. Even though this does not make him a predator, he behaves like one towards his neighbours (*Friedländer*, XXV/2.1:678). But this original state of human nature also provides the first impetus for overcoming it. For hostile behaviour requires that they leave the state of nature and enter a civil state, for only through laws can this natural hostility be kept in check and suppressed. Every form of law-abiding behaviour thus presupposes civil society: "But if this were to cease, refinement would also cease, and all men would be such evil. This malignity lies in the nature of all men" (*Friedländer*, XXV/2.1:679; see also *Idea*, VIII:20). In accordance with the later published writings, human desires and hostilities are instrumentalised by *providence*[28] for this purpose,[29] in order to make humans, who are naturally inclined to inertia, become active and populate the whole world (*Friedländer*, XXV/2.1:679; 691; see also *Idea* VIII:21). Kant provides further examples of the means that nature employs to enforce its intentions, which are ultimately based on a form of natural dialectic. We already know this dialectic from one of Kant's earliest published texts, the *Universal Natural History and Theory of the Heavens*, in which, however, this model takes the form of two antagonistic natural forces, namely attraction and repulsion.

Let us note that Kant believes as early as 1775 that nature, through evil, ultimately aims at good and thus promotes the perfection of the human species. In this way, human predispositions can fully develop including "concepts of justice and morality, and the development of the greatest perfection of which people are capable" (*Friedländer*, XXV/2.1:680; see also 682). Conversely, it also follows that their natural laziness and inertia would have prevented humans from taking upon themselves the sometimes arduous work of civilisation. Mere necessity compels intrinsically lazy humans to act. Otherwise "no civil constitution would have emerged. This latter is the source of the development of talents, of the concepts of justice and all moral perfection" (*Friedländer*, XXV/2.1:681). In the civil state, however, the human being is disciplined: According to Kant, this means that human nature itself is transformed, or, one can also say, that it will suffer "violence to animality" (*Friedländer*, XXV/2.1:684). For the civil state is the starting point for the perfection of morals (*Sitten*) and the perfection of all human talents (*Friedländer*, XXV/2.1:684). Kant makes it clear that "nature's end was the civil society, and the human being is determined to make himself perfectly happy and good as a member of the entire society" (*Friedländer*, XXV/2.1:690). But Kant is well aware that even a complete development of human talents (which presupposes this civil constitution) does not yet involve the moralisation of the human being. At this point, another observation becomes crucial:

> Since as it is however, human beings became ever more refined through civil constraint, and cultivated themselves more and more, thus the constraint of propriety emerged among them, where human beings compel themselves among one another with regard to taste, modesty, refinement, courtesy, and decorum. For everything [which is a matter of] decency in the good life is not produced by any civil constraint; the authorities do not concern themselves with this at all …. If one just does not overtly offend someone, then the authorities are not at all concerned with the rest. However, because of propriety, human beings compel themselves among one another with regard to the rest; they refrain from much because it does not agree with the opinion of others.
> (*Friedländer*, XXV/2.1:692)

Civilisation in the civil state and the development of the human rational predispositions that result from it thus promote a kind of *external morality* that concerns the form of external human interaction. This kind of civilisation, or the formation of moral forms of behaviour, plays a decisive role in achieving the final end of the historical development, namely, the moralisation of the human species:

> Yet the human race is still progressing ever further in perfection. What kind of additional constraint could still be conceived of here? This is the moral constraint, which consists in every human being

fearing the moral judgment of the other, and thereby being necessitated to perform actions of uprightness and of the pure moral life. Human beings have established the constraint of propriety among one another, to which all are subject, and where everyone pays attention to the other's opinion with regard to propriety. Yet human beings have just such a right also to pass judgment about the moral conduct of the human being. First, the concepts of morality must be purified, and respect for the moral law must be instituted; the heart would then already change.

(*Friedländer*, XXV/2.1:692)

Morality is thus promoted by the formation of morals (*Sitten*) which is why the moral character of the human being slowly emerges from morals (*Friedländer*, XXV/2.1:693). What brings about this transition, however, is the compulsion of conscience. We can now see that conscience becomes the decisive moral authority that every human being inherently possesses. But we can also see that it needs to be trained or better cultivated (*Friedländer*, XXV/2.1:693). This cultivation presupposes a kind of external propriety or decency combined with a moral education of the human being (*Friedländer*, XXV/2.1:695). Both in turn require a sufficiently advanced development of the human predispositions or talents (as Kant calls it here), which, however, can only take place in an order established by law. Finally, as Kant claims in his lectures, this development must be complemented by religion (*Friedländer*, XXV/2.1:694–695).

At this point we can clearly see that almost all the decisive ideas of Kant's later philosophy of history were already fully developed by 1775. Thus we also find all three final ends mentioned later in his texts from the 1780s and 1790s and proposed by scholars as the targeted ends of history. In the published *Anthropology*, however, Kant summarises all three ends under one title: the "vocation of the human being":

The sum total of pragmatic anthropology, in respect to the vocation of the human being and the characteristic of his formation, is the following. The human being is destined by his reason to live in a society with human beings and in it to cultivate himself, to civilize himself, and to moralize himself by means of the arts and sciences. No matter how great his animal tendency may be to give himself over passively to the impulses of ease and good living, which he calls happiness, he is still destined to make himself worthy of humanity by actively struggling with the obstacles that cling to him because of the crudity of his nature. The human being must therefore be educated to the good ....

(*Anthropology*, VII:324–325)

Although we cannot, simply on the basis of the lecture notes, determine with certainty the systematic order in which we must combine the three proposed ends of history, it is clear that an account of history must somehow encompass all three ends, which together constitute the vocation of the human being. However, how exactly these three ends relate to each other and how they are to constitute the final end necessary for the formation of a historical system have not been completely settled at this point. Moreover, given that all three ends date back to the mid-1770s, one must wonder how this conception can be reconciled with Kant's critical systematics and the strict separation between morality and nature. We address these questions in the next section.

10.3 The Place of History in Critical Philosophy

Let us now turn to the Third Chapter of the Transcendental Doctrine of Method in the *Critique of Pure Reason*. In this short and very densely argued section of the *First Critique*, Kant is concerned with what he calls the architectonics of pure reason. The whole passage is complicated by the fact that Kant seems to be pursuing different goals with his argument. We cannot deal with the details and problems of this passage here.[30] Two things are particularly important for our investigation. First, this is the passage in the *First Critique* where Kant provides his most comprehensive account of science defined as "the unity of the manifold cognitions under one idea" (*CPR*, A832/B860).[31] Second, he refers to the vocation of the human being as the final end of reason, which determines all subordinate ends and thus makes possible a complete systematic unity (*CPR*, A840/B868). Although Kant makes no mention of history or historiography in this context, as we shall see in this section, there are important links to the preceding discussion, both in terms of terminology and content. This will not only enable us to integrate Kant's approach to history, outlined above, into his critical systematics, but it will also show us that history fully meets the requirements of a science that Kant develops from the very nature of reason itself.[32]

To begin, by architectonics Kant understands "the art of systems" (*CPR*, A832/B860) and systematic unity is that "which first makes ordinary cognition into science" (*CPR*, A832/B860). In line with Gatterer's and Schlözer's reflections on the nature of science, we can read that ordinary knowledge, left to itself, remains merely an indeterminate "aggregate" and only becomes "science" through systematic arrangement.[33] This is where the architectonic of reason becomes relevant. For as Kant explains, this systematisation of cognitions requires an architectonic skill. Architectonic unity is thus treated as a synonym for systematic unity, and the latter is mentioned as a prerequisite for scientificity. This means, however, that architectonic is rooted in human nature, or, differently put, human reason itself is inherently architectonic, since "it considers all cognitions as belonging to a possible system" (*CPR*, A474/502). Ideas play a decisive role in the formation of scientific systems. By idea, as Kant explains here, he means a concept of reason that encompasses the form of the whole (*CPR*, A832/B860).[34] However, the idea requires a schema to fulfil this function. The schema to which Kant refers at this point, at least at first sight, does not seem to have much in common with the schemata from the transcendental analytics. However, both provide unity in a given manifold. What is decisive for the role of a schema in the architectonic is that it contains some kind of internal purposiveness which is directed by an end.

Kant sees two possible options. First, the end is taken from experience and accords to an accidentally assumed intention of the scientist, in which case we get a technical unity. If, on the other hand, the schema is grounded in an idea of reason, that is, if the end arises a priori from the nature of reason itself, then we get an architectonic unity (*CPR*, A833/B861). As Kant makes clear, true science cannot arise from accidental ends, but presupposes an architectonic unity.[35] One aim of this architectonic end is to unite all the sciences into a whole and thus provide the blueprint for the totality of sciences (*CPR*, A835/863). This is the role Kant ascribes to philosophy, as he makes clear in what follows. For in contrast to the other sciences, which for Kant have no actual scientific character (thus he calls the mathematician or the natural scientist mere artists of reason), philosophy points directly to the essential end of reason. Understood in this way, philosophy is in accordance with a "cosmopolitan concept" (*CPR*, A839/B867) which prescribes a perfect systematic unity that can only be based on the "entire vocation of human beings" (*CPR*, A840/B868).[36] Kant points out this is accomplished by morality. This suggests that the final end must be a moral end and that it will be revealed to us through practical reason (*CPR*, A816/B844). Accordingly, we have a good reason to assume that already in the architectonics Kant alludes to the highest good as the final end of reason and thus to the end point that shapes the systematic unity of the entire sciences.[37] That Kant must indeed have the highest good in mind is finally decided by his discussion in the *Critique of the Power of Judgement*, which is crucial for our systematic question. At this point, the internal relations between the subordinate and superordinate ends also become clear.[38]

The discussion relevant to our problem is contained in paragraphs 83 and 84, which deal with the "ultimate end of nature" and the "final end of creation."[39] In the preceding paragraphs Kant has set out the problem that the actual existence of internal purposiveness of natural objects makes it necessary to consider nature as a whole also as a purposively organised system.[40] However, the external order of ends in nature is not determined. Therefore, one cannot readily say what in nature is to be regarded as an end and what as a means. Accordingly, we cannot know a priori whether the grass is the fodder for the sheep or whether the sheep are the reason why the pastures do not become scrubland. In order to arrive at a clear order here, one must determine a final end.[41] What is required is in any case that the presupposed order of external ends can only be justified on condition that a final end of nature can be determined. Otherwise, this order cannot be achieved, since the order in the external relationships of ends and means can be arbitrarily reversed. Kant's central claim here is that only the human being is a suitable candidate for the position as the final end of nature, for only he is endowed with intellect and thus capable of freely setting ends for himself. With regard to this last end, all other natural ends can then be systematically arranged.[42]

However, what is to be promoted in the human being by nature is initially unclear and is to result from his freely chosen ends. There are two possible options: Either the end itself is to be satisfied by nature, or it must be the aptitude to achieve various ends. The former would be the final end to which any human aspires, namely, happiness. By the second end, Kant understands human culture.[43] Since the former not only contradicts our experience but also human nature itself which can never be completely contented, we must assume that nature promotes the development of human culture. This makes culture the final end of nature. The following remarks are for the most part consistent with the argumentation in his lectures. Thus we can read that there are different natures in man, his animal nature and his rational nature, which can also come into conflict. Kant also emphasises that the promotion of the development of man's rational predispositions is through the means of inequality among humans, which leads them to overcome the state of inertia and inactivity. Finally, also in line with the lecture notes, Kant states that the condition under which alone this complete development of predispositions can be achieved must be a civil constitution that maximises freedom.[44] In this way, civil society becomes the condition for the full development of human rational predispositions, which in turn promote fitness for all sorts of ends. This makes the civil constitution the necessary condition for the completion of the full development of all rational predispositions, and thus a final end, subordinate to the first, which nature itself aspires to for the human being.

But where does that leave morality as the final end of history? Up to this point, we have only dealt with the final end of nature. However, nature itself is conditional in that it depends on the subjective forms of intuition. If we are looking for a final or unconditional end, this cannot be the end of nature, but must be what Kant calls the end of creation. For only that which does not depend on any other condition in the order of ends can exist as the unconditioned final end.

It should be noted that only at this point does it become clear that the human being, who until now has only been assumed to be the end of nature, actually occupies this place. For only the human being is a being who himself can only be thought of as an end, because only he is subject to the moral law that makes him the inhabitant of a noumenal realm of reason. The end to which man aspires, however, insofar as he participates in pure practical reason, differs from the end he pursues insofar as he is a sensible being. The latter, as we have seen, is happiness. The former is what Kant calls the ultimate end of a pure rational will (which is practical reason itself), namely the highest good. In summary, it follows:

> the human being is the final end of creation; for without him the chain of ends subordinated to one another would not be completely grounded; and only in the human being, although in him only as a subject of morality, is unconditional legislation with regard to ends to be found, which therefore makes him alone capable of being a final end, to which the whole of nature is teleologically subordinated.
> (*CPJ*, §84, V:435–436)

Now everything falls into place. We can see that Kant adheres to the various ends that we know from the lecture notes, but they are based a priori on basic transcendental philosophical assumptions, which is why they also provide a priori guidance for the ordering of historical events. In order to be able to comprehend nature as a systematically ordered whole, it is necessary to determine a final end, for only in this way can an order be established in the series of conditional end-means relationships. We encountered the same idea in the *Critique of Pure Reason*, where Kant spoke of an architectonic unity that requires a final end grounded in the nature of reason itself. In the *First Critique*, however, Kant calls this final end the vocation of the human being. As we know from the *Anthropology*, the vocation of the human being comprises three distinct ends including the complete moralisation of the human species. This is probably also the reason why Kant claims in the *Critique of Pure Reason* that this required final end is an object of morality. More precisely, as we can now see, what Kant must mean by the final end is the highest good, which consists of two parts: complete moralisation and happiness distributed in relation to morality. Since the human being is obliged to promote the realisation of this good, he must, as far as is in his power, work towards the moralisation and future happiness of the human species.[45] Consequently, the highest good is the final end on which the series of subordinate, conditional ends depend. And only with regard to the highest good can a systematic unity be established in nature, which is understood as a system of ends. In this way, the highest good also forms the keystone or capstone of Kant's critical systematics, since, although it is itself an object of practical reason, it is necessary for the theoretically required unity and complete determination of the laws of nature. Clearly, philosophy of history plays a special role insofar as it conceives of precisely this transition as a historical development.[46] That is, the movement from nature to morality is temporalised by Kant and in this respect also concerns the history of human actions as effects of a noumenal causality. For the purposes of our investigation, then, we can conclude that, for Kant, history can be the subject of a science that itself fulfils the criteria for supreme scientificity and systematics that Kant established in the Transcendental Doctrine of Method in the *Critique of Pure Reason*. Moreover, history, and consequently the philosophy of history, is of extraordinary importance for Kant's critical systematics as a whole, since the latter organises and deals with the necessary transition between the metaphysics of nature and the metaphysics of freedom.[47]

10.4 Conclusion

The initial question of our investigation was whether history can be the subject of a systematic science for Kant. We have seen that this question presupposes the discussion of three closely related issues. First, we examined what exactly historical science understands by a systematic claim. Using the example of the Göttingen historians Gatterer and Schlözer, both of whom had a considerable influence on the developing science of history in Germany, we have seen that the assumption of a definitive end to history is above all decisive for the formation of a historical system, since only in this way can events that are in themselves disjointed be systematically ordered. In the second section, we argued that while there are no direct references to the historians of his time in Kant's published works, we can see from his unpublished lectures that Kant's thoughts on history originally emerged from an anthropological background, which brings him much closer to the developing historiography of his time. It is also clear from this lecture that Kant was already discussing several ends of history in the 1770s which taken together constitute the vocation of the human being. Against this background, we have seen in the last section that Kant takes up the idea of the vocation of the human being in the *Critique of Pure Reason* and uses it as a model for a central aspect of his scientific architectonics. This already indicates that history can fully meet Kant's scientific-systematic requirements insofar as it is dedicated to the vocation of the human being as the final end of practical reason. This assumption was then confirmed by Kant's discussion of the transition of internal to external purposiveness in nature and the special function that history plays in this context in the *Critique of the Power of Judgement*. In summary, then, we can conclude that Kant laid a theoretical foundation for a future systematic science of history which is based on anthropological assumptions within the framework of critical philosophy.[48]

Notes

1 As Sturm (2009, 307) argues, there are fundamental systematic considerations associated with a pragmatic interest in history that link Kant's approach to the emerging historiography of his time. For a similar assessment, see Rohbeck (2006, 80). Kuehn (2009, 93), on the other hand, claims that Kant and his contemporary accounts of universal history "share only the name."
2 The assumption of a final end of history is one of the most controversial claims contemporary scholars see in Enlightenment accounts of history. Many assume that this idea is a theological assumption. Accordingly, Schröder (2011, 35) claims that Kant's conception of history represents a secularised form of earlier theological accounts of history from Augustine to Bossuet. See also Despland (1973). Influential in this regard is the work of Karl Löwith, who did not deal explicitly with Kant, but whose analysis has also been applied to Kant. Löwith, however, had already stated that the philosophy of history is based on a "systematische Ausdeutung der Weltgeschichte am Leitfaden eines Prinzips, durch welches historische Geschehnisse und Folgen in Zusammenhang gebracht und auf einen letzten Sinn bezogen werden" (2004, 11). On this problem more generally, see Kleingeld (1995, 179–82).
3 Many interpreters still believe that Kant's philosophy of history occupies a subordinate position within Kant's philosophical project, motivated by an external cause and need not be taken seriously from a systematic perspective. This is partly because they focus exclusively on the short texts published in the 1780s and 1790s. The first text, in particular, the *Idea for a Universal History from a Cosmopolitan Perspective*, plays a special role. In this text, Kant himself tells us what prompted him to write it, namely an anonymous remark published in the *Gothaische Gelehrten Anzeigen* in February 1784 (see the introduction to Akademie-Ausgabe VIII:468). However, others claim that this text must be understood as a reaction to Herder. See, e.g., Schröder (2011, 31). According to Kuehn (2009, 83), Kant was reacting to Spalding, Abbt, Mendelssohn, and implicitly to Ferguson (Kuehn 2009, 88). In this paper I will take a broader approach and also consider the unpublished material on history from the 1770s.
4 While Yovel (1980) argues that history occupies a central place in Kant's critical systematics, he nevertheless claims that his published texts of the philosophy of history of the 1780s and 1990s are not compatible with basic assumptions of critical philosophy.
5 As Sturm (2009, 307) points out, historiography is considered a prime example of a *pragmatic* discipline in the German-speaking world of the 18th century. Kleingeld (1995, 29) wonders "ob das von Kant beschriebene Systematisierungsproblem in der zeitgenössischen Universalgeschichtsschreibung nicht bereits gelöst worden war." Kleingeld refers to Iselin and also mentions Ferguson's influence on Kant.
6 The personal relationship between the two historians was problematic. Schlözer was initially a student of Gatterer, but then became a rival within the history faculty. See Stagl (1998).
7 For a more comprehensive discussion of the relationship between pragmatic and systematic, see Sturm (2009, 305–65).

8 On the term "universal history" and the discipline within historiography, see Stagl (1998, 525). Anderson-Gold (2009, 457) notes that, correspondingly, 18th-century philosophy of history was primarily concerned with finding principles by which history could be understood "as a whole."
9 Gatterer (1767a, 84). See also Köster (1775, §31, p. 50), who claims that "philosophical often means as much as systematic" ("Philosophisch heißt oft so viel als systematisch"). Accordingly, Köster explains that historical research can be a science if and because it proceeds "philosophically" in the sense of "systematically" (Köster 1775, §32: "Wir nennen eine Abhandlung oft philosophisch, wenn sie gleich nicht eigentlich Sätze aus der Philosophie systematisch genommen, betrifft: sondern wenn dieselbigen so untereinander geordnet und verbunden sind, daß immer das Folgende aus dem Vorhergehenden Licht erhält." See also §34: "Doch glaube ich, daß immer der Nahme einer philosophischen Geschichte, in der jetzt angeführten Bedeutung vorzüglich der Universal-Historie gebühret."). According to Anderson-Gold, historical science thus follows natural sciences (2009, 457). However, there are crucial differences in the importance attached to teleology in the construction of the system.
10 "Man kann sich die Weltgeschichte aus einem doppelten Gesichtspuncte vorstellen: entweder als ein *Aggregat* aller Specialhistorien, deren Sammlung, falls sie nur vollständig ist, deren blosse Nebeneinanderstellung, auch schon in seiner Art ein Ganzes ausmacht; oder als ein *System*, in welchem Welt und Menschheit die Einheit ist, und aus allen Theilen des Aggregats einige, in Beziehung auf diesen Gegenstand, vorzüglich ausgewählt, und zweckmäßig geordnet werden."
11 Schlözer (1772 §7, p. 14): "This form, which lies in the selection as well as the connection of the events, is determined by the end (§.1.2) of world history .... "
12 See also Anderson-Gold (2009, 457) "The particular type of lawfulness thought appropriate to explanations of human behavior is the lawfulness that unfolds from the pursuit of a common or collective purpose attributed to human nature. Such lawfulness, it was assumed, would give to history a direction that not only orders the phenomena of the past, but also would make history intelligible through the projection of an overall purpose. Contemporary historians do not generally accept the claim that history has an overall purpose and this claim is part of what distinguishes a philosophy of history from historiography proper, or the empirical study of past events." As we can see from Schlözer's example, at least the historians of Kant's time agree with the philosophers of history that one needs a final end in order to make historical events comprehensible.
13 That this is also the way Kant conceives of providence is explained by Deligiorgi (2006, 454).
14 These concern not only the related terminology, which is particularly striking in the important distinction between aggregate and system, but also methodological and scientific-theoretical claims. Rohbeck (2006, 84) believes that Kant adopted the distinction between aggregate and system from Schlözer. Sturm (2009, 325) also points to the similarity. However, it is difficult to establish a direct link between the Göttingen historians and the philosopher from Königsberg. Unlike Herder, for example, who was in close contact with Gatterer and critically reviewed Schlözer's book *Presentation of his Universal-History* (*Vorstellung seiner Universal-Historie*), or Schiller, who explicitly refers to Schlözer in early writings on history, we cannot find any such reference to Schlözer or other contemporary historians in Kant's published texts. For the relationship between Herder and Schlözer, see Stagl (1998). I discuss Schiller's relationship

to Schlözer in Hahmann (2023). Although we know from the list of books from Kant's *Nachlass* that he possessed works by Gatterer and Schroekh, these works are not relevant to historical-scientific systematics. See Kleingeld (1995, 14) (note 5) on the historiographical books that, according to Warda, were in Kant's possession. As Kleingeld correctly points out, it is hardly possible to draw conclusions about Kant's actual readings from this, since Kant often borrowed his material from libraries and publishers.

15 Sturm (2009, 360) notes that the historical approaches of Gatterer, Schlözer, Schroeckh, and others cannot offer a convincing theory of teleology presupposed from a systematic standpoint, so that the assumed end point of history remains underdetermined.

16 In a letter to Marcus Herz (1773), Kant claims that in his lectures on anthropology he intends to "disclose through it the sources of all the sciences" (Letters, X:145). I think this is almost nowhere as clear as in relation to the science of history.

17 For a brief summary of the discussion on the question of the vocation of the human being, focusing on the development of Kant's philosophy of history, see Kuehn (2009).

18 Given that one of Kant's first published texts dealt with natural history (*Universal Natural History and Theory of the Heavens*, see also Schröder 2011, 40), we can see that Kant had an early and continuing interest in historical questions. It should be noted that the *Critique of Pure Reason* was published in 1781. Shortly after the *Idea* was presented to the public, the second edition of the *First Critique* followed in 1787, and the *Critique of Practical Reason* appeared in 1788. Consequently, it must seem that Kant's account of history also results from his new transcendental philosophical approach (which again would be in line with Kant's self-assessment and emphasis on the a priori guideline in the *Idea*). This observation has long puzzled Kant scholars, who, in view of the completely different style in which the short texts are written and also in view of an argumentation that does not resemble the more scholastic line of argumentation found, for example, in the *Critique of Pure Reason*, try to find the systematic place of Kant's philosophy of history within his broader critical approach. See, e.g., Beck (1963, vii).

19 With this I contradict Rauscher's view, according to which Kant's historical-philosophical texts "do not bear any relation to Kant's pre-critical thought" (2001, p. 46). Similarly, Despland (1973, 9–10). See however, Yovel (1980, 154–5) who claims that Kant's philosophy of history is of pre-critical origin. Schröder (2011, 33) refers to Kant's notes and reflections, which are said to contain an early version of his argumentation.

20 And not in his philosophy of religion, as Vogt (2016, 204) claims with reference to Despland (1973).

21 Allison (2009, 27). See also Flach (2005, 168); Kaulbach (1975, 83); Pollmann (2011, 76; 82–84). Cf. Kant, *Idea* VIII:22, 27–28; *SF*, VII:91; *Anthropology*, VII:331; *CPJ*, V:432

22 Kleingeld (1999). That morality could be the last end of history is, however, explicitly rejected by Pollmann (2011, 84).

23 There is, of course, a crucial difference between the moral perfection Kant has in mind in the *Second Critique* and that which is decisive for the philosophy of history. The latter concerns only the human species, whereas the former addresses the individual. For a discussion of how Kant's assertion of the perfection of the human species as opposed to the individual relates to the debate

between his contemporaries Spalding, Abbt, and Mendelssohn on the question of the vocation of man, see Kuehn (2009, 83).
24 For a good discussion of the many problems this assumption raises, see Kleingeld (1999). We will eventually reach a similar conclusion to Kleingeld. However, my discussion differs from Kleingeld's in that the other ends are systematically understood in their contribution to the vocation of the human being and in that I develop this claim from Kant's pre-critical lecture notes.
25 In quoting Kant, I follow the convention introduced at the beginning of this volume. Furthermore, I use the following abbreviations: Anthropology (*Anthropology from a Pragmatic Standpoint*), CPJ (*Critique of the Power of Judgment*), CPR (*Critique of Pure Reason*), Idea (*Idea for a Universal History with a Cosmopolitan Aim*), Common Saying (*On the Common Saying: That May Be True in Theory, But It Is of No Use in Practice*). I also use the standard abbreviations for referring to Kant's lecture notes.
26 See, e.g., Deligiorgi (2006, 452). Allison (2009, 28) has noted that both the purposively constituted nature cited by Kant as the starting point of his reasoning and the natural end associated with it that drives the historical process seem far removed from the nature elaborated in the *Critique of Pure Reason*, which is essentially a nature understood according to the model of Newtonian physics.
27 In his later writings, Kant refers to these incompatible qualities as unsocial-sociability. For the historical origins of this idea, see Schneewind (2009); Wood (2009, 115–6).
28 Kant's use of the term "providence" is controversial, see, e.g., Kleingeld (2001) and Hahmann (2017). In his *Anthropology*, Kant briefly explains how he conceives of the work of providence in history: "the human being expects these only from *Providence*; that is, from a wisdom that is not *his*, but which is still (through his own fault) an impotent *idea* of his own reason" (VII:328).
29 See also *Anthropology* VII:322: "But in comparison with the idea of possible rational beings on earth in general, the characteristic of the human species is this: that nature has planted in it the seed of *discord* and has willed that its own reason bring *concord* out of this, or at least the constant approximation to it. It is true that in the **end**, *idea* concord is the **end**, but in *actuality* the former (discord) is the **means**, in nature's plan, of a supreme and, to us, inscrutable wisdom: to bring about the perfection of the human being through progressive culture, although with some sacrifice of his pleasures of life."
30 For a more detailed discussion, see Gava (2014). For the historical context of Kant's concept of architectonic, see Manchester (2003). For a discussion with a focus on the role played for history, see Kleingeld (1995, 16–18).
31 It should be noted that Kant's thoughts on scientificity and systematicity also reach back into his pre-critical thought. See also Sturm (2009, 359) who points out that Kant's concept of the inner systematicity of the sciences emerged in the 1770s. On the historical predecessors of Kant's views on scientific systematicity, see Manchester (2003, 195–8) who claims that "Lambert was undoubtedly the catalyst which eventually led to Kant's attention to architectonic." For Crusius as a possible inspiration, see Gava in this volume.
32 Sturm (2009, 359) points to lecture notes from Kant's logic that sugest that Kant indeed applied this concept of science to history: "Indeßen muß doch die Historie, wenn sie eine Wißenschaft seyn soll ein System seyn, als System braucht sie eine Methode, die auf Regeln der Vernunft beruht" (*Warschauer Logik*, taken from Sturm). See also *Wiener-Logik* XXIV:831: "Ein System ist, wo alles einer

Idee untergeordnet ist, die aufs Ganze gehen, und die Theile bestimmen muß. Z.E. Einer [100] kann viele Historien wissen, ohne eine Wissenschaft davon zu haben. Denn er hat die Form nicht. Er hat sich keinen Abriß vom Ganzen gemacht, und nach einer Idee alles geordnet. Diese Idee also macht die systematische Form. Wir können uns das System als ein aggregat oder Ganzes der Erkenntniß denken. Die Totalität ist nun die Bestimmung des Ganzen, und diese liegt in der Idee. Bey einem aggregat aber ist nichts bestimmt. Denn ich weis nicht, was noch dazu kommt. Bey einem System aber ist alles schon bestimmt." ("A system is, where everything is subordinated to an idea, which must cover the whole, and determine the parts. For example, one [100] can know many histories without having a science of history. For he does not have the form. He has not made an outline of the whole, and ordered everything according to an idea. This idea, then, makes the systematic form. We can think of the system as an aggregate or a whole of knowledge. The totality is now the determination of the whole, and this lies in the idea. In the case of an aggregate, however, nothing is determined. For I do not know what else is added. But in the case of a system, everything is already determined." my translation)

33 See also *CPJ*, V:427. A variation of this thought can be found in the *Anthropology*, VII:328: "The education of the human race, taking its species as a *whole*, that is, *collectively (universorum)*, not all of the individuals (*singulorum*), where the multitude does not yield a system but only an aggregate gathered together."

34 On the concept of "idea" in Kant's philosophy of history, see Allison (2009, 24).

35 That this claim ultimately amounts to the counterintuitive assertion that what are commonly regarded as sciences, such as mathematics, physics, etc., ultimately turn out not to be sciences at all, is discussed by Gava (2014).

36 Gava (2014, 380) refers to V-Met-L2/Pölitz, XXVIII: "philosophy is the only science that has a systematic connection, and it is that which makes all the other sciences systematic."

37 This is also suggested by Gava (2014, 377), who notes that Kant sometimes equates the final end with the highest good, with reference to V-Met-K3/Arnoldt, XXIX:948; V-Lo/Dohna, XXIV:698.

38 Allison (2009, 25) claims that the *Critique of the Power of Judgement* "provides the lens through which the earlier work must be examined."

39 For the important difference between these two ends, see Geismann (2006).

40 *CPJ*, V: 378–9: "It is therefore only matter insofar as it is organized that necessarily carries with it the concept of itself as a natural end, since its specific form is at the same time a product of nature. However, this concept necessarily leads to the idea of the whole of nature as a system in accordance with the rule of ends, to which idea all of the mechanism of nature in accordance with principles of reason must now be subordinated (at least in order to test natural appearance by this idea)." As Allison (2009, 34–36) notes, Kant does not justify the necessity of this transition with any argument.

41 Allison (2009, 35) points out that one must also assume something that is an end of nature in order to avoid an infinite regress of means and ends.

42 This is true, as is well known, with the important qualification that it is only a matter of the regulative (or reflective) and not the constitutive use of the power of judgement.

43 We can leave aside the fact that Kant distinguishes here, as in the lectures, between two forms of culture, a culture of skill and a culture of discipline. See *CPJ*, V:431–432.

44 Allison (2009, 41) emphasises that this is the crucial parallel to Kant's remarks in the *Idea*, which is why §83 is also the place where Kant integrates the philosophy of history into the critical system.
45 Especially for the duty to promote the good of future generations, see *Common Saying*, VIII:309: "For I rest my case on my innate duty, the duty of every member of the series of generations – to which I (as a human being in general) belong and am yet not so good in the moral character required of me as I ought to be and hence could be – so to influence posterity that it becomes always better (the possibility of this must, accordingly, also be assumed), and to do it in such a way that this duty may be legitimately handed down from one member [in the series of] generations to another. It does not matter how many doubts may be raised against my hopes from history, which, if they were proved, could move me to desist from a task so apparently futile; as long as these doubts cannot be made quite certain I cannot exchange the duty (as something liquidum) for the rule of prudence not to attempt the impracticable (as something illiquidum, since it is merely hypothetical); and however uncertain I may always be and remain as to whether something better is to be hoped for the human race, this cannot infringe upon the maxim, and hence upon its presupposition, necessary for practical purposes, that it is practicable."
46 For another interpretation of the relationship between historical development and the highest good, which sees an even closer relationship between the two, see Vogt (2016, 205). Yovel (1980, 29) even goes so far as to call the highest good the "idea of history." On Kant's conception of the highest good, see Hahmann (2023).
47 See also Wood (2006, 245); Schröder (2011, 32); Irrlitz (2002, 399).
48 Similarly, Wood (2006, 247) emphasizes that it was one major aim of the *Idea* to ground the empirical inquiry into human history.

Bibliography

Allison, Henry E. 2009. "Teleology and History in Kant: The Critical Foundations of Kant's Philosophy of History." In *Kant's Idea for a Universal History with a Cosmopolitan Aim. A Critical Guide*, edited by Amélie Oksenberg Rorty and James Schmidt, 24–45. Cambridge: Cambridge University Press.
Anderson-Gold, Sharon. 2009. "Kant and Herder." In *A Companion to the Philosophy of History and Historiography*, edited by Aviezer Tucker, 457–67. Oxford: Blackwell.
Beck, Lewis White. 1963. "Editor's Introduction." In *Kant: On History*, edited by Lewis White Beck, translated by Lewis White Beck, Robert E. Anchor, and Emil L. Fackenheim, vii–xxvii. Indianapolis, New York: Bobbs-Merrill.
Deligiorgi, Katerina. 2006. "The Role of the Plan of Nature in Kant's Account of History from a Philosophical Perspective." *British Journal for the History of Philosophy* 14: 451–68.
Despland, Michel. 1973. *Kant on History and Religion*. Montreal, London: McGill-Queen's University Press.
Flach, Werner. 2005. Zu Kants geschichtsphilosophischem "Chiliasmus" In *Phänomenologische Forschungen*, edited by Karl-Heinz Lembeck, Karl Mertens, and Ernst Wolfgang Orth, 167–74. Hamburg: Meiner.
Gava, Gabriele. 2014. "Kant's Definition of Science in the Architectonic of Pure Reason and the Essential Ends of Reason." *Kant-Studien* 105: 372–93.

Gatterer, Johann Christoph. 1767a. *Allgemeine Historische Bibliothek*, Bd. 1, Halle.
Gatterer, Johann Christoph. 1767b. "Vorrede. Von der Evidenz in der Geschichtskunde." In *Die allgemeine Welthistorie die in England durch eine Gesellschaft von Gelehrten ausgefertiget worden. In einem vollständigen und pragmatischen Auszuge*, edited by Friedrich Eberhard Boysen, Vol. 1, 1–38. Halle.
Geismann, Georg. 2006. "'Höchstes Politisches Gut' – 'Höchstes Gut in einer Welt'. Zum Verhältnis von Moralphilosophie, Geschichtsphilosophie und Religionsphilosophie bei Kant." *Tijdschrift voor Filosofie* 68: 23–41.
Hahmann, Andree. 2017. "Kants kritische Konzeption der Vorsehung im Kontext des höchsten Gutes." *Archiv für Begriffsgeschichte* 58: 71–89.
Hahmann, Andree. 2023. "Schiller's Philosophy of History." In *The Palgrave Handbook on the Philosophy of Friedrich Schiller*, edited by Tim Mehigan and Antonino Falduto, 371–87. London: Palgrave.
Hahmann, Andree. 2023. "Kants Konzeption des höchsten Gutes." *Philosophisches Jahrbuch* 130(1): 21–45.
Irrlitz, Gerd. 2002. *Kant-Handbuch: Leben und Werk*. Stuttgart, Weimar: Metzler.
Kaulbach, Friedrich. 1975. "Welchen Nutzen gibt Kant der Geschichtsphilosophie?." *Kant-Studien* 66: 65–84.
Kleingeld, Pauline. 1995. *Fortschritt und Vernunft. Zur Geschichtsphilosophie Kants*. Würzburg: Königshausen & Neumann.
Kleingeld, Pauline. 1999. "Kant, History, and the Idea of Moral Development." *History of Philosophy Quarterly* 16: 59–80.
Kleingeld, Pauline. 2001. "Nature or Providence? On the Theoretical and Moral Importance of Kant's Philosophy of History." *American Catholic Philosophical Quarterly* 75: 201–19.
Kuehn, Manfred. 2009. "Reason as a Species Characteristic." In *Kant's Idea for a Universal History with a Cosmopolitan Aim. A Critical Guide*, edited by Amélie Oksenberg Rorty and James Schmidt, 68–94. Cambridge: Cambridge University Press.
Köster, Heinrich Martin Gottfried. 1775. *Ueber die Philosophie der Historie*, Giessen: Krieger.
Löwith, Karl. 2004. *Weltgeschichte und Heilsgeschehen. Die theologischen Voraussetzungen der Geschichtsphilosophie*, Stuttgart, Weimar: Metzler (engl. translation: Löwith, Karl, *Meaning in History: The Theological Implications of the Philosophy of History*, Chicago: University of Chicago Press, 1949).
Manchester, Paula. 2003. "Kant's Conception of Architectonic in Its Historical Context." *Journal of the History of Philosophy* 41: 187–207.
Pollmann, Arnd. 2011. "Der Kummer der Vernunft. Zu Kants Idee einer allgemeinen Geschichtsphilosophie in therapeutischer Absicht." *Kant-Studien* 102: 69–88.
Rauscher, Frederick. 2001. "The Nature of "Wholly Empirical" History." In *Kant und die Berliner Aufklärung: Akten des IX. Internationalen Kant-Kongresses*, edited by Ralph Schumacher, Rolf-Peter Horstmann, and Volker Gerhardt, Bd. 4, 44–52. Berlin, New York: De Gruyter.
Rohbeck, Johannes. 2006. "Universalgeschichte und Globalisierung. Zur Aktualität von Schillers Geschichtsphilosophie." In *Schiller und die Geschichte*, edited by Michael Hofmann, Jörn Rüsen, and Mirjam Springer, 79–92. München: Fink.
Schneewind, J. B. 2009. "Good Out of Evil: Kant and the Idea of Unsocial Sociability." In *Kant's Idea for a Universal History with a Cosmopolitan Aim. A*

Critical Guide, edited by Amélie Oksenberg Rorty and James Schmidt, 94–111. Cambridge: Cambridge University Press.

Schlözer, Ludwig August. 1772. *Vorstellung seiner Universal-Historie*. Göttingen: Gotha.

Schröder, Wolfgang M. 2011. "Freiheit im Großen ist nichts als Natur." In *Immanuel Kant. Schriften zur Geschichtsphilosophie. Klassiker Auslegen*, edited by Otfried Höffe, 29–44. Berlin: Akademie Verlag.

Stagl, Justin. 1998. "Rationalism and Irrationalism in Early German Ethnology. The Controversy between Schlözer and Herder, 1772/73." *Anthropos* 93: 521–36.

Sturm, Thomas. 2009. *Kant und die Wissenschaften vom Menschen*. Paderborn: Mentis.

Vogt, Peter. 2016. "What Was 'Geschichtsphilosophie'?" *Journal of the Philosophy of History* 10: 195–210.

Wood, Allen. 2006. "Kant's Philosophy of History." In *Immanuel Kant. Toward Perpetual Peace and Other Writings on Politics, Peace, and History*, edited and with an introduction by Pauline Kleingeld, translated by David L. Colclasure, with essays by Jeremy Waldron, Michael W. Doyle, Allen W. Wood, 243–62. New Haven: Yale University Press.

Wood, Allen. 2009. "Kant's Fourth Proposition: The Unsociable Sociability of Human Nature." In *Kant's Idea for a Universal History with a Cosmopolitan Aim. A Critical Guide*, edited by Amélie Oksenberg Rorty and James Schmidt, 112–28. Cambridge: Cambridge University Press.

Yovel, Yirmiahu. 1980. *Kant and the Philosophy of History*. Princeton: Princeton University Press.

11 Systematicity with a Worldly Orientation?
On Kant's Theory and Practice of Gazing with an "Eye of Philosophy"

Huaping Lu-Adler

Kant often invokes the imagery of Cyclops, the mythical one-eyed giant, to describe a way of knowing. The missing eye, he argues, is that of true philosophy. His account of what this philosophical eye comes down to is multifaceted. An analysis of this account helps to foreground three layers of systematicity, namely the systematic unity of a given science, a systematic interconnection among the sciences, and systematicity with a worldly orientation, whereby all sciences are referred to the final (moral) ends of humanity (Section 11.1). To illustrate the first two layers, I turn to Kant's essays on race, where he shows how one can see "race," as he defines this concept, only from the standpoint of a philosophical investigator of nature (Section 11.2).

I then inquire about "what comes of" this theory, following Kant's account of the worldly orientation of philosophy. I clarify the relevant sense of "humanity" (*Menschheit*). As Kant uses this concept in his pure moral philosophy and anthropology, it refers either to rational beings completely *in abstracto*, regardless of their earthly nature, or to the "whole" of the human species, which is emphatically different from an aggregative sum of individual humans. In either case, for the Kantian philosophical eye to fix its gaze on humanity's final ends, it needs to look away from the conditions of actual human beings here and now. I develop this point in connection with Kant's account of abstraction and attention (Section 11.3), before illustrating it with a study of his views on the place of certain human races – on colonial slavery in particular – along the arc of history (Section 11.4).

11.1 "Cyclopic Learnedness" and the Missing "Eye of Philosophy"

In various contexts, Kant characterizes a cyclopic thinker as someone who possesses a great deal of historical knowledge but no *philosophical knowledge*, philosophizes without a *critique*, or thinks egoistically without taking up the *universal standpoint of humanity*. The italicized

Systematicity with a Worldly Orientation? 255

phrases represent different aspects of the missing "eye of true philosophy" (VII:227). They also roughly correspond to the three layers of systematicity that I mentioned above.

To elaborate, I begin with Kant's distinction between historical and rational knowledge (I will use "knowledge" and "cognition" interchangeably). There are two versions of this distinction. *Subjectively*, I know something historically if I hold it to be true simply because someone has told me so; my knowledge is rational if I have drawn it "from principles of reason" myself. *Objectively*, knowledge is historical if it is possible only based on experience – as in history and geography; it is rational if it is possible only as a "cognition *ex principiis*, which has been drawn from grounds *a priori*" – as in philosophy and mathematics (XXIV:797, 830; see A836–837/B864–865; VII:28; IX:22, XXV; XXIV:52–3, 99–100; XXIV:697, 704, 733).

It is important to gain historical knowledge in the objective sense, Kant argues, because it provides materials for growing our rational knowledge. He speaks disapprovingly of "the brooder who shuts himself up in his room and wants to attain much rational cognition": such a person cannot go far in his pursuit without having "first acquired historical cognition by means of experience from books, or from contact, from association with other people" (XXIV:49, translation amended; see IX:159; IX:77–78; XXIV:245; XXIV:870). But Kant is also concerned about the other extreme, *Vielwissen* or *pansophia* (knowing a lot about everything). This can be either *polyhistoria* or *polymathia*. A polyhistor possesses a vast amount of historical knowledge in the objective sense. A polymath, by contrast, has extensive rational knowledge by memorizing, for instance, entire systems of philosophy.[1] A polymath who has thus learned philosophy but not how to *philosophize* is a polyhistor in the subjective sense (XXIV:797). Mere polyhistory, either subjectively or objectively construed, is "*cyclopic* learnedness, which lacks one eye, the eye of philosophy" (IX:45).

This cyclopic learnedness is not necessarily blameworthy. After all, Kant surmises, some people may have only enough talents for this kind of learnedness while lacking "the faculty of *the power of judgment* suitable for choosing among all this knowledge in order to make appropriate use [*zweckmäßigen Gebrauch*] of it" (VII:184). Nonetheless, *Vielwissen* can lead to the illusion of philosophical understanding and make one arrogant (IX:45). Kant therefore finds it necessary to keep it in check.

> Gigantic learnedness is cyclopic, one-eyed, when it consists merely in historical knowledge and the other eye – reason, philosophy – is missing. A genius sometimes indicates vast learnedness, but it must be regulated; ... Historical knowledge makes one puffed up; philosophy humbles.
>
> (XXV:1315, translation amended)

What makes philosophy humbling? Well, true *philosophizing* is an extremely demanding exercise of the mind. One cannot just imitate others or haphazardly accumulate bits of knowledge. Rather, one must meet at least two requirements.

The first requirement is implicit in Kant's account of objectively rational knowledge: one must identify, *a priori*, the grounds or principles in reference to which given cognitions can be brought into a unified system. In this respect, mere cyclopic learnedness is opposed to a way of thinking that attends to how things are *interconnected* (VII:226–227). Cognitions never add up to an interconnected system through accumulation. An idea of the whole must precede the parts. Even *before* one proceeds to gather (objectively) historical knowledge of the world through travel, for instance, one must form a "conceptual whole" by which to learn about the world; otherwise, one will only have an "aggregation" of information, but never a "system" (IX:158; see IX:93).

Second, one must reflect on and keep in sight the *ends* to which one can align all of one's cognitions. A polyhistor may possess a gigantic amount of knowledge, but the *worth* of knowledge lies somewhere else. "The real magnitude [*Größe*] of knowledge," Kant says, "is based on the extensiveness of its application, not on the quantity [*Menge*] thereof" (XXV:1228, translation amended; see XXIV:68–69; IX:49). This should humble anyone who takes pride in having a sheer quantity of knowledge:

> Philosophy can tear down pride and place his true ends before his eyes. Learnedness without philosophy is cyclopic learnedness. Philosophy is the second eye and sees how to attune the collected cognitions of the one eye to an overarching end.
> (XXIV:818, translation amended; see R2020–2022, XVI:198–199; R2025, XVI:200–201)

Provided "all cognitions stand in a certain natural connection with one another," here is one way of setting an overarching end for oneself: "one makes one principal science his end ... and considers all other cognitions only as means for achieving it." In so doing, one "brings a certain systematic character into his knowledge." One cannot decide on the principal science willy-nilly, however. The decision must reflect "a well ordered and purposive [*zweckmäßigen*] plan," whereby the sciences can be connected according to certain *ideas of reason* "in a whole of cognition that interests humanity" (IX:48–49). That is, the ends to which one must attune all of one's cognitions cannot just be merely contingent or personal ends, but *necessary* and *universal* ones (A839/B867n.).

This reference to the necessary and universal ends of humanity is presumably what Kant has in mind when he claims that the missing "eye of true philosophy" is that by which "reason suitably [*zweckmäßig*] uses this mass of historical knowledge" (VII:227). The true philosophy here is a worldly concept (*Weltbegriff*) of philosophy, as opposed to a merely scholastic one (*Schulbegriff*). Philosophy in the latter sense is "a system of cognition that is sought only as a science without having as its end anything more than the systematic unity of this knowledge." Worldly philosophy, by contrast, is "the science of the relation of all cognition to the essential ends of human reason" (A838–839/B866–867). The "true worth of our use of reason" lies precisely in this teleological orientation (XXIV:799). To this end, we must know "1. the sources of human knowledge, 2. the extent of the possible and profitable use of all knowledge, and finally 3. the limits of reason" (IX:25).

Thus, Kant also characterizes the missing philosophical eye as "critique" (R5081, XVIII:82) or "critical philosophy" (XXIV:625). This reflects his view that pure reason in its dogmatic use – a use, not to be confused with dogmatism, that aims at the systematic unity of a science (Bxxxv) – must "appear before the critical eye of a higher and judicial reason … with modesty" (A739/B767). The requisite critique promises "a scientific and fully illuminating self-knowledge" of human reason, to strike down a speculative reason's arrogant wish "to fly with its own wings." Such self-knowledge is the hallmark of "philosophy in a genuine sense." In that connection, "mathematics, natural science, even the empirical knowledge of humankind [*Menschen*], have a high value [*Werth*] as means, for the most part to contingent but yet ultimately to necessary and essential ends of humanity [*Menschheit*]" (A849–850/B877–878). By contrast, the cyclopic scholar makes an "egoist of science" for lacking this self-knowledge of human reason and hence failing to see sciences from the universal standpoint of humanity. Without being able to take up this standpoint, he also has "no measure of the magnitude of [his] cognition" (R903, XV:395).

In sum, to see the world with a true philosophical eye is to bring three layers of systematicity into one's knowledge. First, one may bring given cognitions into the system of a single science (for example, physics) by grounding them on a unifying principle or set of principles. Second, one may connect multiple sciences and bring them into a larger system by making one of them the principal science to which the others are subordinated as means. Third, one may bring all sciences under the final, universal ends of humanity. This results in what I call "systematicity with a worldly orientation." It means that "even in theoretical cognition" one must eventually consider, on behalf of "reason," the question "What comes of it?" (VII:227). This question was already implicit in Kant's self-reported Rousseauian turn during the early 1760s:

> I feel a complete thirst for knowledge and an eager unrest to go further in it as well as satisfaction at every acquisition. There was a

time when I believed that this alone could constitute the honor of mankind [*Menschheit*], and I had contempt for the rabble who know nothing. *Rousseau* brought me around. This blinding superiority disappeared, I learned to honor human beings, and I would find myself far more useless than the common laborer if I did not believe that this consideration could impart to all others a value [*Werth*] in establishing the rights of humanity [*Menschheit*].

(XX:44)

In short, Kant came to recognize that the true worth of knowledge lies in advancing the final ends of humanity.[2] What does this mean *in practice*, though? A case study of Kant's racialization of perceived human differences from the standpoint of a system-loving natural philosopher (Section 11.2) in conjunction with his normative silence about colonial slavery as an institution (Section 11.4) can help to answer this question in a particularly illuminating way. We will see that the Kantian "humanity," on the final ends of which the worldly philosopher fixes his gaze, is importantly different from the sum of all human beings *in concreto* and that this can have chilling practical implications (Section 11.3).

11.2 Seeing Race with a Philosophical Eye: A Case Study

Kant develops a self-consciously groundbreaking theory of race in three dedicated essays: "Of the Different Races of Human Beings" (1775/7), "Determination of the Concept of a Human Race" (1785), and "On the Use of Teleological Principles in Philosophy" (1788).[3] In all three, he looks at race from the perspective of a philosophical investigator of nature (*Naturforscher*), who seeks to understand phenomena by looking into their law-governed natural causes.

Kant begins his 1785 essay on race by acknowledging the abundance of relevant travel reports, which constitute a crucial source of (objectively) historical knowledge.[4] He does so only to highlight the need for philosophical investigation, though.

> The knowledge which the new travels have disseminated about the manifoldnesses in the human species so far have contributed more to exciting the understanding to investigation on this point than to satisfying it.
>
> (VIII:91)

Specifically, Kant seeks a classification of all humans that is not a mere "school system for memory," which divides creatures based on similarities and brings them "under titles"; it is rather a physical or "natural system for the understanding," which "divide[s] the [human] animals according to *relationships* in terms of generation [*Erzeugung*]" and thereby brings them "under laws" (II:429). Only the latter can give us a "science" by indicating "the ground for the manifold variations" (II:443).

The variation that Kant finds most significant is that of skin color, which he argues is the only contingent characteristic that is *unfailingly hereditary* across generations of humans. Since only this kind of characteristic can establish a race, and since exactly four basic skin colors can allegedly be proved by his method, he divides all humans into four principal races. These races are subspecies of the same human species. Their formation involves two kinds of causes. First, there is the purposive cause, which consists of the various germs (*Keime*) and natural predispositions (*Anlagen*) in the original human phylum (*Stamm*), lying ready to be developed. Second, there must be certain natural causes that provide the material conditions for such developments. Kant sees *air* and *sun* as "those causes which most deeply influence the generative power and produce an enduring development of the germs and predispositions, i.e., are able to establish a race" (II:435–436). How the original germs and predispositions would develop during the first period of human existence depends on the particular climates in which the most ancient dwellers had to adapt and preserve themselves. There are exactly four climates that are radically different in terms of air and sun: humid cold, dry cold, humid heat, and dry heat. Four distinct skin colors developed as a result of long periods of adaptation to those climates, respectively – white, copper-red, black, and yellow. These developments became irreversible and unfailingly hereditary, which is why each skin color marks a race (II:432–441; VIII:93–94, 99–104; VIII:173–177).

Here we have a clear example of the first contrast between a mere cyclopic learner and a philosophical investigator. The former may amass extensive historical knowledge about how human beings around the globe differ from one another physically by reading all the available travel literature, without ever understanding or even inquiring about the cause of their perceived differences. The philosophical investigator, by contrast, seeks to bring the reported phenomena into a *scientific system* by deriving them from a single ground in accordance with certain laws of nature. Furthermore, such an investigator proceeds with antecedently determined principles and concepts, including the very concept of race, insofar as "one finds in experience what one needs only if one knows in advance what to look for" (VIII:91; see VIII:161). Kant suggests that only a philosophically minded observer of nature, who is not content with mere descriptions of phenomena but always seeks to uncover their law-governed causal underpinnings, can see "race" in it (VIII:163).

Now, given all the aspects of Kant's contrast of the learned Cyclops and the true philosopher that I teased out in Section 11.1, two questions suggest themselves. How does his science of race relate to other sciences? And what comes of it if we take into account the final ends of humanity?

To answer the first question briefly, the topic of race offered Kant an exceptional opportunity to test different explanatory models regarding the (re)generation of organisms – for example, an explanation by mere "physical-mechanical causes" versus one that presupposes "preformed" germs and predispositions (II:434–435). In the process, Kant was compelled to clarify the relation between (what we call) biology, of which his scientific theory of race is an integral part, and Newtonian physics, on the one hand, and between natural science and metaphysics, on the other. As someone who started his career in natural science with an overall commitment to Newtonianism, Kant was intrigued by the challenge posed by organisms. Their generation, he proposed in 1763, cannot be explained as "a mechanical effect incidentally arising from the universal laws of nature"; rather, one must grant them a supernatural origin, but only the sort that gives organisms suitability to "generate their kind ... in accordance with universal laws" (II:114–115). As Kant would later put it in the *Critique of the Power of Judgment* (1790), one must find an explanatory model (*Erklärungsart*) that, "with the least possible appeal to the supernatural, leaves everything that follows from the first beginning to nature" and therefore still "leaves natural mechanism an ... unmistakable role" to play by empirically knowable laws of nature. Therefore, "reason would still already be favorably disposed to this explanation" even if it had no recognizable advantage over the alternatives "in the matter of experiential grounds for [its] proof" (V:424). It was in his essays on race, especially the ones published in 1785 and 1788, that Kant first ironed out the key methodological issues to warrant this assertion of the *a priori* endorsement by reason. For example, which of the many competing "maxim[s] of reason" has priority in guiding investigations of the phenomenon of heredity (VIII:96–97)? And how should one apply teleological principles in such investigations without blurring the boundaries between natural science and metaphysics (VIII:150–160, 178–183)? Kant is certain that his take on these issues reflects the universal standpoint of speculative reason (regarding its use in natural science). He can therefore confidently claim that "a philosophical *jury* ... composed of mere investigators of nature [*Naturforschern*]" would rule in his favor in his disputes with the critics (VIII:179).[5]

By this appeal to a jury of *Naturforschern*, Kant is responding to the invocation of an *ächte philosophische Jury* by Georg Forster (1754–1794), who published an essay in 1786 critiquing Kant's 1785 essay on race.[6] Notably, Forster was implicitly calling for a jury of *moral* philosophers, following a skeptical reflection on the *practical* impact that a theory of race like Kant's might – or might not – have on colonial slavery (1991[1786]:165–6). Kant's insistent appeal to a jury of *natural* philosophers instead is puzzling. To compound the problem, far from condemning the cruelty of slavery as Forster did, Kant sides with the anti-abolitionist merchant James Tobin against the abolitionist Reverend James Ramsay in their public debate over whether to liberate "Negro" slaves and use them as free laborers (VIII:174n.).[7]

This observation brings me to the other question I raised above: what comes of Kant's science of race if we take into account the final ends of humanity? Given his argument for the *unity* of the human species (monogenism) and given that he worked out the core theses of his *universalist* moral theory during the 1780s, would it not be natural for him – as his progressive readers today might assume – to use every opportunity to condemn colonial slavery? Moreover, if we take seriously his claim, as I explained in Section 11.1, that the true worth of even a theoretical use of reason lies in serving the final (moral) ends of humanity, should we not expect him to use his monogenetic theory of race to articulate and promote the idea of a world in which *human beings of every race* are treated with dignity, an idea he could then use to reject the institutionalized racial oppression in the form of colonial slavery? Kant did none of that, while managing to insert racist claims about all three non-white races now and then.[8] We cannot simply chalk this up to cognitive dissonance or any other mysterious cognitive-psychological malfunction on Kant's part. Rather, once we have properly understood the concept of *Menschheit* as it figures in his pure moral philosophy and worldly concept of philosophy, we will see that his failure to condemn colonial slavery in unmistakably moral terms is a proverbial feature, not a bug, of how he looks at things with a philosophical eye.

To appreciate this point, we may consider Kant's account of abstraction and attention, two complementary mental acts that work together to ensure a sharp focus of the philosophical gaze on what matters in the end. A critical distinction between *Menschheit* and *Menschen* will allow Kant to make both grand philosophical claims about *Menschheit* in abstraction and blatantly racist claims about some classes of *Menschen* without contradicting himself.

11.3 Abstraction and Attention: How to Focus One's Philosophical Gaze on the Ends of *Menschheit*

Here is one way in which "Menschen" and "Menschheit" can differ: the former, when used as a general term, denotes all individual humans, whereas the latter does not.[9] This is partly because the concepts expressed by these terms are formed differently. The general concept of *Menschen* is the result of "logical abstraction": starting with given (lower) concepts, one compares and reflects on them, abstracts from certain differences among them, and thereby arrives at a higher concept that represents what they have in common; this process may be iterated until one reaches the highest concept, from which nothing further may be abstracted. As a matter of logical necessity, what is true of a higher concept is also true of all the lower concepts (IX:98–99; see XXIV:255–256). Accordingly, what is true of *Menschen* in general must be true of all the humans falling under it.

The same cannot be said of *Menschheit*. This concept or the concept of the human being (*der Mensch*), even in the least abstract sense, signifies the *human species* or *human race* as a whole,[10] not an aggregate of individual humans.[11] Kant has this in mind when he speaks of humanity's destiny: unlike other animals, "with the human being only the *species* ... can work its way up to its destiny" (VII:324; see IX:445; XXV:1196; XXV:839–840; XXV:1416–18). The reason is that the destiny of the human species is defined by its rationality, which separates it from all the other creatures on earth. Kant elaborates this point in the "Idea for a Universal History with a Cosmopolitan Aim" (1784):

> *In the human being* (as the only rational creature on earth), *those predispositions whose goal is the use of his reason were to develop completely only in the species, but not in the individual.* Reason ... needs attempts, practice and instruction in order gradually to progress from one stage of insight to another. ... nature perhaps needs an immense series of generations, each of which transmits its enlightenment to the next, in order finally to propel its germs [*Keime*] in our species to that stage of development which is completely suited to its aim. And this point in time must be, at least in the idea of the human being, the goal of his endeavors, because otherwise the natural predispositions [*Naturanlagen*] would have to be regarded for the most part as in vain and purposeless.
> (VIII:18–19; see VII:329–330; R1471a, XV:650)

The "idea" of the human being – as an essentially rational creature who is endowed with certain *Keime* and *Naturanlagen* that can only be developed and perfected over indefinitely many generations – cannot be formed based on empirical observations about actually existing humans or how the human history has been unfolding so far. Rather, it can only be a rational cognition *a priori*, insofar as it serves as the *architectonic* to guide one's reflections on the material conditions – such as government and education – that are needed for the species eventually to reach its destiny.[12] This idea serves to focus one's *attention* on what matters in the end,

Systematicity with a Worldly Orientation? 263

in *abstraction* from actual experiences of life as it unfolds in front of one's eyes, with all its disappointments, setbacks, and even sheer evils.

Kant characterizes attention and abstraction as two complementary ways "to become conscious of one's representations." Abstraction is "not the mere failure and omission" of attention. It is rather "a real act of the cognitive faculty of stopping a representation of which I am conscious from being in connection with other representations in one consciousness" (VII:131). Abstraction is more difficult because "sensibility always gets in my way" (XXV:1239; see XXV:489; XXIX:878). Thus, the ability to abstract "demonstrates a freedom of the faculty of thought and the authority of the mind, *in having the object of one's representations under one's control*" (VII:131). This cognitive control is most difficult when it comes to philosophical cognitions regarding "principles or fundamental concepts and fundamental rules," because these require "attentiveness to … only one kind of object of thoughts" in abstraction from all sensible representations (XXV:1482).

Accordingly, Kant holds that philosophical cognitions are entirely beyond the horizon of the "common understanding," which is a faculty for cognition *in concreto*. "Speculative understanding," by contrast, is a faculty for cognition *in abstracto* (XXIV:754). The former faculty always "demands an example from experience" for every concept or rule it tries to understand, so much so that it "has a use no further than the extent to which it can see its rules confirmed in experience." Only speculative understanding can "have insight into these rules *a priori* and independently of experience" (IV:369–370; see VII:139–140). To this end, one must *ignore* the empirical applications of a rule, so that "one gets it *in abstracto* and can better consider it in the universal as a principle" (IX:45). This act of ignoring (*ignoriren*), which is distinct from ignorance (*Unwissenheit*), is deliberate: one *chooses* "not [to] take notice of some circumstances that contribute nothing" to one's purpose. That is, "we ignore, i.e., abstract from some things that are known, but are put aside because they do not pertain to the end" (XXIV:837).

But how does one decide which things "pertain to the end" and which do not, if "what one ignores must not be detrimental to the principal end" (XXIV:837)? Kant cautions against either abstracting *too little* or abstracting *too much* in one's social life. If "empirical people do not sufficiently abstract from certain secondary things to attend to principal things," the problem with "speculative minds" is that they tend to "abstract too much": "They do not think at all about how things are *in concreto*; rather they consider them only *in abstracto*." Either extreme can make one "unhappy" (XXV:1240; see VII:131–132; XXV:488–489). Might there be an analogous problem in the case of the learned Cyclops versus the person who views the world with the Kantian philosophical eye? The learned

Cyclops abstracts too little, dispersing his attention equally among a vast quantity of knowledge without a guiding idea or unifying principle. Now, might the philosophical mind run into the opposite problem by abstracting too much? I am referring not just to philosophy in the scholastic sense, but Kant's worldly concept of philosophy. By steadily training one's philosophical eye on the final ends of humanity, which can at best be realized in a remote future, might one risk completely losing sight of problems that require urgent attention here and now?

To clarify this question, let me return to Kant's notion of *Menschheit*. In the most abstract sense, the Kantian human being is a mere (finite) *rational being*, in abstraction from its earthbound nature. This is the concept of humanity in Kant's pure moral philosophy. To set moral concepts and laws in their universality (*im Allgemeinen*),[13] he states, is to "set [them] forth ... *in abstracto*" and trace their "origin completely a priori in reason." That is, they "cannot be abstracted from any empirical and therefore merely contingent cognitions." Rather, they must be derived "from the universal concept of a rational being as such [*überhaupt*]." Accordingly, the proposition that moral concepts and laws hold for all "human beings" comes with a significant caveat: these must be considered as mere (finite) rational beings, in total abstraction from "the nature of the human being" and from "the circumstances of the world in which he is placed." Nor can the relevant concept of "rational being" be derived through logical abstraction from lower concepts representing different classes of rational beings (the earthly ones along with those on other planets). Rather, it must be a concept "of pure reason" (IV:389, 409, 411–412).

The system of morals, which must first be set forth *a priori* as a pure philosophy, therefore "needs anthropology for its *application* to human beings" (IV:412). Even in this case, the expression "human beings" still does not denote all concretely embodied individuals. What Kant takes into account is only "the particular *nature* of human beings." The anthropology that investigates this nature

> would deal only with the subjective conditions in human nature that hinder people or help them in *fulfilling* the laws of a metaphysics of morals. It would deal with the development, spreading, and strengthening of moral principles (in education in schools and in popular instruction), and with other similar teachings and precepts based on experience.
> (VI:217; see IV:388–389; XXVII:244–245; XXIV:599)

This passage resonates with Kant's account of "the character of the species" in his writings and lectures on anthropology. This account shows how "the human being, as an animal endowed with the *capacity of reason* (*animal rationabile*), can make out of himself a *rational animal* (*animal rationale*)" in accordance with "the idea of possible rational beings on earth in general" (VII:321–322). Anthropology is the science – insofar as it is "systematically formulated" – that seeks to "know the human being according to his species as an earthly being endowed with reason." The part of this knowledge that considers what the human being "as a free acting being makes, or can and should make of himself" (VII:119) is called "pragmatic."

> The sum total of pragmatic anthropology, in respect to the vocation of the human being and the characteristic of his formation, is the following. The human being is destined by his reason to live in a society with human beings and in it to *cultivate* himself, to *civilize* himself, and to *moralize* himself by means of the arts and sciences.
> (VII:324–325; see XXV:436; XXV:471–472; XXV:735; XXV:1211, 1367–1368)

Clearly, "the human being" here refers to the human species (when contrasted with non-rational animals) or the human race (when contrasted with non-terrestrial rational beings) as a whole, but not individual humans. To be specific, Kant is interested in understanding the character of the earthbound human being as such (*überhaupt*), in order to establish the hope – as "an idea which is possible" – that "the human race ... will attain the greatest degree of perfection [in moralization]" even though it may take an indeterminably long time (XXV:675–697). Explicating the "character of human species" in this way is the same as analyzing "the concept of human nature in general" (XXV:838). This captures what is unique about Kant's conception of pragmatic anthropology, which is "not a description of human beings, but of human nature" (XXV:471).

Kant has reasons to abstract from the actual conditions of individual humans – so that he can train his philosophical gaze firmly on the final (moral) destiny of humanity. This destiny is an idea of reason that represents a hoped-for future state, which the human species can only approximate through indefinitely many generations. If culture, civilization, and moralization are, in that order, the "three kinds of progress that nature has laid in human beings" and if "in moralization we have done almost nothing," Kant claims that "we have reason to hope for it" (XXV:1197–1198; see VII:324–325; VIII:26; IX:451). He recognizes that the current state of humanity is so full of ills that "Rousseau was not so wrong when he preferred to it the condition of savages" (VIII:26). He is nevertheless hopeful that humanity *as a species* "can attain" highest moral perfection (XXVII:471). Maintaining this hope requires a deliberate act of abstraction, an act of *looking away* to focus well on the final ends of humanity.

What does Kant choose to look away from so that he can hold onto the hope for a future of moralized humanity? These include the appearances of human "doings and refrainings on the great stage of the world" that make one suspect that "everything in the large is woven together out of folly, childish vanity, often also out of childish malice and the rage to destruction." Confronted with this seemingly "nonsensical course of things human," the Kantian "philosopher" can only "try whether he can discover an *aim of nature*" in it, in abstraction from the aims of the human actors in them (VIII:17–18). Someone like Rousseau may lament "the malevolence of human nature, which can be seen unconcealed in the free relations of nations" (VIII:355). After all, the political states seem to be forever at war with one another, as they "apply all their powers to their vain and violent aims of expansion" (VIII:26). The philosopher reasons, however, that wars (or threats thereof) and, more generally, human beings' "antagonism in society" or "unsociable sociability" can serve as the "means nature employs in order to bring about the development of all its predispositions." Nature thereby "awakens all the powers of the human being, brings him to overcome his propensity to indolence, and, driven by ambition, tyranny, and greed," to develop all his talents bit by bit (VIII:20–21). So, Kant asserts without irony:

> Thanks be to nature, therefore, for the incompatibility, for the spiteful competitive vanity, for the insatiable desire to possess or even to dominate! For without them all the excellent natural predispositions in humanity would eternally slumber undeveloped. The human being wills concord; but nature knows better what is good for his species: it wills discord.
>
> (VIII:21)

The Kantian philosopher can thus remain hopeful that, for all the ills now besetting the human being, he has "a still greater ... moral predisposition to eventually become master of the evil principle within him"; it is just that this predisposition is "at present dormant" (VIII:355). Humanity will remain in this condition "until ... it will labor its way out of the chaotic condition of the present relations between states" (VIII:26).

Such a resolute orientation toward the hopeful *future* of humanity can simultaneously obscure one's perception of *present* human relations. This has to do with how abstraction and attention work together. Abstraction happens when "I extract from [*entziehe*] all the other representations in the vicinity so much clarity that they become completely obscured and only the one [representation] remains"; this allows me to "keep away other representations [so] that their impressions do not act on my consciousness," whereby I can be exactly attentive to the one representation that I must focus on (XXV:1239). In the case of the Kantian philosopher, this means that he may have to *look away* from present moral wrongs so that he can hold fast to the idea of humanity with its promise of a future moral state.

Kant knows how to see the world in this way without cognitive dissonance. To him, it is all about what *principle* to use in a given case. To understand the characteristics of the female sex, for instance, Kant chooses as his principle something that "does not depend on our choice but on a higher purpose for the human race." Accordingly, the question is "not what we *make* our end, but what *nature's end* was in establishing womankind." In this way, what appears to be "the foolishness of human beings, must still be wisdom according to nature's purpose." Now, one of nature's ends for humanity is "the preservation of the species" (since the species can approximate its moral end only through continuously reproduced generations of biological humans). To this end, Kant claims, nature plants a "weakness" in the woman, namely "fear of *physical* injury and *timidity* before similar dangers," through which "this sex rightfully demands male protection for itself"; nature can thereby ensure the fulfillment of the pledge that it has "entrusted to woman's womb," namely "to propagate and perpetuate" the species (VII:305–306). A similar logic (that *nature* may use some parts of humanity to serve *its* end for this species), as I shall explain next, may help to make sense of Kant's normative silence about colonial slavery as an institution.

11.4 Considering Races with a Worldly Orientation: The Above Case Study Continued

The separation between what *we* make *our* end and what *nature* does to serve *its* end provides an important clue to understanding the complex relations between different parts of Kant's philosophical system. In particular, it helps to shed some light on how his theory of race interacts with his moral philosophy, political philosophy, and philosophy of history. In this regard, an especially puzzling fact is that Kant made no unequivocal moral judgments about colonial slavery. This is sometimes seen as an example of certain "cognitive failings" on Kant's part, as a moral philosopher who wrote so forcefully about the Categorical Imperative while expressing racist as well as sexist views (Allais 2016). Overall, commentators tend to assume that Kant's racist remarks directly contradict his universalist moral teachings (if these are interpreted literally); this assumption leaves them with a binary choice, either to conclude that Kant's moral teachings are not genuinely universal (if we are to take his racism seriously) or to insist that his moral teachings are unquestionably universal and outweigh his racist prejudices.[14]

These opposing approaches share two basic mistakes. First, the *strictly universal* claims of Kant's pure moral philosophy, which concern the human being only as a (finite) rational being *in abstracto*, are mistaken as *general* claims about actual humans *in concreto*. According to Charles Mills, for instance, a genuinely universalist reading of Kant's moral philosophy would make him simultaneously believe two contradictory claims – the unqualified universalist belief that "all biological humans, including all of the races, are normatively equally human/full persons," and the particularist racist belief that "the races of blacks and Native Americans are natural slaves." The contradiction here is so "flagrant," Mills contends, that it "would have been noticed by anyone of the most minimal intelligence, let alone one of the smartest minds" (2014, 146). Mills has misrepresented Kant's moral universalism, however, by referring it to the domain of *biological humans*. Mills's critics have hardly noticed this problem.

Second, there is also a tendency to equate Kant's claims about the human species as a *whole* with claims about an *aggregate* of individuals. This is most evident in Robert Louden's attempt to locate a middle ground where a responsible interpreter can prioritize what Kant is "logically committed" to by his moral theory, while acknowledging that, when it comes to race and sex, what he "personally" believes contradicts those logical commitments. Louden goes so far as to suggest that the racist and sexist Kant, on account of the "many private prejudices and contradictory tendencies" exhibited in his writings, might "not accept these logical implications of his own theory." Louden nonetheless insists that "Kant's theory is fortunately stronger than his prejudices" and that we as "philosophers" should focus on the theory and take care not to "inflate" the prejudices. The theory itself, Louden claims, implies a belief representing "the core value of universality in Kant's ethics," namely that the *entire human species* must eventually share in the destiny of the species: moral perfection." That is, Kant's moral universalism, properly understood, "refers to the hard and painful work of ... figuring out what changes in human institutions and practices need to be made so that *all members* of the species will be brought into the moral community" (2000, 105–6, italics added). The italicized parts of these remarks suggest an attempt to read Kant's theory as inclusive and egalitarian concerning all races and sexes. Unfortunately, they also reveal a division fallacy, the inference that what is true of the whole must also be true of individual parts. Louden makes this inference more explicit in the following claim:

> because [Kant] believes that the entire species progresses in perfection, he must also accept that the *entire species* is destined to eventually work its way through the preparatory steps of culture and civilization to moralization. It therefore cannot be the case ... that *women or people of color* will always remain mere passive citizens in the realm of ethics.
>
> (2000, 105, italics added)

As I pointed out above, however, within the Kantian framework what is true of the human species as a whole is not thereby true of all individual humans. If Kant's pure moral philosophy dictates that no human being qua rational being be used *by other human beings* as mere means for *their* end, his overall system may allow certain hegemonic human relations to be used *by nature* as means to serve *its* end for humanity as a species.

I have explained elsewhere how this way of thinking gives Kant ample room to exclude, without self-contradiction, the "savages" represented by Amerindians from participating in humanity's cultural, civilizational, or moral progress (Lu-Adler 2022b). Now I briefly turn to the case of "Negroes," who Kant portrays mainly as the objects of colonial slavery.[15] The same tropical climate that occasioned the formation of a truly black skin, Kant writes, has also "result[ed] in the Negro, who is well suited to his climate, namely strong, fleshy, supple," because that kind of climate favors "the robust growth of animals in general" (II:438). Accordingly, "one uses Negroes [as opposed to the red slaves (Amerindians)]" for "field labor" (II:438n.). If the "Negro" is naturally "lazy, soft and trifling" due to "the abundant provision of his mother land" (II:438), he is at the same time full of affects and passions, sensitive, and "afraid of beatings," wherefore he can be trained (*abrichten*) and acquire a "culture of slaves" (XXV:1187; see R1520, XV:877–878; XXV:450–451). Given this depiction of the "Negro" race as entirely lacking *self*-control and as needing to be *driven* to labor through external coercion, Kant is unsurprisingly skeptical about Ramsay the abolitionist's proposal to "use all Negro slaves as *free* laborers." He speculates that their lack of an "immediate drive to activity," due to the minimal needs in their native climate, must have been so interwoven with their natural predispositions that it "extinguishes just as little as the externally visible [black skin]" long after their ancestors were transported to the new world (VIII:174n.).

Although Kant expressed occasional qualms about *certain practices* of colonial slavery in the 1790s, he never came to condemn the very *institution* of colonial slavery, let alone endorse the abolitionist movement that, as he knew, was gaining momentum in England.[16] All in all, he seemed less interested in defending whatever rights the enslaved might have than he was concerned about the specter that how colonial slavery was practiced in the late eighteenth century could have disastrous consequences for the Eurocentric global order (VIII:359; XXIII:174). This case further illustrates how Kant, in reality, looked at things with a philosophical eye. If he belatedly became somewhat critical of colonial slavery, it was not thanks to any new moral epiphany. It was rather because the late Kant could no longer see colonial slavery as something that could facilitate humanity's progress – for example, by dramatically increasing material production and promoting an intricate global trading system – without such advancements being overshadowed by its corrosive political effects. In other words, if Kant voiced no objection to colonial slavery at all in the 1780s because he might be convinced of its overriding benefits to progress (even if he could still deem it a moral ill on its own), its practice seemed objectionable in the 1790s by the same logic: now it might jeopardize any prospect of forming a perpetually peaceful union among sovereign European states.[17]

This again reflects Kant's consistent focus on the final ends of *humanity*. To fix his philosophical gaze sharply on such ends, he chose to *look away* from the present injustices done to some *human beings*. We can grant for the sake of argument that Kant might have privately regarded colonial slavery as morally wrong all along (he never explicitly made any unequivocal moral judgment to this effect). It is just that whether he would openly object to it – to how it was practiced to be precise – hinged on how he, as a philosopher resolutely guided by the rational *idea* of human progress, would evaluate it in light of the political realities at a given point in history. He could do this by using as his principle "not what we *make* our end, but what *nature's end* was" (VII:305). In these terms, he could, coherently, regard colonial slavery both as morally wrong (a *human being* ought not to use another as mere means to *his* end) and as an arrangement conducive to *nature's* end – until it no longer seemed to serve this end.

11.5 Conclusion

If you read only Section 11.1 of this chapter, you might find yourself nodding along with Kant. Of course, you might think, as aspiring philosophers we should

1 seek principled interconnections among our cognitions and first bring them into the system of a single science (for example, physics), then
2 see how this science may be connected with other sciences (for example, how is physics related to biology?), and finally
3 ask ourselves why we are pursuing any of those sciences in the first place and recognize that, the more they serve to improve the conditions of humanity (for example, when we can use physics and mathematics to construct more earthquake-resilient cities), the worthier they are.

But you read on. You saw that Kant developed a theory of race precisely because he beheld the reported phenomenon of human differences with a philosophical eye. Had he been a mere polyhistor, no amount of travel literature he consumed would have added up to a "science" of race like the one he developed in his three dedicated essays on race (Section 11.2). You also saw that the notion of humanity operative in Kant's worldly concept of philosophy does not refer to the sum of all human beings *in concreto*; therefore, what is said of the former is not necessarily true of the latter. To train his eye steadily on the final ends of humanity as such, the Kantian philosopher sometimes finds it necessary to look away from the present sufferings of some human beings (Section 11. 3). This might well be how Kant looked at colonial slavery, as he never showed any true *moral* regard for the enslaved as human beings with intrinsic dignity (Section 11.4). Does any of this mean that we should fundamentally reassess Kant's very account of systematicity, with its insistence on seeing the world with a "philosophical" eye? You may find it difficult to pin down a straightforward answer to this question.

Notes

1 See IX:45; XXIV:519, 522; XXIV:619, 625; XXIV:713–714; XXIV:818; R2025, XXVI:200; R2016, XXVI:196.
2 For a detailed analysis, see Velkley (1989).
3 On the history of these essays, see Mikkelsen (2013, 18–32). For a study that situates them both within Kant's overall philosophical system and against the relevant historical backdrop, see Lu-Adler (2023, 111–240).
4 On the place of travel reports – as a form of historical knowledge – in Kant's natural philosophy, see Lu-Adler (2022a).
5 For further explications of Kant's theory of race in connection with his evolving natural philosophy, see Mensch (2013, 95–104); Sanford (2018); Lu-Adler (2023, 111–61).
6 On Forster versus Kant, see Goldstein (2019, 92–104).
7 On the historical significance of the Ramsay-Tobin debate, see Swaminathan (2016).
8 For a partial, assorted list of Kant's racist remarks, see Mills (2005, 173–5). As I argue in Lu-Adler 2023 (especially 76–107), we should not interpret those remarks simply as expressions of personal prejudice; rather, we should attend to their ideology-forming and world-making power as pronouncements by a prominent philosopher and lifelong educator like Kant.
9 Kant does not always use these terms in the same way. Sometimes, they appear interchangeable. Here I only focus on the sense in which they differ.
10 Humanity is a "species [*Gattung*]" of creatures on earth, when it is differentiated from other species of animals on account of its essential *rationality*. It is a "race [*Rasse*]" of rational creatures *on earth*, however, when it is compared with the "rational beings on other planets, as a multitude of creatures arising from one demiurge" (VII:331).
11 The difference between *whole* and *aggregate* matters here: while a whole always precedes its parts, it is the other way around for an aggregate.
12 According to Kant, "the plan for education must be made in a cosmopolitan manner," which has as its "final end the best for the world and the perfection to which humanity is destined" (IX:448).
13 Kant is talking about "strict universality," as opposed to merely "empirical" or "assumed and comparative universality (through induction)." What is strictly universal is necessary, permitting "no exception at all"; it is therefore "not

derived from experience, but is rather valid absolutely *a priori*" (B3–4). This strict universality essentially differs from mere generality, which is a matter of degree (A843/B871).

14 For a detailed analysis of this false dichotomy and a critique of the contradiction thesis that underpins it, see Lu-Adler (2023, 33–75).

15 Kant locates what he calls "true Negroes" in the West-African region of Senegambia (II:441–442). His rationale is that the air in this region is so "phlogistized" that only those with the blackest skin can survive there (VIII:103; VIII:169–170n.). For this reason, I retain Kant's use of "Negro" as a term with a special meaning to him. It is also worth highlighting that West Africa was the epicenter of the transatlantic slave trade. So, it is especially telling that Kant characterizes the "Negro" race as uniquely and naturally fit for slavery.

16 As I explain in Lu-Adler (2025), we need not invoke Kant's anti-black racism to explain his failure to support abolitionism. Rather, this failure may tell us something about certain limitations of his political philosophy, which was systematically articulated only in the 1790s, in that they might have made it *theoretically difficult* for him to figure out exactly what to *do* about the institution of colonial slavery as an entrenched political reality. If this is a plausible interpretation, then we have yet another case against Louden's claim that Kant's theory is stronger than his racist prejudice.

17 I argue for this interpretation in Lu-Adler (2022c). For a careful and illuminating study of Kant's views on slavery in its various guises, see Jorati (2024, 280–307). This study makes clear that not all of Kant's later remarks about slavery were about colonial slavery: the majority of them were about penal slavery, which he strongly endorsed, and about contractual slavery, which he rejected as self-contradictory.

Bibliography

Allais, Lucy. 2016. "Kant's Racism." *Philosophical Papers* 45: 1–36.
Forster, Georg. 1991 [1786]. "Noch etwas über die Menschenrassen" (1786). In *Georg Forsters Werke 8* (Berlin: Akademie–Verlag, 1991), 130–56.
Goldstein, Jürgen. 2019. *Georg Forster: Voyager, Naturalist, Revolutionary*. Chicago: University of Chicago Press.
Jorati, Julia. 2024. *Slavery and Race: Philosophical Debates in the Eighteenth Century*. New York: Oxford University Press.
Louden, Robert. 2000. *Kant's Impure Ethics*. Oxford: Oxford University Press.
Lu-Adler, Huaping. 2022a. "Kant's Use of Travel Reports in Theorizing about Race—A Case Study of How Testimony Features in Natural Philosophy." *Studies in History and Philosophy of Science* 91: 10–19.
Lu-Adler, Huaping. 2022b. "Kant on Lazy Savagery, Racialized." *Journal of the History of Philosophy* 60(2): 253–75.
Lu-Adler, Huaping. 2022c. "Kant and Slavery—Or Why He Never Became a Racial Egalitarian." *Critical Philosophy of Race* 10(2): 263–94.
Lu-Adler, Huaping. 2023. *Kant, Race, and Racism: Views from Somewhere*. Oxford: Oxford University Press.
Lu-Adler, Huaping. 2025. "Slavery and Kant's Doctrine of Right." *Journal of Modern Philosophy* 6: 1–26.
Mensch, Jennifer. 2013. *Kant's Organicism: Epigenesis and the Development of Critical Philosophy*. Chicago: University of Chicago Press.

Mills, Charles. 2005. "Kant's *Untermenschen.*" In *Race and Racism in Modern Philosophy*, edited by Andrew Valls, 169–93. Ithaca: Cornell University Press.
Mills, Charles. 2014. "Kant's Racism Redux." *Graduate Faculty Philosophy Journal* 35: 125–57.
Mikkelsen, Jon, tranlator and ed. 2013. *Kant and the Concept of Race: Late Eighteenth–Century Writings*. Albany: SUNY Press.
Sanford, Stella. 2018. "Kant, Race, and Natural History." *Philosophy and Social Criticism* 44: 950–77.
Swaminathan, Srividhya. 2016. "(Re)Defining Mastery: James Ramsay versus the West Indian Planter." *Rhetorica: A Journal of the History of Rhetoric* 34: 301–23.
Velkley, Richard. 1989. *Freedom and the End of Reason: On the Moral Foundation of Kant's Critical Philosophy*. Chicago: University of Chicago Press.

Part III
The Systematicity of Philosophy

12 The Systematicity of Natural Science
Logical and Real

Eric Watkins

12.1 Introduction

It is widely acknowledged that Kant has a strict conception of natural science – most cognition will not satisfy the criteria required to qualify as natural science – and that part of what makes it so strict is that systematicity plays an ineliminable role in it. But how exactly does Kant understand systematicity in the context of natural science, and what exactly is it about his understanding of systematicity that makes his conception of science so strict?

In the context of eighteenth-century German philosophy, it was commonplace to require systematicity in science, with Leibniz, Wolff, Baumgarten, and Meier all explicitly endorsing it. However, given their rationalist orientation, it would be natural to interpret systematicity as a purely logical requirement, one that is satisfied if one can place the cognitions that are to constitute science in a series of syllogisms. Now if we took Kant's understanding of systematicity to involve a purely logical requirement, that would not help much in explaining why his conception of science is so strict, for he would simply be endorsing a demand that many others had stated long before: Put your scientific claims in the form of syllogisms, where the premises either need no support or are supported by higher, more general principles that themselves either need no support or are supported by even higher principles, until you reach self-evident most fundamental principles (such as definitions and axioms, like the principle of contradiction or the principle of sufficient reason), at which point you have reached the first principles of science.

In this chapter, I argue that understanding systematicity as a purely logical requirement that concerns only the logical relations between cognitions in syllogisms does not fully capture Kant's conception.[1] Instead, he takes the kind of systematicity that is required for natural science to include a grasp of the systematic unity of the objects of nature itself in addition to the kind of logical unity that is provided by syllogisms. In short, for Kant, systematicity in natural science must be both logical and real. For the logical use of reason, which concerns our mere forms of thought and the logical conditioning relations on which syllogisms are based, does not capture what reason's most fundamental interest is in natural science. In addition to its logical use, reason has a real use, one that concerns the real conditioning relations among objects in nature and that is ultimately oriented toward what is fundamental therein, both metaphysically and in the order of explanation. What's more, it is precisely this aspect of Kant's conception of systematicity that helps to explain the strictness of his conception of science, for the standards that must be met to attain cognition of the systematic unity of nature are so demanding that it turns out to be impossible for us ever to achieve it and we are left with the task of approximating it as best we can.

To argue for these claims, I begin, in Section 12.2, by introducing Kant's views on systematicity as they are presented in several central passages of the *Critique of Pure Reason* and the *Metaphysical Foundations of Natural Science*. In Section 12.3, I consider how one might articulate a logical conception of systematicity in terms of the logical use of reason, logical conditioning relations, and syllogisms before arguing, on both textual and philosophical grounds, that this conception does not capture Kant's views. I then show, in Section 12.4, how one can understand the "systematic unity attaching to the objects themselves" (A651/B679), which one might call "real systematicity," in terms of the real use of reason, real conditioning relations, and compete explanations that invoke unconditioned conditions.[2] In Section 12.5, I defend this view against the objection that establishing logical systematicity would automatically satisfy the demand for real systematicity, by arguing that it is not possible to derive real systematicity from logical systematicity, because real systematicity is an independent and substantive requirement on natural science.

12.2 Systematicity Introduced

To understand Kant's conception of systematicity and how it is to be understood in the context of natural science, three passages are of particular note, two from the *Critique of Pure Reason*, and a third from the *Metaphysical Foundations of Natural Science*. The first passage is located in the Appendix to the Transcendental Dialectic:

> If we survey the cognitions of our understanding in their entire range, then we find that what reason quite uniquely prescribes and seeks to bring about concerning them (*darüber*) is the **systematic** in cognition, i.e., its interconnection (*Zusammenhang*) based on one principle. This unity of reason always presupposes an idea, namely that

of the form of a whole of cognition, which precedes the determinate cognition of the parts and contains the conditions for determining *a priori* the place of each part and its relation to the others. Accordingly, this idea postulates complete unity of the understanding's cognition, through which this cognition comes to be not merely a contingent aggregate but a system interconnected in accordance with necessary laws.

(A645/B673)

The second passage, which expands on Kant's notion of a system, is found in the Architectonic of Pure Reason:

I understand by a system, however, the unity of the manifold cognitions under one idea. This is the rational concept of the form of a whole, insofar as through this the domain of the manifold as well as the positions of the parts with respect to each other is determined *a priori*. The scientific rational concept thus contains the end and the form of the whole that is congruent with it. The unity of the end, to which all parts are related and in the idea of which they are also related to each other, allows the absence of any part to be noticed in our awareness (*Kenntnis*) of the rest, and there can be no contingent addition or undetermined magnitude of perfection that does not have its boundaries determined *a priori*. The whole is therefore articulated (*articulatio*) and not heaped (*coacervatio*); it can, to be sure, grow internally (*per intus susceptionem*) but not externally (*per appositionem*), like an animal body, whose growth does not add a limb but rather makes each limb stronger and fitter for its end without any alteration of its proportion.

(A832/B860)

At first glance, in both of these passages, Kant seems to articulate what a system is as follows. A system is a whole consisting of parts, the parts are the understanding's cognitions, and the form is the principle of organization that relates those parts to each other such that they form a unified whole. Thus, a systematic whole is composed of those cognitions that are united in virtue of a certain form, that is, in virtue of standing in certain relations. Kant indicates further that the cognitions must be related "in accordance with necessary laws" and that reason is the faculty that prescribes this form to the parts and thereby seeks to bring about its unity in accordance with necessary laws. Moreover, the idea of the unity of the end precedes the parts and determines the place of each part within the whole and it does so a priori. In virtue of its form the whole is not a contingent aggregate or heap, but rather articulated, and any changes that it might undergo would have to be not external additions, but rather the result of internal growth, that is, growth that makes the whole stronger and better suited to its end.

Though these passages provide much useful information, they leave open a number of questions.[3] (1) What exactly is the form of the whole, that is, what specifically are the kinds of relations that relate the parts into a single whole? And what are the necessary laws that govern those relations that relate the parts into a whole? (2) What is the idea of the end of the whole? And how is the form related to that end? (3) How does the idea of the unity of the end determine its domain a priori and in such fashion that all the parts are related both to each other and to the whole in such a way that the absence of any one of them could be immediately noticed? (4) How is change of any kind possible when we are talking about a domain that has been determined a priori, and what could it mean to speak of this change as an internal growth? Even if these passages are suggestive, they do not explicitly answer these questions and thus leave Kant's position underdetermined in several important respects. Below, we investigate two different conceptions of systematicity by considering two different sets of answers to these questions.

Kant clarifies several aspects of his conception of systematicity as it applies to natural science in particular in the Preface to the *Metaphysical Foundations of Natural Science*.[4] After reiterating that science requires systematicity, understood "as a whole of cognition ordered according to principles" (IV:467), Kant divides the doctrine of nature (*Naturlehre*) into the historical and the rational doctrine of nature, depending on whether the connections that make the cognitions into a whole are empirical or rational. In the case of the *historical* doctrine of nature, where the connections are contingent, they consist either in similarities (which form the basis for the doctrine of natural description) or in times and places (which are the subject matter of the doctrine of natural history). Either way, the result is a systematically ordered set of facts about nature. In the case of the *rational* doctrine of nature, where the connections are necessary, Kant distinguishes those doctrines whose grounds are wholly a priori, which amounts to natural science properly so-called (for which he takes physics to be the paradigm case), from those that are based on empirical laws, which is natural science improperly so-called (for which he takes chemistry and psychology to be paradigm cases).

Though all of these doctrines include systematicity as a requirement, systematicity has significantly different meanings in each case. For the cognitions that constitute each doctrine are connected into a systematic whole by means of significantly different kinds of connections (contingent vs. necessary) and on the basis of different kinds of principles (empirical vs. a priori). Specifically, whereas the cognitions that constitute natural description, natural history, and the improperly called sciences are related either contingently or by empirical principles, those cognitions that constitute science properly so-called stand in necessary relations that are grounded in a priori principles, and what systematicity requires in each case is different. We return below to what it amounts to in the case of natural science, in particular.

Immediately after dividing up the doctrine of nature in this way, Kant claims that apodictic certainty is a criterion for science properly so-called (IV:468). Now it might seem as if taking apodictic certainty to be a criterion for science introduces a provocative new idea into the discussion, one that would also make his conception of science extraordinarily strict. For even the best supported laws of (empirical) physics could seem to fall far short of such a standard. However, by "apodictic certainty," Kant does not mean indubitable knowledge, as Descartes did, but rather a fallible consciousness of the necessities that ground what is systematic in our cognition.[5] Given the way that Kant introduces natural science, the idea here is not new; he is simply pointing out that cognition can be science proper only if it is grounded in principles whose necessity we can be conscious of a priori, that is, only if it is systematic in the robust way in which he has characterized natural science. The a priori principles of whose necessity we can be conscious are supposed to emerge as the argument of the *Metaphysical Foundations of Natural Science* unfolds.

12.3 The Logical Use of Reason, Logical Conditioning Relations, and Logical Systematicity

Given the views of his immediate predecessors and his own claims, it can be tempting to think that Kant's conception of systematicity is purely logical in character.[6] According to such a conception, the parts that constitute the whole would be cognitions, they would have to be related to each other logically in the form of syllogisms, and science would be systematic if and only if its content can be fully expressed in a series of interconnected syllogisms. This basic thought is tempting because syllogisms possess precisely those features that the passages from the first *Critique* discussed above seemed to indicate are required for systematicity. For the premises and conclusions of a syllogism are cognitions, a syllogism unites its premises and conclusion into a whole by relating them according to logically necessary rules of inference, and since syllogisms are iterative, one can connect syllogisms to each other and thereby create a whole of cognitions. To understand this thought more precisely, however, we need a more detailed grasp of Kant's conception of syllogisms as it is presented in general logic.

Kant's conception of syllogisms can be seen to rest on two doctrines in particular. The first, which he shares with Leibniz, Wolff, and Meier, is that the truth of the premises in a syllogism is based on the containment relations in which concepts stand to each other. Consider the following syllogism (the Animality Syllogism):

All animals are mortal.
All humans are animals.
Therefore, all humans are mortal.

In this syllogism, the major premise attributes mortality to animality, the minor premise attributes animality to humanity, and the conclusion then attributes mortality to humanity. The truth of the major and minor premises is determined here by the fact that, according to standard definitions, the concept <animality> stands in containment relations with the concepts <mortality> and <humanity>. The conclusion follows from the premises because one can subsume the subject term of the minor (humanity) under the predicate term of the major (mortality), on the grounds that anything attributed to one concept must also be attributed to all the concepts that it contains. More broadly, if (nearly) all of our concepts stand in containment relations with each other (e.g., by being interdefinable), it stands to reason that one can construct a series of interrelated syllogisms that are supported by these containment relations.[7] In the tradition of Leibniz, Wolff, Baumgarten, and Meier, syllogisms are thus based on containment relations between concepts, since the containment relations in which concepts stand to each other guarantee the truth of the syllogism's premises.[8]

The second doctrine of note, which is relevant to the logical form of syllogisms rather than to the truth of their premises, is Kant's distinctive conception of reason. Kant conceives of reason most generally as a spontaneous faculty of principles (A299/B356), one that seeks not only the *conditions* for what is *conditioned*, but also the *totality* of conditions and hence the *unconditioned*. He applies this conception of reason to logical inference by describing the logical form of syllogisms in terms of conditions. Accordingly, he views the conclusion of a syllogism as a cognition that is *conditioned*, the premises as those cognitions that are the *conditions* of the conclusion, syllogisms as expressing the different kinds of logical conditioning relations in which the premises stand to the conclusion, and finally, those premises that stand in need of no further justification as *unconditioned*.

Further, because Kant maintains that the relational forms of judgment express the basic kinds of conditioning relations and he holds that there are three different kinds of relational forms of judgment (categorical, hypothetical, and disjunction), he views syllogisms as belonging to one of three kinds. A categorical syllogism is based on a major premise whose

The Systematicity of Natural Science: Logical and Real 283

logical form is that of a categorical judgment (as was the case in the example given above), while a hypothetical syllogism is based on a major premise that is a hypothetical judgment, and a disjunctive syllogism is based on a disjunctive judgment. Specifically, hypothetical syllogisms contain a hypothetical judgment in the major and then a minor that either affirms the antecedent or denies the consequent of the major, giving rise to *modus ponens* or *modus tollens*. By contrast, disjunctive syllogisms have a major premise that is a disjunctive proposition and then a minor that either affirms one of the disjuncts or denies all but one of the disjuncts, such that the conclusion expresses either a denial of any of the remaining disjuncts or an affirmation of the sole remaining disjunct. In all three cases, something in the minor is subsumed under something in the major in such a way that the conclusion necessarily follows from these conditions. Their differences arise from the different kinds of conditions asserted in the major premise. Syllogisms thus express what one might call logical conditioning relations between cognitions according to which the truth of the conclusion both depends on and is explained by the truth of the premises.[9]

Recall, however, that on Kant's account, reason seeks *all* of the conditions for whatever is conditioned. Thus, in its logical use, once reason has discovered the premises, or conditions, from which a given cognition follows as a conclusion in a syllogism, it must then consider whether the conditions of that syllogism are themselves conditioned, that is, whether its premises need to be derived from elsewhere. If so, then reason must seek a further syllogism, a so-called prosyllogism, that can deliver the premise in question as a conclusion. The repeated application of the logical use of reason thus generates a series of syllogisms that reaches increasingly universal conditions, and that thereby seeks "to bring the greatest manifold of cognition of the understanding to the smallest number of principles (universal conditions), and thereby to effect the highest unity" (A305/B361). That is, it seeks a logically structured body of cognitions, with a small number of increasingly universal conditions (at the top) that serve to entail and explain a plurality of increasingly particular cognitions (at the bottom). In this way, reason in its logical use aims for comprehensiveness by seeking a single system of syllogisms that would contain *all* of the conditions from which a given conditioned cognition would logically follow. If, as we saw above, (almost) all of our concepts are interdefinable, a systematic whole of containment relations between concepts could provide the basis for a systematic whole of cognitions related in a series of syllogisms.

Recall further that reason's ultimate goal is to seek not only the totality of conditions, but also the unconditioned.[10] In the case of the logical use of reason, the unconditioned is simply the highest, or most universal cognitions that are not conditioned by, or derived from, any others. They are thus first logical principles, such as axioms and those definitions that neither depend on nor are logically derivable from any other cognition.[11] Reason is interested in such first principles, because once it has reached an unconditioned principle, there is no further condition, that is, no further dependence relation and no further explanation left for it to seek and thus its desires are satisfied and its end is fully realized; the unconditioned alone can offer reason a satisfactory "resting place" (A584/B612), from which it can view its task as completed.[12]

With this description of Kant's account of syllogisms in hand, we can see how one might answer the questions about systematicity that had been left open earlier. (1) The logical conditioning relations that unite cognitions in a syllogism constitute the form of the whole, and the laws of logic are the necessary laws that are the ground of such relations.

(2) The idea of the end of the whole is that of having a complete logical explanation, which is expressed in a series of syllogisms that are grounded in first fundamental principles (e.g., axioms) that entail and explain the truth of all the remaining cognitions that are the parts of the whole. That is, having a series of syllogisms that terminate in logically unconditioned first principles allows one to understand how a small set of highly general self-evident principles can entail and explain the truth of a large number of specific cognitions. The logical conditioning relations that are the form of the whole contribute to this end because they make it possible for the first fundamental principles to support all the others in this way.

(3) The fundamental principles and the logical conditioning relations that support the series of syllogisms jointly determine the domain of the parts that constitute the whole a priori because these principles and relations determine a priori whether any given cognition is or is not entailed by them and thus whether or not they are parts of the whole. What's more, this account allows one to see why any missing element should be obvious, for one can determine a priori which propositions do and which do not follow logically from taking a cognition or set of cognitions as first principles.

(4) We can even see how to make sense of Kant's metaphor of "internal growth." For if we can find further syllogisms that relate more of our cognitions to each other, a more powerful series of syllogisms emerges. For example, if one thinks about the containment relations that obtain between the concepts <animality>, <mammality>, and <humanity>, one could "grow" the Animality Syllogism given earlier into the following syllogism (the Extended Animality Syllogism):

P1 All animals are mortal.
P2 All mammals are animals.
C1 Therefore, all mammals are mortal.
P3 All humans are mammals.
C2 Therefore, all humans are mortal.

Though the Animality Syllogism suffered from no defect, it is clear that the "growth" that is generated by the Expanded Animality Syllogism is an improvement. By drawing on Kant's conception of syllogisms and logical conditioning relations in these ways, one can see how a purely logical conception of systematicity can appear attractive, both textually and philosophically.

In the context of natural science, however, this conception of systematicity cannot be entirely correct. One indication of this is evident if we look carefully at the first passage quoted above. According to the logical conception of systematicity, the parts that constitute the whole must be cognitions, since only cognitions (as a kind of judgment) can stand logical relations in syllogisms. However, Kant's reference to "cognition *of* the parts" (emphasis added) suggests not that the cognitions *are* the parts that constitute the whole, but rather that the parts are the *objects* of the cognitions. If we take into consideration a passage from the Appendix to the Transcendental Dialectic, where Kant describes the systematic unity of cognition, we see further that he takes systematic unity to be "not merely something subjectively and logically necessary, as method, but objectively necessary" (A648/B676), a description that does not fit well with the logical conception. He goes on to illustrate this claim with the example of causal powers, where what is unified is not the cognitions of the effects of causal powers in syllogisms, but rather the causal powers themselves. These remarks all suggest that the parts of a systematic whole are not (only) cognitions, but rather (also) the objects found in nature (broadly construed so as to include substances, causal powers, accidents, states, representations, etc.).

Kant concludes his discussion there by remarking: "In fact, it cannot even be seen how there could be a logical principle of rational unity among rules unless a transcendental principle is presupposed through which such a systematic unity attaching to the objects themselves is assumed a priori as necessary" (A650-651/B678-679). Here, Kant is claiming that the logical principle of systematic unity is incoherent on its own and that it can make sense only in conjunction with a "transcendental" principle of systematic unity, one that pertains to "the objects themselves." While Kant's claim that such a logical principle requires a transcendental principle is controversial and would need considerable elaboration and defense, it is an additional textual indicator that Kant thinks that systematicity is not merely logical, but must also involve objects.[13]

What's more, the problem is not merely textual. For, philosophically, it is clear that logical systematicity reflects only one small part of our interest in natural science. For the fundamental intent in science is to determine the real relations that obtain between objects in nature, rather than to exploit containment relations between our concepts for the sake of identifying the logical relations that obtain between our cognitions. And it attempts to arrive at cognition of the way that nature actually is, whereas logic is interested in the different logical systems that pertain to our concepts. The fact that many logical frameworks could in principle be adequate for our scientific descriptions of nature reveals that science must go beyond logical systematicity. The passages cited in the previous paragraphs give a clear indication that on Kant's view natural science requires a systematic unity in the objects of nature themselves.

12.4 The Real Use of Reason, Real Conditioning Relations, and Real Systematicity

But how is the real systematicity that underlies the unity of nature to be understood? To see how to answer this question, it is helpful to return to Kant's account of reason and its interest in conditioning relations. Above, we noted that in its logical use reason focuses on the logical conditioning relations that obtain between cognitions in a syllogism. But Kant thinks that there is also a real use of reason, one that focuses on what one might call the real conditioning relations between objects (continuing to take "object" here in a broad sense).[14] Though Kant thinks of real conditioning relations as structured according to the relational categories, just as logical conditioning relations depend on the relational forms of judgment, it can be useful to consider some specific instances to get a proper sense of what a real conditioning relation is. For example, Kant thinks that a real whole is conditioned by its parts; an event is conditioned by what causes it; one moment in time is conditioned by the moment immediately preceding it; one region in space is conditioned by the regions in space that are next to it; a thought is conditioned by the thinking subject that thinks it; the reality (or material content) of possibility is conditioned by something actual.

Despite significant differences, these examples are all instantiations of a univocal notion of a real conditioning relation, which is, at its core, an explanatory asymmetrical metaphysical dependence relation. That is, what is conditioned depends on and is explained by its conditions, but not vice versa. Accordingly, a real whole depends on and is explained by its parts, but not vice versa; an event depends on and is explained by its causes, not vice versa; a thought depends on and is explained by the thinking subject, or I, that thinks it, not the other way around; and so on. When reason asks why the world is the way it is, real conditioning relations constitute the materials for appropriate answers because of their explanatory character.

Further, just as is the case with its logical use, so too in its real use reason seeks both the totality of conditions and the unconditioned. Thus, when reason seeks to understand why an event has happened, it considers not only what its cause is, but also whether that cause is itself caused.

The Systematicity of Natural Science: Logical and Real 287

For if that cause is caused, appealing only to the event's most proximate cause provides only a partial explanation of the event. In its search for a complete explanation of what has happened, reason pursues the series of causal conditions back until it has located all of them, whether that leads, via an infinite regress, to infinitely many causes or to a terminal member of the series, an uncaused cause. A parallel story can be told for wholes and their compositional conditions. Reason first seeks the parts into which a given whole can be divided. But if those parts themselves have parts, reason would have provided only a partial explanation of the whole in appealing only to the first set of parts. As a result, in seeking a complete explanation, reason seeks the parts of those parts, and the parts of those parts, etc., until it has located all of the parts. If the regress has no terminal member, then reason seeks infinitely many ever-smaller parts (infinitely many compositional conditions). If the regress has terminal members, then it seeks simples (uncomposed compositional conditions). In both cases reason is seeking all relevant conditions so that it can offer a complete explanation of what exists, and it does so by seeking out all the conditions on which what exists depends. But note that when reason has identified *all* the conditions of something conditioned, there are no more conditions, and thus no further explanation is possible. In short, when reason has reached the totality of conditions, it has thereby also reached the unconditioned and its explanatory task is complete. On the basis of this understanding of Kant's conception of real conditions and the unconditioned, we can understand why Kant characterizes the objects of traditional metaphysics – God, freedom, and the soul – as unconditioned conditions. They are precisely the kinds of objects that could serve as an ultimate or complete explanation; the explanatory buck stops with them.

Now, in the first *Critique* Kant is primarily interested in whether metaphysics is possible, that is, in whether we can have cognition of the unconditioned objects of traditional metaphysics. What is important for current purposes, however, is that the real use of reason, along with its notion of a real conditioning relation, makes it possible for us to specify how real systematicity is to be understood by answering the questions about systematicity that were raised but left open by the passages in the first *Critique* discussed above. (1) The parts that constitute the whole are the real objects found in nature, the real conditioning relations in which they stand to each other constitute the form of the whole, and the laws of nature are the necessary laws that are the ground of the real conditioning relations.

(2) The idea of the end of the whole is that of a complete scientific-metaphysical explanation, which is expressed in cognitions of all the real conditioning relations that obtain between objects in nature and are grounded in the unconditioned conditions of those conditioned objects that we encounter in the world.[15] That is, having cognition of the

unconditioned conditions of the different conditioned objects that one encounters in the world would allow one to understand how these objects depend on and are explained by what is metaphysically and explanatorily fundamental. The real conditioning relations that are the form of the whole contribute to this end because they make it possible for the entities that are in fact fundamental to be part of a complete explanation of all those entities that are derivative.

(3) The ideas of unconditioned objects and the real conditioning relations in which they stand in to the objects that they condition assign each conditioned object its place in the sense that they constitute an explanatory framework which can then be filled in over time with further cognitions of the various conditioned objects that the unconditioned conditions ground.

(4) We can even see how to make sense of how the changes that could take place would still be internal to the whole. For prior to cognizing every member of the series of real conditioning relations, reason has an idea of what the conditions of a conditioned object would have to be like such that when experience reveals what these conditions are, the "reveal" is not a change from without (a *Deus ex machina*), but rather something that already lies implicit within the explanatory framework that was initiated by the experience of the conditioned object and projected by reason's desire for cognition of the unconditioned conditions of the object. Though these remarks amount to no more than a cursory sketch that would need to be spelled out in detail, they express a basic outline of Kant's understanding of the systematic unity of nature.[16]

This conception of real systematicity is consistent with the passages cited above, but it also puts us in a position to understand why Kant would be motivated to include it as a condition on natural science. Natural science seeks knowledge of nature and of why it is the way it is. By discovering the real conditioning relations that obtain among objects in nature, natural science is able to *explain why* nature is the way that it is. Indeed, it is a striking feature of our scientific ambitions, as Kant understands them, that science will not happily settle for anything less than a *complete* explanation of everything we find in nature. For this reason, it will stop at nothing short of the *unconditioned*. However, since, according to Kant, we cannot have cognition of either the totality of conditions or the unconditioned, the best that reason can do is to cognize as many real conditions as it can, thereby coming ever closer to cognition of the totality of conditions, even if it can never fully realize this end. Reason's ideas thus function as regulative principles for science, by projecting the order of what a complete system of nature would be like. By understanding the systematicity requirement in terms of real conditioning relations in this way, one can see how and why real systematicity is a perfectly reasonable aspirational requirement for natural science.

12.5 Logical and Real Conditioning Relations

If we distinguish between logical and real systematicity in this way and accept that natural science requires both, it would be natural to wonder how real systematicity could be embodied in science.[17] At this point, one might be tempted to think that if Kant is right to require that the results of science be put in syllogistic (or axiomatic) form, as is required by logical systematicity, then perhaps satisfying the requirements of logical systematicity might also satisfy the requirements of real systematicity. For even if the two kinds of requirements are distinct, they could still be parallel in important respects, and if we were able to state the results of science in a series of interrelated syllogisms that terminated in axioms, how could that not be enough to establish real systematicity, since one's conclusions are in that case fully justified by self-evident highest principles? And if that were not sufficient, what more could one do? All would be well, so the thought goes, if in achieving logical systematicity you get real systematicity for free.

It turns out that significant obstacles stand in the way of showing that satisfying the requirements of logical systematicity automatically entails satisfaction of the requirements of real systematicity, and identifying these obstacles will help us to appreciate more fully one important aspect of Kant's understanding of real systematicity. Recall, from our description of syllogisms above, that the logical conditioning relations that pertain to syllogisms are such that the major premise must be more general than the conclusion in the sense that the concept <animality>, which appears in the premise "All animals are mortal," is more general (i.e., has greater scope) than the concept <humanity>, which appears in the conclusion, "All humans are mortal." And if reason demands that one seek a further syllogism that delivers the major of this syllogism as its conclusion, it is clear that its major will have to appeal to an even more general concept, such as <embodied living being>. Accordingly, the further reason pursues the conditions of a given cognition in a series of syllogisms, the more general are the cognitions that stand in the major premises of these syllogisms. However, as we have also seen, when reason considers the real conditioning relations that obtain between objects, these objects will always be fully particular. For the conditioned object that we cognize will always be particular, but so will its conditions, and the conditions of its conditions as well. For example, the parts of a whole are particulars just as the whole they compose is, and the cause of an event is just as particular as the event it causes. There is thus a fundamental mismatch between the increasing generality of what is related by the logical conditioning relations in syllogisms and the thoroughgoing particularity of what is related by real conditioning relations in the world. Given this mismatch, it is difficult to see how there could be an automatic mapping relation between logical and real systematicity.

Similarly, on the basis of logical conditioning relations alone, there is no way of either anticipating or inferring what the content of the real conditioning relations must be. For example, the logical forms of judgment neither specify nor require that there must be compositional or causal conditioning relations among objects in the world. It is a contingent feature of our world that it contains compositional or causal relations, given that we can imagine other, very sparsely populated possible worlds that consist of, say, only one simple immaterial being (e.g., God), in which case there would be neither composition nor causal relations. What's more, even if one could establish that causal conditioning relations were an essential feature of any world that we could experience, it is clear that the specific causal relations that obtain in such a world cannot be derived from the logical relation expressed in the hypothetical form of judgment alone. For the fact that our world has objects that not only move, but also are accelerated, is a contingent fact, and the fact that only accelerations and not motions are causally conditioned entities cannot be derived from the existence of logical conditioning relations among cognitions. These facts make it clear that the differences between the content of the logical and real conditioning relations are such that it is impossible to show that satisfying the requirements of logical systematicity would also automatically satisfy the requirements of real systematicity.

If logical and real conditioning relations are thus fundamentally different (despite similarities in some of their formal properties) and there is therefore no direct or straightforward way to infer the presence of real systematicity in nature from those relations that establish logical systematicity, it is natural to ask how we can establish real systematicity at all. If the real conditioning relations that are required for real systematicity are not expressed by logical conditioning relations in syllogisms, how can they be expressed in our cognitions at all?

To see our way forward here, consider the following syllogism (the Featherless Biped Syllogism):

P1 All featherless bipeds are mortal.
P2 All humans are featherless bipeds.
C1 All humans are mortal.

The Featherless Biped syllogism is formally valid and its premises are true. Further, the concept <featherless biped> is more general than <humanity>, so the requirement that the major premise in a syllogism is more general than its conclusion is satisfied here as well. However, in scientific contexts it is clear that we (strongly) prefer the Animality Syllogism over the Featherless Biped Syllogism, even though purely logical grounds cannot justify favoring one over the other. What justifies this preference?

Intuitively, we think that any genuine explanation of *why* human beings are mortal must appeal to our animality (or perhaps to our being embodied living beings), since our animal nature is an essential, or at least more explanatorily basic feature of us, rather than to our lack of feathers, which seems to be an accidental feature of human beings that does not capture the causal conditioning relations in virtue of which human beings die. In short, to justify our preference for the Animality Syllogism over the Featherless Bipedal Sylloogism, we have to appeal to the real conditioning relations of objects (the fact that our animality, and not our featherless bipedality, causes our mortality), but this shows that under certain circumstances, syllogisms *can* express real conditioning relations and when they do, they can be genuinely explanatory of objects in nature.

But if a syllogism can express a real conditioning relation, how is it able to do so, given the fundamental differences noted above between logical and real conditioning relations? Now Aristotle argues that syllogisms can be explanatory but only if their premises are "prior in nature" to their conclusions (and not only "priori in knowledge"). Kant's position can be understood as developing a broadly Aristotelian answer here by focusing on real conditioning relations, and their asymmetrical metaphysical dependence, in particular. To do so, it will be instructive to consider briefly the premises of a syllogism that is genuinely explanatory. As we saw above, for Kant's rationalist predecessors, the premises in a syllogism are supposed to be analytic truths, justified by the containment relations that obtain between the concepts employed in the major and minor premises. But for Kant, outside of general logic, not all principles are analytic. Indeed, he thinks that the principles of metaphysics, in particular, must be synthetic a priori cognitions, that is, cognitions that are precisely not justified on the basis of an analysis of the containment relations that obtain between the relevant concepts. Instead, in synthetic a priori cognitions, the concept of the predicate is "added" to that of the subject concept (though not because of experience). The suggestion here is that this "adding" is based on a grasp of the relevant real conditioning relations among the objects rather than a grasp of any conceptual containment relations. Thus, in addition to being formally valid, to be genuinely explanatory a syllogism must be based on premises whose truth is supported not solely by the analysis of concepts, but rather by a grasp of the real conditioning relations that obtain among objects in the world. And it is by adding this further requirement regarding real conditioning relations to syllogisms that they can express real systematicity and be explanatorily significant in science.

Given this, much will turn on how our concepts can grasp the real conditioning relations that obtain in the world. As readers of the first *Critique* are well aware, this is a task that one cannot avoid, but must rather face head-on, lest our most fundamental concepts end up being empty figments of the brain. While the Transcendental Deduction of the Pure Concepts of the Understanding takes up this challenge for the categories, the pure principles that are established in the *Metaphysical Foundations of Natural Science* must accomplish similar work for several highly abstract concepts of matter, thereby contributing to a (special) metaphysics of nature.

In light of the fundamental differences between logical and real conditioning relations, it is now clear that constructing a series of formally valid syllogisms that satisfy the requirements of logical systematicity does not necessarily or automatically satisfy the requirements for real systematicity. Constructing an interrelated series of syllogisms *can* nonetheless bring about a grasp of real systematicity. For that to happen, however, our judgments must grasp the real conditioning relations that obtain in the world, since these dependence relations carry with them the kind of explanatory dimension that reflects science's most fundamental interest.[18]

12.6 Conclusion

The tremendous achievements made in formal logic over the past 150 years have shown that formal reasoning can be an incredibly powerful tool, one that should not be overlooked when we turn to natural science and its attempt to give expression to our understanding of the world. These advances might be taken to suggest that one should turn to logical constraints to express what science can justify. Given this, it can appear tempting to hold that when Kant considers what requirements reason can demand in natural science, his remarks about systematicity should be taken to mean that he is requiring that the demands of logical systematicity in particular must be satisfied, where these demands can be explained in terms of logical conditioning relations among cognitions. Above, I have argued that the nature of the systematicity that Kant requires for natural science cannot be exclusively logical, for he maintains that science also requires real systematicity, which can be explained in terms of real conditioning relations. This interpretation not only makes sense of those passages in which Kant refers to a kind of systematicity that cannot be logical, but is also motivated philosophically insofar as real conditioning relations can be used to explain in what ways objects in nature depend on each other, which is a core goal of natural science. An added benefit of this interpretation is that it helps to explain the strictness of Kant's conception of natural science. While requiring logical systematicity for natural science may not be an entirely modest requirement, requiring real systematicity in addition raises the stakes significantly. For it is extremely burdensome to determine not only what dependence relations obtain among objects, but also, in line with reason's demands, the *totality* of those relations, such that we have a *complete* scientific explanation. While one might lament the stringency of Kant's conception of natural science – it makes science unrealizable – one might instead applaud its ability to explain what it is that makes science so difficult, namely not so much discovering the appropriate logical form, but rather identifying a complete and thus truly satisfying explanation of the world in its entirety.[19]

Notes

1 I do not attempt to settle the question of whether Leibniz, Wolff, Baumgarten, and Meier all accepted a purely logical conception of systematicity. On the one hand, if all truths are analytic and are thus based on containment relations between concepts, which are purely logical, then it can seem that all truths are based on purely logical considerations. On the other hand, Leibniz, at least, seems to acknowledge the Aristotelian idea that some truths are prior "in nature" and not logically prior or prior in our knowledge. For discussion of Leibniz's complex views, see Di Bella (2005, 67–93).
2 All translations are my own, though I have frequently consulted the Cambridge Edition of Kant's Works.
3 For discussion of a range of further questions, see Sturm (2009, 2020, 1–28) as well as Gava (2014, 372–93, 2023).
4 For a helpful discussion of the Preface to the *Metaphysical Foundations of Natural Science*, see Pollok (2001). Pollok suggests (2001, 56, n. 79) that Kant's conception of systematicity that pertains to science properly so-called is not based on reason in the narrow sense, but is rather due to the table of categories. While I agree that the table of categories is a crucial structuring device for the *Metaphysical Foundations*, I argue below that systematicity is also based on reason in the narrow sense.
5 See, for example, Kant's discussion of apodictic certainty at *Logik Wien* (XXIV:830).
6 For interesting discussions of systematicity, see especially (van den Berg 2014, 17–34; Willaschek 2018, 53–56). Hein van den Berg provides an account of systematicity that starts out purely logical (by focusing exclusively on concepts and their containment relations), but then goes on to argue for the necessity of fundamental propositions that ground all other propositions that are known, where these grounding relations involve the real use of reason and thus are plausibly taken to require what is akin to real systematicity. Willaschek distinguishes between the logical use of reason *in abstracto* (which would be the concern in formal logic, which abstracts from all relation to an object) and its use *in concreto* (which includes an interest in the objects of cognition and the epistemological concerns that would support the cognition). It is unclear whether Willaschek thinks that systematicity is only logical, or whether the logical use of reason *in concreto* requires additional constraints that would amount to real systematicity.
7 On Kant's view, any "highest concept" (*conceptus summus* (IX:97), e.g., that of "something" (XXIV:911), "thing" (XXIV:755), or "possible thing" (XXIV:259)) will not contain any other concept in itself.
8 We return to this issue below.
9 Note that on this view, syllogisms are not simply truth-preserving inferences for Kant, but are also supposed to explain *why* the conclusion is true.

10 Though Kant argues that the existence of the totality of conditions entails the existence of something unconditioned, there is at least a distinction of reason between the two concepts. Indeed, Kant expresses some ambivalence about whether reason's ultimate goal is thoroughness (in finding the totality of conditions so as not to leave unexplained anything that can be explained) or finality (in finding what is unconditioned, such that reason can rest, since it has reached the limits of intelligibility). Nor is this matter settled by the fact that Kant sometimes emphasizes that reason's distinctive accomplishment is comprehension rather than understanding (A311/B367).
11 Kant refers to "all rational and unprovable cognitions" as axiomata in the *Jäsche Logik* (IX: 231-232).
12 See also, e.g., the Canon of Pure Reason, where reason is said to be driven by a propensity of its own nature "to find peace for the first time only in the completion of its circle, in a self-subsisting systematic whole" (A797/B825).
13 For insightful discussion of Kant's conception of real systematicity and certain puzzles that arise for it, see Guyer (2005, 57-68).
14 For a fuller account of Kant's understanding of real conditioning relations, see Watkins (2019b, 1133-40, 2019c).
15 It is not the case that only the cognitions (and not their objects) might be systematic, because the real conditioning relations tracked by the cognitions must be systematic, given the nature of real conditioning relations.
16 For discussion of how real conditioning relations are central to teleological judgment, see Watkins (2019a, 174-88).
17 Béatrice Longuenesse also notes that Kant is interested in determining how logical relations "reflect" the real relations among objects. See Longuenesse (1998, 151).
18 To be clear, the fact (if it is one) that at least one of the premises in a syllogism would need to be a synthetic a priori principle for the conclusion of that syllogism to be a synthetic a priori proposition, does not immediately explain how such a syllogism can reflect a grasp of real conditioning relations. Some connection between the a priori principles and the real conditioning relations would need to be established.
19 For comments on earlier versions of this paper, I thank Gabriele Gava, Clinton Tolley, Hein van den Berg, Marcus Willaschek, and audience members at the conference held at the Goethe University in Frankfurt in the summer of 2019.

Bibliography

Di Bella, Stefano. 2005. "Leibniz's Theory of Conditions: A Framework for Ontological Dependence." *The Leibniz Review* 15: 67-93.
Gava, Gabriele. 2014. "Kant's Definition of Science in the Architectonic of Pure Reason and the Essential Ends of Reason." *Kant-Studien* 105: 372-93.
Gava, Gabriele. 2023. *Kant's Critique of Pure Reason and the Method of Metaphysics*. New York: Cambridge University Press.
Guyer, Paul. 2005. *Kant's System of Nature and Freedom*. New York: Oxford University Press.
Longuenesse, Béatrice. 1998. *Kant and the Capacity to Judge*. Princeton: Princeton University Press.
Pollok, Konstantin. 2001. *Kants Metaphyische Anfangsgründe der Naturwissenschaft: Ein kritischer Kommentar*. Berlin: Felix Meiner Verlag.

Sturm, Thomas. 2009. *Kant und die Wissenschaften vom Menschen*. Paderborn: Mentis Verlag.
Sturm, Thomas. 2020. "Kant on the Ends of the Sciences." *Kant-Studien* 111: 1–28.
Van den Berg, Hein. 2014. *Kant on Proper Science*. New York: Springer Verlag.
Watkins, Eric. 2019a. *Kant on Laws*. New York: Cambridge University Press.
Watkins, Eric. 2019b. "Kant on Real Conditions." In *Proceedings of the 12. International Kant Congress Nature and Freedom*, edited by Violette L. Waibel and Margit Ruffing, 1133–40. Berlin: De Gruyter.
Watkins, Eric. 2019c. "What Real Progress Has Metaphysics Made Since the Time of Kant? Kant and the Metaphysics of Grounding." *Synthese*. https://doi.org/10.1007/s11229-019-02180-2
Willaschek, Marcus. 2018. *Kant on the Sources of Metaphysics: The Dialectic of Pure Reason*. New York: Cambridge University Press.

13 What Is a System of Moral Philosophy for?
Systematicity in Kant's Ethics[1]

Stefano Bacin

13.1 The Necessity of a System of Morals

The idea of a system takes centre stage in the final moment of the development of Kant's practical philosophy from the very beginning. The Preface of Kant's *Metaphysics of Morals* states at the outset that "[t]he critique of practical reason was to be followed by a system, the metaphysics of morals" [*Auf die KpV sollte das System, die MS, folgen*] (VI:205). Although it has hardly attracted specific attention in the scholarship, it is an extremely important statement.[2] At a first glance, the initial sentence of the *Metaphysics of Morals* follows up to the systematic project outlined in the Architectonic of Pure Reason. To a closer look, though, that statement raises many questions concerning Kant's project in moral philosophy that would require closer scrutiny. Remarkably, the *Groundwork* is not mentioned at all, although that work was presented as the first installment of the project of a metaphysics of morals. The *Groundwork* could play a role parallel to the *Critique of Pure Reason* as a "propaedeutic to the system" (B25) and had in fact anticipated the further development of the system with a few hints (see IV:391, IV:421n). Conversely, the second *Critique* only vaguely mentioned a future "system of the science" succeeding the "system of the critique" (V:8) and never referred to a metaphysics of morals. Yet, Kant connects the new work to the *Critique of Practical Reason*.[3] In terms reminiscent of the Architectonic of Pure Reason, that initial statement includes a significant innovation, since the systematic project in the Architectonic did not include a second *Critique*, which was not foreseen at that stage.[4]

The initial sentence of the Preface to the *Metaphysics of Morals* thus rapidly sounds less uncontroversial than it might appear. Conspicuously, Kant takes the same angle in the Preface to the "Doctrine of Virtue," which opens quite similarly:

> If there is a philosophy (a system of rational cognition from concepts) of any subject, then for that philosophy also there *must* be a

DOI: 10.4324/9781003166450-17

system of pure rational concepts independent of any conditions of intuition, that is, a metaphysic.

(VI:375; my emphasis)[5]

This reprise of the general Preface gives the second of the two issues that I have pointed out further prominence.[6] The *Critique of Practical Reason* is here not mentioned anymore, but the emphasis is on the systematic nature of the entire enterprise. Whereas at the beginning of the first volume of the *Metaphysics of Morals* Kant stresses the connection between the preliminary investigation unfolded in a critique and the proper system (according to the outline prominent in the first *Critique*), in the "Doctrine of Virtue" the main focus is on the appropriate completion of a project that is systematic throughout. In spite of the differences, the close parallelism between the two statements emphasizes the significance of the angle that Kant takes in both cases. His first and arguably most important instruction to the readers of the new work is that the main task of the *Metaphysics of Morals*, in both its parts, is to provide a *system*, which is thus presented as the most appropriate way to achieve the purpose of practical philosophy.

These opening statements are so prominent and emphatic that they should be sufficient to show how central the thought of systematicity is not only in Kant's theoretical philosophy, but in his practical philosophy as well. As Paul Guyer has rightly stressed, "[t]he idea of systematicity is clearly central to Kant's moral philosophy."[7] Yet, the issue is hardly discussed with regard to that part of his general project.[8] Among other factors, this neglect might have to do with the fact that Kant's view of systematicity yields in practical philosophy a perspective that is significantly different from the angle from which the possibility of a system of morals is considered in current debates in ethics. That arguably makes these features more difficult to construe as philosophically relevant. A big part of the work that has been done on Kant's moral philosophy in the last decades has centred on its philosophical resources in dialogue with current perspectives. Now it might be suspected that Kant's emphasis on the systematic nature of his project cannot be reclaimed to a fruitful debate, since it would only be historically relevant, even the by-product of a time, between the eighteenth and the early nineteenth century, in which systematicity was held to be an essential requirement for any theoretical enterprise. However, this worry should be overcome.[9] Systematicity proves to be an integral component of Kant's overall philosophical project, as recent scholarship has come to appreciate again, primarily with regard to his theoretical philosophy. As I shall show, quite the same holds true for his practical philosophy.

In the following I shall examine the strong necessity of a system that is expressed by the modal verbs in the opening of both parts of the *Metaphysics of Morals*. Why does Kant maintain that a system of practical philosophy *has to* follow? What is the task of a genuinely *systematic* treatment in practical philosophy? More specifically, I shall focus on the role of systematicity in the ethics of his "Doctrine of Virtue," where Kant's distinctive view can helpfully be contrasted with the predominant understanding of systematicity in current debates in moral philosophy. An underlying claim of this chapter is that Kant's development of a system of ethics must be interpreted in light of his own conception of a system.[10] First, I shall consider how Kant's conception of systematicity in this domain is construed as to the relation of the elements of the system with each other and the grounding of the system. Second, I shall clarify how Kant understands the outline and the scope of the system of ethics. Third, I shall explain how far that systematicity can have practical significance. This will prove to be of special importance in clarifying the task of Kant's system of moral philosophy.

13.2 Systematicity, Not Systematization: The Priority of a Principle

If in current moral philosophy there is talk of systematicity at all, what is mostly discussed is in fact a coherentist conception, according to which the coherence of certain moral beliefs with (most or all) others would provide justification for them. As the most widespread variant of this general approach has it, the method of reflective equilibrium, the main task is to systematize intuitions or basic moral beliefs into a tenable set.[11] The coherence of that set is meant to have justificatory force with regard to the items of the set. Seen in this light, a system is "a network of credibility transfers that can raise the level of the whole set of beliefs."[12] This is what is questioned by critics of that systematizing ambition in moral philosophy, which argue that mere coherence cannot be the sole source of justification, because it pushes a theoretical construction onto the substance of ethical life.[13] If endorsed or rejected, systematization of moral beliefs is considered as a possible source of validation of moral principles.

Kant operates with a decidedly different conception of systematicity in moral philosophy. As the opening statement of the general Preface to the *Metaphysics of Morals* shows, the system at issue is not self-standing, but follows a previous, crucial step of investigation, which Kant there calls a critique of practical reason and has the primary task of establishing a general principle of morality. The system of morals is not built by directly operating on ordinary, available beliefs, but only once a principle has been validated, which in turn shall provide the proper grounding for a system.[14] There is room for a system only once the investigation has led to an authoritative principle. In the Preface to the *Metaphysics of Morals*, only two pages after the initial statement of the work, Kant accordingly characterizes the doctrine of virtue as a "system that connects all duties of virtue by one principle" (VI:207), thereby spelling out the fundamental connection of the genuine system with a fundamental principle as its first defining feature.

More distinctively, this way to understand the systematic approach to what Kant calls the doctrinal part of practical philosophy also conforms to his own conception of systematicity. Kant thus applies to practical philosophy the general idea that a system as a body of scientific knowledge necessarily follows from a principle, namely that "the systematic of cognition [*das Systematische der Erkenntniß*]" lies in "its interconnection based on one principle [*Zusammenhang derselben aus einem Princip*]" (A645/B673; cf. IV:467, V:151). This underlying thought is at odds with the current notion that a system is put together by securing the consistency of its elements, possibly even without any general principle, or in order to establish principles in the first place.

To this extent, however, Kant does not merely elaborate on his own account of systematicity, but he also follows a widespread paradigm in early modern moral philosophy. That the fundamental principle of natural law yields a system of obligations is a prevalent assumption in eighteenth-century natural law.[15] Although often overlooked, this represents a significant part in the history of the conceptions of a system in early modern philosophy, which unfolds parallel to the usages in metaphysics and natural philosophy. The general project of providing such a system as a guide to the conduct of rational subjects was first articulated by Pufendorf. Centring his revision of natural law on the connection between the fundamental principle and the resulting system, Pufendorf had observed:

> When I decided to bring natural law to the rightful form of a discipline, whose parts should be mutually consistent and derive from one another in an evident way, *my first concern was to establish a solid foundation, that is a fundamental proposition*, which should comprehend and summarize in itself all its precepts, *from which all further rules could be derived with an easy and evident demonstration*, and in which they all could then be easily resolved.
> (Pufendorf 2002, 142; my emphasis)

If the word "system" is not used here, the thought is in fact at the centre of Pufendorf's conception. The comprehensive set of rules that are to be derived from the principle is precisely what would have soon to be called a system. Also, bringing about a system is the way to develop natural law as a proper science. Accordingly, following Pufendorf's paradigm, most expositions of natural law since the early eighteenth century aimed at providing a complete collection of norms that instantiate the general principle with regard to the different matters and circumstances and can thereby offer the needed guidelines to the human beings.

What we can call the Pufendorfian paradigm is thus significantly different from a coherentist strategy. While the latter sets off from a set of beliefs, the traditional approach of eighteenth-century natural law develops a system of obligations from a fundamental proposition that is not itself under scrutiny, since it has previously been established as valid and true, that is, as a principle. The justification, thus, is not attained by virtue of the coherence among elements of the system, but by virtue of the relation of derivation from the principle. The principle can be said to "comprehend and summarize in itself all its precepts" precisely because it provides the foundations for their normative significance.

The convergence with the Pufendorfian paradigm represents an important aspect of Kant's connection with early modern natural law theory that goes beyond the aspects that have been appreciated so far.[16] Like eighteenth-century natural law, also Kant's general strategy in practical philosophy proceeds in two main steps, whose connection is of the utmost importance: (1) Establishing a principle, drawing on which alone it is possible (2) to develop a system of obligations. Their connection validates both steps: the principle makes a system of duties possible, and conversely developing such a system confirms the validity of the principle.

When Kant maintains that the critique of practical reason was to be followed by a system of practical philosophy (see VI:205, VI:375), thus, he does not envisage a system of obligations constructed by a series of logical relations of consistency among its elements, but a system of duties deriving from a fundamental moral principle. Systematicity here is not about systematization, unlike what is mostly assumed in the recent debates. The opening passages of both volumes of the *Metaphysics of Morals* state that a system of moral philosophy is needed because it accounts for the entire scope of moral demands on the basis of the principle of morality, from which it follows. In this general strategy, there is no direct continuity between ordinary moral thinking and moral science. The continuity is only indirectly secured by the principle itself, from which the system can unfold. Unlike in current views, systematicity in Kant's practical philosophy does not have to do with making elements of ordinary moral thinking consistent with one another.

13.3 A System of Ethical Demands via a System of Ends

If Kant follows the Pufendorfian paradigm of a doctrine of duties in the crucial aim to cover the entire scope of morality, his way to unfold a proper system differs significantly. On the traditional model, the system is based on a principle that is general enough to apply to as many cases as possible.

Accordingly, the comprehensiveness that the system aims at is reached by classifying the demands that can be derived from the principle. In this perspective, "system" denotes the largest possible collection of items (cases and duties) that can be accounted for. Notably, this approach is the target of further criticisms by the opponents of systematization efforts in ethics, who lament precisely that "[t]heory looks characteristically for considerations that are very general and have as little distinctive content as possible, because it is trying to systematize and because it wants to represent as many reasons as possible as applications of other reasons."[17] In this respect, Kant parts company with the traditional way to develop that Pufendorfian paradigm.

At a first glance, Kant would seem to follow that traditional paradigm also in structuring its system through a taxonomy of duties. When he comes to discuss the outline of the "Doctrine of Virtue," however, he presents the reader not with one, but with two divisions. First, he draws one concerning "the subjective relation between a being that is under obligation and the being that puts him under obligation." This would tentatively yield a fourfold division in duties to the self, duties to others, duties to "subhuman beings," and duties to "superhuman beings" (VI:413). Then, Kant adds an "objective" division that mirrors the general outline of a critical investigation into one of the uses of reason, that is, the division, familiar to the readers of the *Critiques*, between a Doctrine of Elements and a Doctrine of Method, which he here specifies further, adapting it to the field at issue.[18] As it becomes clear, the two divisions are to be "taken ... together [*zusammen verbunden*]." The latter, "objective" division roughly corresponds to the main outline of the published text, in which the former division concerning the subjects that put the agent under obligation has merely the task to give a tentative overview of the further specifications of duties that are dealt with in the Doctrine of Elements. Moreover, the fourfold division of subjects, which per se would be rather uncontroversial is eventually not carried out by Kant in the terms in which he initially presents it, since God and non-human living beings are left out from the morally relevant relations (see VI:488). While one division is basically confirmed by the development of the treatise, the other one is stated only to be significantly revised.

The leading thread for the formal unity of Kant's treatment of ethics is thus given by the general division of doctrine of elements and doctrine of method, combined with (part of) the distinction between different moral relations. However, the genuine organizing principle of the system presented in the "Doctrine of Virtue" is in fact provided by a different source, that is, the two "ends that are at the same time duties," one's own perfection and the happiness of others, which Kant presents in the Introduction to the "Doctrine of Virtue" even before explaining the criteria according to which the work is outlined (cf. VI:385f.).[19] In fact, most of the first half of the Introduction to the "Doctrine of Virtue" (§§ II–V) centres on the thought of an obligatory end and its implications. Kant confirms the systematic primacy of those ends when he writes that "[e]thics can ... be defined as the *system of the ends* of pure practical reason" (VI:381; my emphasis).[20] I suggest that Kant's treatment of ethics can be systematic precisely because it unfolds from the consideration of *all* possible obligatory ends.[21] Their systematic role is thus crucial to understand Kant's final solution to the issue of a system of ethical demands.

Kant had arguably first attempted to assure genuine systematicity for his normative theory drawing on the classification of the possible relations between rational agents. For instance, in a note from the drafts for the later *Metaphysics of Morals*, he considered that the "completeness of the division of laws" could be connected with "exhaustiveness in specifying the cases that are under the laws that result a priori from the possible relationships among human beings [*Ausführlichkeit in Specificirung der Fälle unter den Gesetzen die sich a priori aus den möglichen Verhältnissen der Menschen ergeben*]" (XXIII:406). This kind of attempt could not suffice, since it imposes no further constraints on the fact that rational agents act in relation to one another. In those terms, the contents of ethical demands could hardly be determined. That attempt, however, displays Kant's awareness that the mere taxonomy of duties cannot suffice to outline a system of ethical demands.[22]

In contrast, the organization via the system of ends yields a proper system of duties, in Kant's view. The two obligatory ends build the "system of ends of pure practical reason," because they comprehend all the ways to determine a general content for possible maxims that can embody the fundamental principle of ethics. Thus, those two ends can provide the outline of the whole that makes the doctrine of ethical demands a properly systematic theory. Strictly speaking, Kant's system of ethical duties cannot immediately follow either from specification or derivation from the one or the other variant of the categorical imperative, in contrast to what is often assumed. Were the system of duties derived in successive, single steps, from the categorical imperative, this would yield rather an aggregate of normative directions with regard to diverse circumstances.[23] The categorical imperative is accordingly first specified in the principle of ethics, as a principle that puts constraints on ends ("The supreme principle of the doctrine of virtue is: Act in accordance with a maxim of ends that it can be a universal law for everyone to have," VI:395). Then, the systematic development of that principle is made possible through the two obligatory ends, which are not mere general characterizations of the different kinds of duties, but the only two possible overarching ends that meet the constraint imposed by the general principle.[24] From Kant's perspective, only they can warrant the systematicity of the normative theory, as they can both account for the classification of duties and for the generation of the content of ethical demands.[25] I shall briefly consider the two aspects in turn.

First, the two obligatory ends vindicate the distinction between self- and other-regarding duties as well as the ethical irrelevance of the relations to God and non-human beings. Since the only two ends that are commanded by the principle of ethics are to be realized in relations to oneself and others, the other possible relations are, morally speaking, not directly significant, on Kant's view. They can be acknowledged to be indirectly significant, insofar as they can be connected with the two obligatory ends (as Kant does with regard to the conduct towards animals: see VI:442–444, 491). Kant can thus cover the entire territory of ethics attaining a system of duties via a system of ends. Here it is important to differentiate (1) ways to adopt and realize those ends, and (2) ways to guarantee the ability to have and realize them. In these terms, Kant's system can account for every duty of virtue and, at the same time, differentiate between perfect and imperfect duties as to their respective obligation. Whereas the former have to do with the strict necessity to safeguard the crucial ability to pursue morally not-discretionary ends at all, the latter require to actually pursue those ends in some of the many possible ways to do it.

More importantly, the system of ends as organizing thread matches the structure of the determination of the will, that is, fits the architectonic purpose of the system at issue. If the purpose is the full determination of the will from principles of pure reason, the obligatory ends as morally required contents of maxims are perfectly adequate. With this aim in view, the systematic arrangement is achieved through a system of non-discretionary contents of possible maxims, that is, obligatory ends. The system of duties is aptly unfolded not as a list of mere constraints on choice, but as a development of basic objects of the will which are to be considered from a first-person standpoint as the content of possible maxims.

Second, according to Kant's view of systematicity, the systematic role of the twofold system of ends does not merely lie in accounting for an overarching order of the doctrine of duties, but in allowing for further development within the system, in order to deal with future issues. If the categorical imperative provides us with a compass, as Kant writes in the *Groundwork* (see IV:404), it is impossible to determine in advance all the places it can lead us to, but it is possible to clarify in which directions it will guide us. The two obligatory ends specify those directions. The system of ends provides a unifying idea of the whole of ethical demands, since it reduces them to their essential contents, without giving a complete, final description of all those demands. The systematic nature of their treatment is not limited by any factual completeness of the doctrine of duties presented in the "Doctrine of Virtue." The generative role of the two obligatory ends allows for, even suggests, the possibility of further duties beyond those examined in Kant's work. If understood in Kant's terms, the systematicity of moral science allows for substantive moral progress, that is, for the acknowledgment of further demands beyond those that were previously acknowledged.

Along these lines, Kant's approach yields a system of ethics that meets the desiderata of the Pufendorfian paradigm while it also satisfies the demanding requirements of Kant's own notion of a system. The doctrine of ethical demands can be comprehensive because its principle is not a ground for the successive derivation of single duties through an isolated application of the principle to particular circumstances. On Kant's view, a doctrine can be properly comprehensive if, and only if, its principle (a) determines *a priori* "the domain of the manifold as well as the position of the parts with respect to each other" (A832/B860) and (b) institutes a whole that "can ... grow internally (*per intus susceptionem*) but not externally (*per appositionem*)" (A833/B861). Both conditions are met by the system of ends produced by Kant's principle of ethics: the two obligatory ends exhaust the entire scope of ethical demands, leaving it nevertheless open to different ways to pursue them, as no traditional taxonomy could have. As Kant is reported to have observed in his lectures, "morals is an inexhaustible field" (Kaehler, 358); now the same holds true in the "Doctrine of Virtue." This feature of morals would appear to resist a systematic treatment only if we follow a traditional notion of system. Kant's own view of systematicity, instead, centres on the thought that a whole with that property is to be at the same unified and open to further development, for a system is necessary and complete, although not concluded.[26] It holds true also in the "Doctrine of Virtue" that its "systematic unity ... is only a projected unity" (A647/B675). Hinging on the two obligatory ends, Kant's system of ethics thus satisfies the Pufendorfian demand of comprehensiveness by providing a system that is both complete in its scope and open-ended in its contents.[27] In Kant's terms, it is a "whole" that is "articulated ... and not heaped together" (A833/B861). Here the two obligatory ends provide the internal articulation according to which the whole of ethical demands can be established and further develop. They provide a general outline of a system, giving it at the same time a crucial plasticity.

13.4 Enhancing Ordinary Moral Thinking: Systematicity and Orientation

Because of its comprehensiveness and structure, a system of cognitions marks a difference from ordinary, i.e. non-scientific cognition (see A832/B861; IV:467). As to a system of morals, its relation to ordinary thinking can be of different sorts, though. Different ways to understand a system diverge in this respect. A system can give order to pre-scientific cognition, thereby justifying its elements, in coherentist accounts. In a traditional deductive account, as in the Pufendorfian paradigm, the system traces back the contents of ordinary moral thinking to the principle that grounds them. Kant's approach diverges from both these paths. The sort of systematicity that is distinctive of his ethics differentiates his approach from that of the traditional versions of the Pufendorfian paradigm, which provide deductive derivations from the principle via logical relations and a taxonomical overview of the demands of morality. Furthermore, Kant's view can be helpfully contrasted also with a different understanding of a system of morals, which does not conform to the deductive outline of the Pufendorfian paradigm. In a remarkable passage of his *Essay on the Active Powers of Man*, Thomas Reid, who subscribes to the project of a comprehensive doctrine of duties, writes:

> A system of morals is not like a system of geometry, where the subsequent parts derive their evidence from the preceding, and one chain of reasoning is carried on from the beginning, so that, if the arrangement is changed, the chain is broken, and the evidence is lost. It resembles more a system of botany, or mineralogy, where the subsequent parts depend *not* for their evidence upon the preceding, and *the arrangement is made to facilitate apprehension and memory, and not to give evidence*.[28]

On Reid's view, the materials of a system of ethics do not require any justification, but have to be laid out in a comprehensive order, without being traced back to a unifying principle. The criteria that govern moral judgement are self-evident and do not need to be backed up by an underlying standard.[29] They are merely fixed points of moral reasoning that must be kept in plain sight, in order to avoid confusion in thinking about moral issues.[30] Unlike Reid, Kant has in view a system of morals that does not aim at mapping already available convictions and beliefs into a more perspicuous arrangement. Rather, as we have seen in the previous sections, Kant's own conception of a system leads him to devise a system of ethics that draws on the foundations provided by a principle and that unfolds by projecting a whole that shall grow out of the two obligatory ends commanded by the principle. The relation of such a system to ordinary moral thinking is accordingly different: while in Reid's account the system is mainly a memorizing tool for ordinary moral thinking, a system that is complete, yet not concluded, is primarily about recognizing a systematic order in morality within which any further demand will find their place. Building on the standard given by reason, the system articulates the shape that ordinary moral thinking should take, if it would fully embody the principle of reason. Ordinary moral thinking will thus be enhanced by a firmer awareness of the fundamental standards that shall guide deliberation even beyond already recognized demands.

Along these lines, the conception of ethics as a system of duties that unfolds from a system of ends strengthens ordinary moral thinking through simplicity and principled completeness. Recently, the "Doctrine of Virtue" has been characterized as "an explanatory grounding project," which aims "to derive (and thus justify) a set of duties but also to explain and thus provide insight into the deontic status of a range of actions."[31] The *Doctrine of Virtue* has thus the task to clarify not only "that certain types of action are required of us," but also "why they are."[32] That explanatory task of the system of ethical duties entails that the contents of the system go beyond "intuitive moral judgement." Here again, crucially, systematicity is not about systematization of already available beliefs of ordinary moral thinking. Characterizing the system of ethics as an explanatory project overall is appropriate, provided that that explanation is understood as the enlightenment of the rational agent about his, or her, capacity to grasp the complex variety of ethical demands. Systematizing those demands as a dynamic whole is rather about strengthening the awareness of the proper determination of rational deliberation and, thereby, cultivating reason in its ordinary use. As any systematic enterprise in Kant's philosophy, the systematic account of obligatory ends and the corresponding duties amounts to a specific mode of self-knowledge of reason.[33] With respect to our starting point, a further clarification of the necessity stated in the opening passage from the preface of the *Metaphysics of Morals* is now available, namely that such a system *has to* follow in order to unfold the normative import of the principle of morality, thereby strengthening its availability to reason.

If a system of ethics aims at strengthening reason in ordinary moral thinking, however, it would seem that Kant eventually comes closer to the current understanding of the role of systematicity in ethics, after all. A crucial task of such a system of moral philosophy would be to provide by a lexical order of moral demands the means to address and solve difficult cases, that is, first and foremost instances of supposed conflict between different demands. As Paul Guyer has put it, "[i]f Kant's classification of duties is a genuine system, then it ought to provide a basis for the resolution of these sorts of conflicts too." Guyer has suggested that "Kant does not explicitly explain how it can, but he does offer hints and materials that can be developed for this purpose."[34] I shall suggest that the systematicity of Kant's ethics is indeed crucially connected to that need, although it addresses it in a different way than the current understanding of a system of morals assumes. I will consider not so much whether Kant's moral theory has the resources to deal with purported conflicts of demands, but the more specific, and somehow more neutral, issue whether the systematicity of that theory plays a role in that regard and whether to deal with those conflicts should be regarded as a task for the system of ethics.

Addressing the need for orientation with regard of cases of purported conflict between different obligations would be a matter for what the previous tradition calls casuistry. Kant's view reverses the traditional perspective, though. When he remarks that "ethics falls into a *casuistry*" "because of the latitude it allows in its imperfect duties," he adds that casuistry so understood cannot properly belong to the system that is to be developed: "casuistry is ... *neither a science nor a part of a science*; ... it is woven into ethics in a fragmentary way, *not systematically* ..., and is added to ethics only by way of *scholia to the system*" (VI:411; my emphasis; cf. XXIII:389). It would then appear that systematicity encounters at this point its boundaries. Yet, Kant's distinctive notion of a system of ethics reveals here its peculiar practical significance. On Kant's view, casuistry has to play a role in orienting judgement within the space of options that an imperfect duty leaves open.[35] It thus focuses on how to comply with one ethical demand (say, the duty of beneficence) in the given circumstances or, as Kant puts it, "to decide how a maxim is to be applied in particular cases" (VI:411). This, however, is not the same as dealing with a supposed conflict of duties, where contrasting grounds of obligation are confronted (see VI:224).

If casuistry is limited in its scope, in Kant's view it is the *system* that does provide orientation in problematic cases. It is not simply because the scientific treatment of ethics argues that "a collision of duties and obligations is not even conceivable" (VI:224), thereby maintaining that difficult cases are in fact only apparently dilemmatic. Ethics can provide orientation in that regard because a systematic account opens up a broader perspective that goes beyond a mere classification of duties. If casuistry can only sharpen *the power of judgment* in applying specific demands, the system thereby strengthens pure practical *reason* and expands its outlook (see VI:411). "Falling into casuistry," that is, going beyond the limits of systematicity, is not an unfortunate weakness of ethics, but a danger or a defeat for moral theory, as it makes unable to see the connections between different duties. To the contrary, properly systematic ethics should aim at taking on the task that traditionally was attributed to casuistry, namely preparing the subject to deal with difficult cases. Instead of practicing the power of judgement, which cannot assure any significant results, and might in fact jeopardize the clarity achieved in the "dogmatics" of duties, moral theory should aim at strengthening reason in the individual agent, making it capable to grasp the complex connection of demands.

Kant's systematic view of ethics does respond to the need of dealing with difficult cases and apparent conflicts, after all. It does "provide a basis for the resolution of these sorts of conflicts," as Guyer puts it, but not through its classification of duties. Ends as systematic standards make a comparative consideration of grounds of obligation easier, as different purported obligations are warranted by the reference to the morally necessary ends. The entire system of duties contributes to facilitate the consideration of perplexing cases, as it clarifies the specific character of each demand. In fact, it is precisely the systematic nature of the treatment of ends and duties that allows rational agents to better explore moral options in problematic circumstances. A mere taxonomic systematization, like that endorsed by Reid, would provide no clue at all to address perplexing cases – just a catalogue of separated options from which the agent should intuitively pick the most appropriate. In contrast, a genuine system that unfolds the complex web of normative demands presents the rational agents with an orientation that is closer to the terms of deliberation.

The systematic connection of demands grounding on the system of obligatory ends provides the rational agent with the better perspective. Systematicity makes the particular understandable in light of its being part of a whole. Alleged moral conflicts never present themselves in isolated cases, but always in already normatively loaded situations, in which previous choices have been made and other demands have been considered.[36] Instead of artificially separating normative claims, Kant's perspective underscores the necessity of focus on their interplay. In fact, cases of this sort present themselves in conditions in which "one obligatory maxim" could be legitimately limited "by another (e.g. general love of one's neighbour by the love of one's parents)." When this happens, as Kant remarks, "the field for the practice of virtue *is indeed widened*" (VI:390; my emphasis). Connections, convergence, or conflict between different grounds of obligation are indeed only one aspect of the more general web of moral demands. The normative relations between grounds of obligation are better construed not by narrowing the focus of moral reasoning, but rather by widening it. Remarkably, one of the clearest way to find the solution to a conflict is to recognize that one option is supported not only by one ground of obligation, but by several grounds of obligation.[37]

The present issue, however, is not Kant's view on how to deal with conflicts of moral demands, but the extent of the role that systematicity can play in this regard. Because this complexity becomes relevant at the level of the individual agent's maxims, then a systematic doctrine of moral demands ultimately does play a role also with regard to perplexing cases. It is not the role of solving conflicts in advance, but that of shaping reason so that it can be prepared and flexible enough to face perplexing cases. The wide obligation of ethical duties also requires that, unlike the doctrine of right, ethics encompasses a part concerning the individual learning and assimilation of the system of duties, which Kant calls a doctrine of method.

Notably, Kant is explicit in regarding the doctrine of method as a part of the system (see VI:412), although it does not present any specific principle or demand. This shows, again, that systematicity is not about systematization of given elements, but about the ongoing implementation of principles of reason according to its fundamental normative purpose. In contrast to the episodic focus on the power of judgement that is characteristic of casuistry, which can easily lead to micrology, a thorough articulation of the system of ends and duties is ultimately, for Kant, what makes a rational agent well armoured to confront perplexing cases from a broader perspective. It is not the power of judgement, but reason in Kant's specific sense that makes an agent able to deal with difficult cases, because the fundamental standard of morality that it provides also issues a complete, yet growing network of demands that are connected through a system of obligatory ends. The systematicity of the overall examination of the entire web of ethical demands thus trumps the isolated attempts at handling particular difficulties concerning specific duties.

Here, again, the significance of the systematicity of ethics displays a further notable difference from the role that systematization would now be taken to have. A systematic account of ethical demands that arranges them in a coherent disposition would handle perplexing cases and possible conflicts by putting forward a lexical order according to which some demands have to be prioritized above others. Differently, Kant suggests that a proper system does not provide a mere order of prioritization, but a reconstruction, as thorough as possible, of the complex web of ethical demands and their connections that makes its systematic structure apparent. In this perspective, the absence of any genuine conflict has a different meaning than in traditional rationalist accounts of morality: a conflict of obligations is not merely ruled out because it counters the overall consistency of ethical demands, but because it would trouble the relations connecting them. The aim of a system in Kant's distinctive sense is not only an exposition of morality free from any contradiction, but the possibility of a thorough determination of the maxims of the individual agents. With respect to the opening passage from the preface of the *Metaphysics of Morals*, here emerges a further reason why a system *must* be unfolded, namely because it provides rational agents with all the orientation reason can offer to deal with the intricacies of morality.

In the first *Critique* Kant had observed that" [t]he greatest systematic unity, consequently also purposive unity, is the school and even the ground of the possibility of the greatest use of human reason" (A694f./B722f.). I suggest that the same holds true for the system of ethics, in which the articulation of ethical demands as forming an open-ended unity amounts to a genuine "school … of human reason." The broader the perspective on ethical demands we can reach, the closer we come to their complete systematic unity, the closer we reach to satisfying the need of reason for totality also in its practical aspect, that is, as the fullest determination of maxims instantiating the principle of morality.

13.5 The Distinctive Systematicity of Kant's Ethics

Kant's emphasis on the systematic nature of the *Metaphysics of Morals* proves to be not merely historically, but also philosophically significant. That both parts of the work begin stressing that the project is systematic in nature is thus not accidental. Specifically, ethics needs to be developed as system because the mere identification and corroboration of a principle is hardly enough with regard to the needs of moral life and the weaknesses of ordinary moral thinking in dealing with it. In this respect, a system of morals *has* to follow the preliminary investigation on the principle of morality.

The distinctive character of Kant's systematic development of ethics revolves around a threefold difference between his conception of systematicity as it is applied in his ethics and a current standard conception, which mostly informs the reference to systematicity, or systematization, in moral philosophy. Each of the features that I have pointed out also contributes to make the relationship between ethics in its systematic development and ordinary moral thinking more precise.

First, in Kant's view, the system of ethics is not about the systematization of moral beliefs, but about the *articulation of moral demands in their connection with the fundamental principle of morality*. Unlike in the current understanding, a systematic treatment of ethics is not instrumental to a coherentist project that aims at justifying moral convictions and beliefs embedded in ordinary moral thinking by constructing them into a coherent set. Here, on the contrary, Kant's project follows the Pufendorfian paradigm of a comprehensive account of the demands of morality that draws primarily on a fundamental principle, which provides the proper ground for the system. A system of ethics is needed to unfold the obligations that are justified by the fundamental principle.

Second, drawing on the fundamental principle of morality, Kant's systematic treatment of ethics is developed as a *system of ends* that yields *an open-ended system of duty types*. While Kant's approach shares the first feature with the traditional approach, here it markedly diverges, as the novelty of Kant's distinctive notion of a system becomes relevant. According to that conception, a system finds its unity in an underlying principle of reason, which generates a set of determinations that is complete in its scope and yet can grow further. The system of ethics that Kant puts forward is thus neither a logically consistent arrangement of demands, nor a concluded collection of ethical duties. The central system-building feature here is the role of the obligatory ends, from which the corresponding system of duties originates, in an ongoing development. In this respect, a system of ethics entails an advancement from ordinary moral thinking insofar as the system presents a dynamic reconstruction, maybe even a revision, of the ordinary moral convictions that is able to show how it could expand.

Third, because of its dynamic character and its relation to ordinary moral thinking, the system of ethics that Kant has in view by virtue of his distinctive notion of systematicity has a different, more substantial purpose that a system of moral demands should have according to other conceptions. Instead of providing merely a more coherent or orderly arrangement of moral beliefs, Kant's system of ethics aims at strengthening reason in its practical use. This becomes apparent, for instance, with regard to intricacies such as supposed conflicts of obligations. In Kant's view, the systematicity of ethics as a comprehensive and dynamic body of demands is supposed to *provide orientation and a broader perspective from which perplexing cases should be considered*. Here the scientific treatment of morality provides crucial support to ordinary moral thinking by emphasizing the holistic character of ethical demands and clarifying their connections.

These three features define Kant's view on the significance of systematicity for moral philosophy and ethics specifically, which constitutes an original rationalist conception that takes its clue from the system that reason unfolds from its own principles. Having his characteristic notion of a system in view, systematicity proves to be an integral component of Kant's distinctive approach to ethics.[38]

Notes

1 In loving memory of my father, an altogether unsystematic man.
2 A rare exception to the general disregard of this passage is Thorndike (2018, 22). However, Thorndike focuses on it to draw a parallel with the project of a transition from metaphysics of nature to physics.
3 See an analogous statement in the drafts for the *Metaphysics of Morals*, XXIII:247.
4 I cannot consider here the further important claim that the system shall take the shape of a metaphysics of morals. What does it mean exactly in 1797 would deserve some clarification. I have argued that Kant's project of a metaphysics of morals undergoes significant changes through the decades. See Bacin (2006, 223ff.).
5 I have revised the Cambridge Edition translation, which here reads: "A philosophy of any subject (a system of rational cognition from concepts) requires a system of pure rational concepts independent of any conditions of intuition, that is, a metaphysic." Kant's own phrasing stresses more strongly the necessity for a system of metaphysics and the connection of *two* systematic tiers: the general systematicity of philosophy as such and the systematicity of metaphysics in particular.
6 Yet, this opening has not drawn much attention either. See, e.g., the recent brief remarks on "philosophy as a system" in Timmons (2021, 27–29), which

merely connect the work to the general project sketched in the Architectonic. The main issue here, though, is not how practical philosophy is a part of the overall system of critical philosophy, but how practical philosophy as such is systematic.
7 Guyer (2005, 243). See also Guyer (2019, 43ff.).
8 An exception is Barbara Herman's emphasis on systematicity in her recent work; see Herman (2021).
9 Herman (2021) should suggest such a change of direction.
10 In contrast to Mark Timmons' otherwise excellent and thought-provoking interpretations collected in Timmons (2017), which, in spite of the general title of the volume, understand "system" in a rather generic sense, without considering the connection with Kant's distinctive notion.
11 See, e.g., the characterization of this approach in Cath (2016, 214f.).
12 Griffin (1996, 123); cf. 16. See also Griffin (2015, 84). As Crisp (2000) shows, Griffin's understanding of a system of morals is equivalent to Sidgwick's, despite their opposite views on the matter.
13 See Griffin (1996, 124–8); Cueni and Queloz (2021).
14 To show that Kant does not follow a coherentist approach in establishing the principle of morality (against, e.g., Gillessen 2016), I should examine the strategy of the *Groundwork* and the second *Critique*, which I shall do in a separate paper. Here, though, it is sufficient to my present purposes to emphasize that for Kant a system depends on a prior principle, instead of justifying it.
15 See especially Scattola (2017, 132–9). See also Scattola (2008, 240f., 2011, 264f.).
16 See, e.g., the traditional account of the impact of natural law on Kant's moral philosophy in Schneewind (1993).
17 Williams (1985, 116f.).
18 Note, incidentally, that this is the only case in his published writings in which Kant employs the Doctrine of Elements/Doctrine of Method division not in a critique, but in a part of the system. (Traces of that division are to be found in the anthropology lectures: see e.g. VII:421.5-8, XXV:1529.)
19 Here I cannot discuss the specific content of the two ends.
20 The same primacy is already articulated in the drafts for the *Metaphysics of Morals*. See, e.g., XXIII:374: "doctrine of virtue as doctrine of wisdom [shall] consider ends that it is presented as a duty to set oneself [*Tugendlehre als Lehre der Weisheit [hat] von Zwecken zu reden die sich zu setzen es als Pflicht vorgestellt wird*]."
21 Gregor misses this crucial point when she suggests that "our obligatory ends are, in a sense, abstractions from" the system of the ends of pure practical reason (1963, 93).
22 See also XXIII:417.14–20.
23 See e.g. XXIX:5: "When the parts come before than the whole, one has an aggregate."
24 Here I find myself in agreement with Baum (1998).
25 Since the distinction of the two obligatory ends is never mentioned in the main text of the "Doctrine of Virtue," but only in the Introduction, Bernd Ludwig has suggested that it could be a later addition to the actual development of the system of ethical duties (see Ludwig 2013, 80, 83). Passages of connected texts prior to the *Metaphysics of Morals* that contain virtually explicit statements of the thought of the two obligatory ends (see, e.g., XXIII:374 and XXVII:543.30ff.) invalidate the suggestion, though. Even as a possible *ex post*

arrangement, however, the distinction makes sense of the whole as a system in Kant's distinctive sense, as it is apt to confer completeness and organicity in a way that would not be available to other criteria.
26 Thorndike (2018, 132f.) insists that the "Doctrine of Virtue" lacks systematicity and *cannot* provide more than "an aggregate of fragmentary precepts." But that would reduce the "Doctrine of Virtue" to a casuistry, in contrast to Kant's remarks to this regard (see VI:411), on which I shall comment in the next section. Also, Thorndike does not consider that, in Kant's view, a system does not have to be concluded, although it must be complete.
27 Herman (2021, especially chap. 9) similarly emphasizes that Kant's is a "dynamic system."
28 Reid (2010, 281, my emphasis). The passage is so remarkable that J.G.H. Feder quotes it at length in his review of Reid's work (see *Philosophische Bibliothek* 2, 1789, 115f.). It might have been known to Kant thanks to that mediation.
29 See Reid (2010, 31, 271).
30 See also Cuneo (2011, 112f.).
31 Timmons (2017, 176) (see Smit and Timmons 2013).
32 Timmons (2017, 178).
33 See Baum (2019).
34 Guyer (2005, 269).
35 On Kant's view on the role of casuistry, see Schuessler (2012, 2021).
36 Here I find myself in agreement with Herman (2021, 78f.).
37 See Timmermann (2013, 52f.).
38 I should like to thank Paul Guyer and Marcus Willaschek for their remarks on the first version of this paper and Eric Watkins for his extensive comments.

Bibliography

Bacin, Stefano. 2006. *Il senso dell'etica: Kant e la costruzione di una teoria morale*. Bologna-Napoli: Il Mulino.
Baum, Manfred. 1998. "Probleme der Begründung Kantischer Tugendpflichten." *Jahrbuch für Recht und Ethik/Annual Review of Law and Ethics* 6: 41–56.
Baum, Manfred. 2019. "Systemform und Selbsterkenntnis der Vernunft bei Kant." In *Kleine Schriften 1*, edited by Marion Heinz, 227–40. Berlin, Boston: De Gruyter.
Cath, Yuri. 2016. "Reflective Equilibrium." In *The Oxford Handbook of Philosophical Methodology*, edited by Herman Cappelen, Tamar Gendler, and John Hawthorne, 213–30. Oxford University Press.
Crisp, Roger. 2000. "Griffin's Pessimism." In *Well-Being and Morality: Essays in Honour of James Griffin*, edited by Roger Crisp and Brad Hooker, 115–28. Oxford–New York: Clarendon Press.
Cueni, Damian, and Matthieu Queloz. 2021. "Whence the Demand for Ethical Theory?." *American Philosophical Quarterly* 58(2): 135–46.
Cuneo, Terence. 2011. "Reid on the First Principles of Morals." *Canadian Journal of Philosophy* 41: 102–21.
Gillessen, Jens. 2016. "Kants ethischer Kohärentismus." *Kant-Studien* 107(4): 651–80.
Gregor, Mary J. 1963. *Laws of Freedom: A Study of Kant's Method of Applying the Categorical Imperative in the Metaphysik der Sitten*. New York: Barnes and Noble.

Griffin, James. 1996. *Value Judgement: Improving Our Ethical Beliefs*. Oxford: Clarendon Press.
Griffin, James. 2015. *What Can Philosophy Contribute to Ethics?* Oxford: Oxford University Press.
Guyer, Paul. 2005. *Kant's System of Nature and Freedom: Selected Essays*. Oxford–New York: Clarendon Press.
Guyer, Paul. 2019. *Kant on the Rationality of Morality*. Cambridge: Cambridge University Press.
Herman, Barbara. 2021. *The Moral Habitat*. Oxford: Oxford University Press.
Ludwig, Bernd. 2013. "Die Einteilungen der *Metaphysik der Sitten* im Allgemeinen und die der *Tugendlehre* im Besonderen (MS 6:218–221 und RL 6:239–242 und TL 6:388–394, 410–413)." In *Kant's "Tugendlehre" A Comprehensive Commentary*, edited by Andreas Trampota, Oliver Sensen, and Jens Timmermann, 59–84. Berlin, Boston: De Gruyter.
Pufendorf, Samuel von. 2002. *Eris Scandica und andere polemische Schriften über das Naturrecht*, edited by Fiammetta Palladini. Berlin: Akademie Verlag.
Reid, Thomas. 2010. *Essays on the Active Powers of Man* (1788), edited by Knud Haakonssen and James A. Harris. Edinburgh: Edinburgh University Press.
Scattola, Merio. 2008. "Die Naturrechtslehre Alexander Gottlieb Baumgartens und das Problem des Prinzips." *Aufklärung* 20: 239–65.
Scattola, Merio. 2011. "Scientific Revolution in the Moral Sciences: The Controversy Between Samuel Pufendorf and the Lutheran Theologians in the Late Seventeenth Century." In *Controversies*, edited by Marcelo Dascal and Victor D. Boantza, 251–76. Amsterdam: John Benjamins Publishing Company.
Scattola, Merio. 2017. *Prinzip und Prinzipienfrage in der Entwicklung des modernen Naturrechts*, edited by Andreas Wagner. Stuttgart-Bad Cannstatt: Frommann-Holzboog.
Schneewind, J. B. 1993. "Kant and Natural Law Ethics." *Ethics* 104: 53–74.
Schuessler, Rudolf. 2012. "Kant die Kasuistik: Fragen zur *Tugendlehre*." *Kant-Studien* 103: 70–95.
Schuessler, Rudolf. 2021. "Kant, Casuistry and Casuistical Questions." *Journal of Philosophy of Education* 55(6): 1003–16.
Smit, Houston, and Mark Timmons. 2013. "Kant's Grounding Project in the 'Doctrine of Virtue'." In *Kant on Practical Justification: Interpretive Essays*, edited by Sorin Baiasu and Mark Timmons, 229–68. Oxford: Oxford University Press. Reprinted in *Significance and System: Essays in Kant's Ethics*, 175–218.
Thorndike, Oliver. 2018. *Kant's Transition Project and Late Philosophy: Connecting the Opus Postumum and Metaphysics of Morals*. London: Bloomsbury.
Timmermann, Jens. 2013. "Kantian Dilemmas? Moral Conflict in Kant's Ethical Theory." *Archiv für Geschichte der Philosophie* 95: 36–64.
Timmons, Mark. 2017. *Significance and System: Essays in Kant's Ethics*. Oxford: Oxford University Press.
Timmons, Mark. 2021. *Kant's Doctrine of Virtue: A Guide*. Oxford: Oxford University Press.
Williams, Bernard. 1985. *Ethics and the Limits of Philosophy*. London: Fontana.

14 Kant's System of Systems

Paul Guyer

14.1 The System of Nature and Freedom

Kant concluded his critical philosophy with a vision of a single system of nature and freedom. This vision is expressed in the Doctrine of Method of the Critique of the Teleological Power of Judgment in the *Critique of the Power of Judgment* by means of the idea that just because we human beings are the "final" ends of morality, we must also consider ourselves the "ultimate" end of nature: because apart from all our particular, material ends, the realization of none of which is of unconditional value, "there remains only the formal, subjective condition, namely the aptitude for setting [ourselves] ends at all and (independent from nature in [our] determination of ends), using nature as a means appropriate to the maxims of [our] free ends in general, as that which nature can accomplish with a view to the final end that lies outside of it," we therefore "have sufficient cause to judge the human being not merely, like any organized end, as a natural end, but also as the **ultimate end** of nature here on earth, in relation to which all other natural things constitute a system of ends in accordance with fundamental principles of reason" (*CPJ*, §83, V:431, 429).[1] That is, Kant sees human beings as part of nature as a system in which we can cultivate our natural abilities under the guidance of and for the realization of the moral law that our own reason imposes upon us in the fullest exercise of our freedom. Kant goes further and says that "even the unifiability of the two ways of representing the possibility of nature," as merely mechanical and as teleological, "may well lie in the supersensible principle of nature (outside of as well as inside us), since the representation of it according to final causes is only a subjective condition of the use of our reason when reason would not judge the objects merely as appearances, but rather demands that these appearances themselves, together with their principles, be related to the supersensible substratum in order to find possible certain laws of their unity, which cannot be represented except by means of ends (of which reason has ones that are supersensible)" (*CPJ*, §82, V:429). That is, we must see the moral law to which we must subject our own actions in nature as stemming from our own supersensible nature, but we must also see nature as a whole, the stupendous theater in which we act, as compatible with our own moral goals, and we can do this only if we see the laws of the whole of nature, for which we are certainly not causally responsible, as having been written by an author also governed by morality, namely God. To be sure, Kant adds, this vision, or chain of inferences, is not valid "for the determining power of judgment, yet [is] for the reflecting power of judgment" (*CPJ*, §83, V:429), that is, we cannot treat the subordination of the laws of nature to the moral law and their divine authorship with an eye to this subordination as a fact demonstrable by theoretical reason, in the way in which a general law of nature such as that every event has a cause or a foundational principle of physics such that every action is met with an equal and opposite reaction is demonstrable, but neither can this vision be disproven by the theoretical use of reason, so we can regard it as a way of looking at nature that we can be allowed to adopt for regulative purposes and even must adopt for practical purposes.

I have described this culminating vision of Kant's philosophy before.[2] Here I dive deeper into Kant's conception of the two systems that he ultimately unifies in this system of systems, that is, the system of nature, on the one hand, and the system of freedom, on the other. In spite of an obvious difference between the two systems, namely that while the realization of the system of nature does not depend upon human choice, *a fortiori*, in Kant's view, on the freedom of human choice, the systematic realization of morality does depend on the free choice of human beings, there are also deep similarities between the two systems.[3] In particular, each system has both a subjective and an objective level, or a level of representation and a level of reality. That is, for Kant not only must our representation of nature in the form of empirical *concepts* be systematic, but the *laws* of nature and further the *objects* of nature must also comprise a system, and likewise not only must our moral *maxims* comprise a system – an intra- as well as interpersonal system – but all moral *agents*, our *ends*, and the *actions* that we perform with the intention of realizing them must also comprise a system, or systems.[4] In the first case, Kant uses the distinction between a "logical" and a "transcendental" level of systematicity to mark these two demands. He does not use this distinction in the case of the moral system of ends. An obvious reason for this is that in the moral case we cannot simply presuppose that the system of ends exists, but we must attempt *to bring it into existence* by our own efforts. Yet we must also presuppose that nature is *receptive* to our efforts to create a system of ends. In the *Critique of Practical Reason* this presupposition takes the form of the postulate of pure practical reason that the laws of nature and the moral law have a common author; in the *Critique of the Power of Judgment* it takes the form of the "hints" of both aesthetic and teleological judgment that nature is receptive to our ends. So in spite of the both substantive and terminological differences between the two cases, there is still much commonality of structure. My primary goal will be to reveal these parallel levels of systematicity within the two systems of nature and morality or freedom.

14.2 The System of Nature

Kant writes of the necessity for a system of *concepts* of nature, of *laws* of nature, and of *nature* itself. The first two are not always clearly distinguished, but there is an important distinction between them.[5] Concepts are our general *representations* of kinds of things in nature, in particular our discursive representations of things made up of marks or characteristics (*Merkmale*) that can be presented in intuition and shared among individual things, thereby allowing the concepts they constitute to designate kinds of things, whereas laws of nature are regularities in things in nature, dispositions that appearances, in the sense of things as they appear to us, actually have.[6] Or, more fully, our representations of laws of nature are representations of such regularities, while the regularities themselves exist in nature (subject to the interpretation of existing in nature itself that would be required by transcendental idealism).[7] Lawlike relationships among the things in nature, or the system of laws of nature, constitute nature itself into a system. As I said at the outset, in the end Kant will argue that for things in nature, more precisely the things of nature, to constitute a system they must not only be bound together by laws but also have an ultimate *purpose* that they can realize *through* their laws. But here I want to clarify the systems of concepts and of laws as Kant conceives them in his theoretical philosophy of nature considered on its own.

Kant does not present the layers of theoretical systematicity straightforwardly in the order of the system of the concepts of nature, the system of the laws of nature, and then the system of nature itself or of the things of nature taken collectively. The first system he expounds is rather the most general system of the laws of nature, namely the "System of the Principles of Pure Understanding" comprised by the Axioms of Intuition, the Anticipations of Perception, the Analogies of Experience, and the Postulates of Empirical Thinking in General (A148–235/B187–287).[8] These are the laws that everything in nature, as represented by us, must have a determinate magnitude in time or space and time (A162/B202); that the real in appearance that corresponds to sensation in us must have a determinate degree of intensive magnitude (A165/B207); that all events in nature are alterations in the states of enduring substance and are subject to laws of causation and interaction (A182/B224, A188/B232, A211/B256); and that the categories of possibility, actuality, and necessity can be applied to things in nature only on the basis of their satisfaction of these laws (A218/B265–266). In the *Critique of the Power of Judgment* Kant retrospectively characterizes these as the "universal laws of nature" or "universal transcendental laws, given by the understanding," in contrast to the "particular empirical laws" of nature (*CPJ*, Introduction, section IV, V:179–180). We will come back to the contrast between universal and particular laws of nature shortly, but first we must note the significance of Kant's use of the term "transcendental" in this context: it is contrasted to "logical," and the contrast is precisely that the logical concerns our own concepts but the transcendental concerns the things to which we apply our concepts, although of course those things as they appear to us in space and time rather than as they are in themselves (I won't keep repeating this qualification) – that is, the transcendental concerns nature as we represent it, not merely our representation of it.[9]

The contrast between the logical and transcendental systems of theoretical philosophy is not immediately apparent in the *Critique of Pure Reason*'s next and most extensive discussion of systematicity, namely the Appendix to the Transcendental Dialectic "On the regulative use of the idea of pure reason" (A642/B670) and "On the final aim of the natural dialectic of human reason" (A669/B697). Kant's argument is that ideas of pure reason, the ideas of the soul as ultimate subject of thought, the world as a complete whole, and God, which have been shown in the preceding Transcendental Dialectic to have no sound "speculative" or constitutive use in the representation of nature, nevertheless have an indispensable regulative use in guiding our inquiry into nature, to which he will subsequently add, in the Canon of Pure Reason in the Doctrine of Method, that they also have an indispensable use in the practical use of reason.[10] But he initially talks about the use of ideas of reason in establishing "the **systematic** in cognition" (A645/B673), or the use of reason as "directed at the systematic unity of the understanding's cognition" (*Verstandeserkenntnisse*) (A647/B675), without specifying whether by "cognition" he means concepts considered as our representations or knowledge of laws considered as real regularities in nature itself. As he continues, however, Kant makes clear that the framework of systematicity, comprised by the requirements of upwardly increasing generality, downwardly increasing specificity, and continuity between its levels, is applied in the first instance to *concepts* of genera and species of things, and thus is a condition on the formation of empirical *concepts* (A653–654/B681), but through these concepts to *things* and their properties and powers, such as "chalky" and "muriatic earths" (A657/B685) and orbits of celestial bodies (A663/B691). This is why "the logical principle of genera," that is, the principle that lower-order species are to be subsumed under increasingly more general, higher-order genera, and conversely the logical principle of species, "presupposes a transcendental principle if it is to be applied to nature" (A654/B682). That is, our concepts must be systematic, but if they are to be applied to nature then nature itself must be systematic, or the systematic relations among our concepts must mirror systematic relations in nature itself. This is particularly clear in Kant's example of orbits, where he argues that the system of curves we can represent in geometry (technically, types of pure intuition rather than concepts, but we can have concepts of types of intuitions)[11] can be used to represent a system of objective relations among the orbits of planets, comets, and stars. The logical system is thus the system of concepts of curves, while the transcendental system is the system of actual orbits. Or, even more fully, our concepts of objects typically include concepts of their dispositions or the particular laws that govern their behavior, and the laws included in our concepts must be the laws of nature itself for science to be true.

Kant does not make the distinction between the systematicity of concepts *qua* mental representations and laws *qua* features of nature that we represent by means of concepts explicit in the *Critique of the Power of Judgment*, but the distinction between a logical and a transcendental principle for the power of reflecting judgment which depends on this prior distinction is deployed there too. And as it turns out, the first draft of the Introduction to the third *Critique* ("First Introduction") focuses its discussion on the systematicity of our classificatory *concepts* of nature, while the published version of the Introduction focuses on the objective content of a system of concepts, namely the system of the *laws* of nature – but there is no conflict, because our concepts of objects include concepts of their properties as governed by laws. The Critique of the Teleological Power of Judgment, the second half of the body of the work, then shows how we can and must consider nature as a whole as a system of objects.

Both versions of the Introduction begin their discussion of systematicity with the distinction between determining and reflecting judgment. This distinction is made in terms of concepts: the "determining power of judgment" is "a faculty for **determining** an underlying concept through a given **empirical** representation," while the "reflecting power of judgment" is "a faculty for **reflecting on a given** representation, for the sake of a concept that is thereby made possible" (*FI*, section V, XX:211). The distinction is thus between making a given concept more determinate with additional information about its objects, on the one hand, and finding a concept for a given object, on the other. In the First Introduction, Kant then states that "The principle of reflection on given objects of nature is that for all things in nature empirically determinate **concepts** can be found," and in particular that they can be found through "a hierarchical order of species and genera," or "**artistically**, in accordance with the general but at the same time indeterminate arrangement of nature in a system, as it were for the benefit of our power of judgment" (XX:213–214). Searching for such a system of concepts, that is, empirical concepts closer to actual objects than, but subsumed under, the most general laws of nature, within such a system prevents "all reflection" from becoming "arbitrary and blind ... undertaken without any well-grounded expectation of its agreement with nature" (XX:212). Kant then goes on to say that what we will represent *through* such a system of concepts is "the suitability of [nature's] particular laws (about which understanding has nothing to say) for the possibility of experience as a system, without which presupposition we could not hope to find our way in a labyrinth of possible empirical particular laws" (XX:214); the understanding "has nothing to say" about these particular laws because it provides only the most general laws of nature as expounded in the first *Critique*, that is, the Analogies of Experience and other synthetic *a priori* principles of judgment. (Kant was clear in the first *Critique* that we cannot simply deduce particular laws of nature from the general principles of the possibility of experience; see A127 and B165.) Kant then concludes this exposition with a repetition of his distinction between a logical and transcendental principle of judgment:

> The principle of the reflecting power of judgment, through which nature is thought of as a system in accordance with empirical laws, is however merely a principle **for the logical use of the power of judgment**, a transcendental principle, to be sure, in terms of its origin, but only for the sake of regarding nature *a priori* as qualified for a **logical system** of its multiplicity under empirical laws
>
> (XX:214)

Here Kant says that the logical principle of systematicity is also transcendental because it is an *a priori* way of regarding nature, presumably due to our own cognitive powers as a whole, but the point remains that he is conceiving of a logical system of empirical concepts as representing an objective system of particular laws of nature. So he clearly has two levels of systematicity in mind.

The published Introduction to the third *Critique* also makes the distinction between determining and reflecting judgment, but from the outset makes it as a distinction about the laws of nature rather than mere concepts: "If the universal (the rule, the principle, the law) is given, then the power of judgment, which subsumes the particular under it ... is **determining**. If, however, only the particular is given, for which the universal is to be found, then the power of judgment is merely **reflecting**" (*CPJ*, introduction, section IV, V:179). This suggests that the determining power of judgment subsumes more particular laws under more general ones and ultimately particular objects under laws, while the reflecting power of judgment must start from particular objects to find laws under which to subsume them or from more particular laws to subsume them under more general laws. But either way, Kant is here thinking of a system of laws, although of course such a system will be represented through a system of concepts, and the concepts of particular kinds of objects will include their powers as predicates and thus imply the particular laws that govern their behavior. Kant formulates the principle of reflecting judgment accordingly:

> Now this principle can be nothing other than this: that since universal laws of nature have their ground in our understanding, which prescribes them to nature (although only in accordance with the universal concept of it as nature), the particular empirical laws, in regard to that which is left undetermined in them by the former, must be considered in terms of the sort of unity they would have if an understanding (even if not ours) had likewise given them for the sake of our faculty of cognition, in order to make possible a system of experience in accordance with particular laws of nature.
> (*CPJ*, Introduction, section IV, V:180)

Here Kant makes it clear that we must conceive of a system of the laws of nature itself. He also adds something missing from the first version of the Introduction, namely the claim that just as we conceive of the most general laws of nature as originating from our own cognitive powers, we must also conceive of the more particular laws of nature, which we cannot ascribe to our own cognitive powers, as originating from an understanding, "even if not ours." Of course this is not intended as a theoretical proof of a divine mind as creator of nature, but as the basis of a principle for our own reflecting judgment. Still, this is the only way, Kant thinks, in which we can think of a system of laws of nature.

In the next section of the published Introduction to the third *Critique* Kant adds another element missing from the preliminary draft, namely the claim that we must think of the particular laws of nature as members of a system of laws and think of that system as if it originated in a mind greater than our own in order to be able to think of these laws as *necessary*. Kant makes this claim in an intricate passage:

> The understanding is of course in possession *a priori* of universal laws of nature, without which nature could not be an object of experience at all; but still it requires in addition a certain order of nature in its particular rules, which can only be known to it empirically and which from its point of view are contingent. These rules, without which there would be no progress from the general analogy of a possible experience in general to the particular, it must think as laws

(i.e., as necessary) because otherwise they would not constitute an order of nature, even though it does not and never can cognize their necessity.

What gives these laws the semblance of necessity that we cannot furnish from our own understanding alone is the place of each in a hierarchical system of laws:

> Thus although it cannot determine anything *a priori* with regard to those (objects), it must yet, in order to investigate these empirical so-called laws, ground all reflection on nature on an *a priori* principle, the principle namely, that in accordance with these laws a cognizable order of nature is possible – the sort of principle that is expressed in the following propositions: that there is in nature a subordination of genera and species that we can grasp; that the latter in turn converge in accordance with a common principle, so that a transition from one to the other and thereby to a higher genus is possible; that since it seems initially unavoidable for our understanding to have to assume as many kinds of causality as there are specific differences of natural effects, they may nevertheless stand under a small number of principles with the discovery of which we have to occupy ourselves, etc. This agreement of nature with our faculty of cognition is presupposed *a priori* by the power of judgment in behalf of its reflection on nature in accordance with empirical laws, while at the same time the understanding recognizes it objectively as contingent, and only the power of judgment attributes it to nature as transcendental purposiveness (in relation to the cognitive power of the subject): because without presupposing this, we would have no order of nature in accordance with empirical laws, hence no guidelines for an experience of this in all its multiplicity and for research into it.
> (*CPJ*, Introduction, section V, V:184–185)

The (long!) last sentence of this quotation makes the same point as the first draft of the introduction, namely that we have to think of the laws of nature as a system for heuristic purposes, in order to have a methodical way to search for them – if we think of the laws as comprising a system, we can see gaps in it, use laws that we have found at one level as models for hypotheses about other laws at that level we have not yet found, and so on. But the earlier part of the quotation suggests a different point: namely, that we have to be able to think of the particular laws of nature as necessarily true even though we cannot appeal to our own legislation of the most general laws of nature for that purpose; so instead we imagine them as if they are members of an actually existing system of laws instituted by an actual mind other than our own, in which the content of particular laws is necessitated by their relation to others. Again, we cannot assert the actual existence of this system and its source on purely theoretical grounds, but neither is it just a heuristic device; it is more like a quasi-explanation of quasi-necessity. But even with this qualification, it is clear that we must think of these laws as existing in nature, outside of our concepts, although of course represented by our system of concepts.

So, I claim, Kant conceives of us as representing systematicity as obtaining, with appropriate epistemological qualifications, at two different levels, at the level of our concepts and at the level of nature itself. In this way we represent nature as "purposive" (*zweckmäßig*) for our understanding. I now want to show that Kant replicates the two-level theoretical structure just described in his practical philosophy, in the form of a distinction between a system of *maxims* or *duties* and a system of *ends* achieved, rather than merely represented, through the former, although with the profound difference that in this case we can presuppose no more than that nature is *receptive* to our efforts to establish a system of ends, which however we must establish by our own decisions and efforts.

14.3 The System of Duties and the System of Ends

Morality concerns both representations and objects, consisting as it does of rules or principles that ought to be adopted by persons for their treatment of persons or, depending on the theory, some wider domain of beings, for example, all sentient beings, all living beings, etc. Moral rules significantly differ from factual or scientific cognitions in their "direction of fit," as it has come to be called,[12] for in Kant's words morality "yield laws that are imperatives, i.e., objective **laws of freedom**, and that say **what ought to happen**, even though perhaps it never does happen, and that are thereby distinguished from **laws of nature**, which deal only with that **which does happen**" (*CPuR*, A802/B830), thus morality commands that we make its objects correspond to our representations rather than our representations conform to their objects. But the same two levels, representations and objects, are involved in both cases: we must not merely *represent* a system of ends but must work *to bring one into existence*, while presupposing that *nature itself allows* us to do so. Here Kant is not thinking of nature as something constituted by ourselves, but as the arena outside of our will, although perhaps including our own predispositions and inclinations, within which we must act.

These two levels, of maxims on the one hand, and objects, namely persons and their ends, on the other, are evident from Kant's summary of his formulations of the categorical imperative in the *Groundwork for the Metaphysics of Morals* in three formulae, the first two of which concern form and matter and then combine in the third to yield a complete specification of the goal of morality, in the form of the sum of its maxims but through them of a system of objects of morality. The first formulation of the categorical imperative, which says that our "maxims must be chosen as if they were to hold as universal laws of nature," concerns, as it says, our maxims, that is, our *representations* of the principles in accordance with which we should act; the second formulation, which says "that a rational being, as an end by its nature and hence as an end in itself, must in every maxim serve as the limiting condition of all merely relative and arbitrary ends," concerns the *objects* of morality, namely the beings affected by our actions in accordance with our maxims, who must be treated as the rational – and therefore free – beings that they really are; and the third formulation of the categorical imperative, the "**complete determination** ... that all maxims from one's own lawgiving are to harmonize into [*zu*] a possible empire[13] of ends as a kingdom of nature" (G, IV:436; not merely "like" an empire of ends, as it is sometimes translated), describes the objective *state of affairs* that is to be brought about by our action in accordance with moral maxims, or the relation that is to be established among the rational beings who are the object of our moral maxims. So, maxims, on the one side, rational beings, and the relation of an empire of ends among them, on the other – representation and reality, although in this case a reality to be straightforwardly constituted by our actions.

But this means that the concept of systematicity has a twofold application in Kant's moral philosophy: on the one hand, Kant demands that our maxims comprise a system; on the other, by demanding that we work to bring about an empire of ends, morality demands that we bring about a systematic relationship among persons and their particular ends. Kant actually describes both of these levels of systematicity in his exposition of the third formulation of the categorical imperative in the *Groundwork*, for this is first stated as "the **principle** of every human will as **a will giving universal law through all its maxims**" (G, IV:432; often called the "Formula of Autonomy"), but Kant then says that "The concept of every rational being as one who must regard himself as giving universal law through all the maxims of his will" leads to the "very fruitful concept ... **of an empire of ends**," and that is defined precisely as "a systematic union of various rational beings through common laws," even more specifically as "a whole of all ends in systematic connection (a whole both of rational beings as ends in themselves and of the ends of his own that each may set himself) ... which is possible in accordance with the above principles" (G, IV:434). That is, if the maxims of each rational being (Kant should say, more precisely, rational *agent*) and of all (interacting) rational agents taken together comprise a system, then what will result will be a systematic union of all those rational beings and of all their particular ends. That is, each rational agent will be able to freely set and pursue her own ends to the greatest extent possible compatibly with the equal freedom of every other rational agent to do the same, so that the rational agents will constitute a system of free agents and their ends will constitute a system of freely chosen ends.

In the *Metaphysics of Morals* for which the *Groundwork* is the groundwork, Kant describes the same two levels of systematicity by describing a system of *duties*, on the one side, and the system of persons and their ends, on the other, that will result from conduct in accordance with this system of duties. Both the system of duties and the system of persons and their ends that will result from the former are themselves divided into two parts, the system of duties being divided into the systems of juridical and ethical duties and the system of the objects of duty being divided into the system of persons connected within law-governed states and ultimately into the worldwide peaceful system of states that "is the entire final end of the doctrine of right within the limits of mere reason" (*MM*, Doctrine of Right, Conclusion, VI:355), on the one hand, and, on the other, the system of ends that are also duties, that is, the self-perfection of each and the happiness of all (*MM*, Doctrine of Virtue, Introduction, section IV, VI:385–6, and §27, VI:450–451) that would constitute the "highest good" that is possible "in the world," as Kant often says (*CPJ*, §83, V:435; *TP*, VIII:279), that is, in *this* world. Of course there will be no room here to describe all these components of Kant's system of moral philosophy in detail, but let me document at least briefly why it is appropriate to understand Kant's thought in terms of systematicity in the first place.

The *Metaphysics of Morals*, as is well known, was originally published in two parts, the *Metaphysical Foundations of the Doctrine of Right* in January 1797, and the *Metaphysical Foundations of the Doctrine of Virtue* in the following August; and since the prefatory material to the work as a whole was included in the first volume, it emphasizes the Doctrine of Right and the principle of its distinction from more than its commonality with the Doctrine of Virtue. However, what the Preface to the work immediately says about the Doctrine of Right also applies to the Doctrine of Virtue, namely that *each* of these parts is to describe a system of duties and that *together* they should describe an exhaustive system of human duties – *human* duties, not duties for all rational beings, because the derivation of specific duties from the general principle of morality that does apply to all rational beings must also take account of the specific nature of human beings and of the nature, the terraquaeous globe, within which they live (VI:217). Thus Kant's Preface begins with the statement that "The critique of **practical** reason was to be followed by a system, the metaphysics of **morals**, which falls into metaphysical and first principles of the **doctrine of right** and metaphysical first principles of the **doctrine of virtue** ... The Introduction that follows presents and to some degree makes intuitive the form which the system will take in both these parts" (*MM*, Preface, VI:205). Thus the two parts of the work together constitute a system of duties. But each part itself also constitutes a system: "For the **doctrine of right**, the first part of the doctrine of morals, there is required a system derived from reason," although precisely because this is a system of duties that must be realized in concrete spatio-temporal conditions and therefore involves empirical concepts, it can never be stated completely (VI:205–206); and in the separate Preface and Introduction to the Doctrine of Virtue, Kant makes it clear that the same requirement of systematicity also holds, for "A **philosophy** of any subject (a system of rational cognition from concepts) requires a system of **pure rational** concepts independent from any conditions of intuition" (*MM*, DV, Preface, VI:375), but also a system that takes into account the empirical conditions of human agency, above all that humans need to perfect their potential abilities, that they need the assistance of others to achieve their ends, and that they have to overcome inclinations contrary to morality, or perfect their virtue, in order to be able to do what morality requires of them. The great principle of division that Kant introduces within the entire system of moral duties is then that some duties – the duties of right – may and indeed must be enforced with external, coercive laws, while others – the ethical duties – cannot and may not be so enforced, so can be enforced only be the virtue of each agent instead.

> In ancient times "ethics" signified the **doctrine of duties** (*philosophical moralis*) in general, which was also called the **doctrine of duties**. Later on it seemed better to reserve the name "ethics" for one part of moral philosophy, namely for the doctrine of those duties that do not come under external laws (it was thought appropriate to call this, in German, the **doctrine of virtue** [*Tugendlehre*]). Accordingly, the system of the doctrine of duties in general is now divided into the system of the **doctrine of right** (*ius*), which deals with duties that can be given by external laws, and the system of the **doctrine of virtue** (*Ethica*), which treats of duties that cannot be so given ...
> (*MM*, DV, Introduction, VI:379; see also *MM*, Introduction, section IV, VI:218–219)

Duties that prohibit interference with the freedom of others can be coercively enforced, Kant argues, because "if a certain use of freedom is itself a hindrance to freedom in accordance with universal laws (i.e., wrong), coercion that is opposed to this (as a **hindering of a hindrance to freedom**) is consistent with freedom in accordance with universal laws, that is, it is right" (*MM*, DR, Introduction, D, VI:231); but duties to promote one's own perfection or the happiness of others, and thereby to enhance their freedom, cannot be coercively enforced on this principle. Further, although people can often be coerced into performing certain *actions*, these duties require the adoption of certain *ends* rather than the performance of specifiable actions in specifiable circumstances, and the adoption of ends, on Kant's non-behavioristic conception of what that is, cannot be coercively enforced.

All of this is a large subject which I cannot pursue further here,[14] but I do want to add to Kant's explicit characterization of the system of our systems of duties one further point where he presupposes that our duties constitute a system without saying so. This is his well-known claim that "a **collision of duties** and obligations is inconceivable" although "two **grounds** of obligation" can "conflict with each other." Kant says that "since duty and obligation are concepts that express the objective practical **necessity** of certain actions and two rules opposed to each other cannot be necessary at the same time," two duties cannot be necessary at the same time so there cannot be a conflict between them; one and only one of them must be necessary at a particular time. But how are we to decide *which* of two conflicting *grounds* of duty, or in later terminology, two conflicting *prima facie* duties, is the necessary one? According to Kant, by the principle that "the stronger **ground of obligation** prevails" (*MM*, Introduction, section III, VI:224). But how are we to know which is the stronger obligation? What Kant does not say is that we know this by knowing the place of each duty within a system of duties with hierarchical relationships, like a system of scientific concepts.

Kant does not say this because it was thought to be so obvious at his time that he did not need to say it. Others did, for example, Georg Friedrich Meier, whose textbook on logic Kant used for decades and whose other works Kant presumably knew. "No true duty," Meier asserted in his *Universal Practical Philosophy*, "and no true moral rule, can contradict another true duty, and another true moral rule." Sometimes, Meier argued, apparent conflicts can easily be resolved simply by attending to the different duties at different times, but sometimes the resolution of apparent conflicts requires a lexical ordering of duties: "In the general theory of the practical disciplines it is demonstrated that we are obligated by a duty only so far as it is possible"; "consequently also only in those cases in which it can be observed without detriment to all other higher duties."[15] Meier then uses the word "systematic" in his description of the moral life as the satisfaction of a coherent set of duties:

> The greatest perfection of practical philosophy [is] that therein the natural duties are connected to one another in the best and most excellent order.... For the sake of the whole end of these disciplines it is not sufficient that one be convinced that an action is a duty; one

> must also know whether a duty is a higher or lower duty, a more important and necessary one or [less] indispensable and necessary; whether it is a chief duty or an ancillary one; whether it must be fulfilled prior to another or subsequent to it? Our entire moral life must be an orderly observation of all our duties. What we must do first we must not postpone, and what we must do foremost we must not do only by-the-by. The virtuous life must not be a disorderly and tumultuous observation of our duties, but a systematic and methodical observation of the laws.[16]

Only by means of a systematic order of duties, in other words, can conflicts among grounds of obligation be resolved.

Now Kant does not say very much about how his system of duties will actually resolve conflicts among grounds of obligation. Perhaps he did not feel the need to, since writers like Meier had presented their own systems of duties in multi-volumed detail,[17] and Kant's real concern was to revise the foundations of the existing system of duties rather than its superstructure, that it, its actual list of duties.[18] No doubt he took for granted and assumed that everyone else does as well that the fulfillment of perfect duties takes precedence over the fulfillment of imperfect duties. This follows from the very nature of such duties: for example, fraud is prohibited under any circumstances, but what is commanded with regard to beneficence is only that we adopt that as a necessary end, not that we perform a beneficent deed in every conceivable circumstance, thus, for example, in any circumstance in which we could perform a beneficent act only if we also committed some fraud, we simply have to defer the practice of beneficence to some other occasion when we would not have to commit fraud. But whether there are any hierarchical relations between perfect duties alone Kant does not say, and about the possibility of hierarchical relations among imperfect duties he says only that while "in wishing I can be **equally** benevolent to everyone ... in acting I can, without violating the universality of the maxim, vary the degree greatly in accordance with the different objects of my love (one of whom concerns me more closely than another)" (*MM, DV*, §28, VI:452), which suggests that I can resolve conflicts among possible benevolent acts by external considerations such as closeness of relationships, for example that when forced to choose I may direct my limited resources for beneficence to my children before strangers.

But there are more ways in which we can understand the systematic structure of Kant's catalogue of duties. We can understand Kant's initial definition of an empire of ends as implying a hierarchy of duties: in a systematic union of both persons as ends in themselves and then of their particular ends, we must *first* treat all persons as ends and never merely as means, and *only then*, that is, only so far as is compatible with that, treat their particular ends as our own. This would in fact give rise to the lexical ordering of perfect and imperfect duties: the perfect duties are those necessary for treating persons as ends rather than merely as means, and the imperfect duties have more to do with promoting the realization of particular ends insofar as they are compatible both with the status of all persons as ends in themselves and with each other. Within the realm of imperfect ends, however, it might also be possible to consider self-perfection, which includes the perfection of the capacities necessary for successful beneficence, as ordinarily having priority over the performance of specific acts of beneficence, which can often if not always be deferred to other occasions. Although, as Barbara Herman has recently made clear, establishing a hierarchy at least within the domain of imperfect duties might not be so simple: there may be circumstances in which responding with gratitude to a previous beneficence might have to come before some other possible act even of well-needed beneficence; cases in which the dire needs of other children might put them ahead even of one's own children in the line for one's possible beneficence, etc.[19] Herman does not say as much about duties to self, but here too the hierarchy might not be as simple as first appears: there may be cases in which the duty to perfect one's talents should come before a specific duty of beneficence to some others, because doing the former might, among other things, allow one to help even more other people in the future; but there might be cases in which the need of others is so dire that one should suspend everything else if one can help at all.

We can also understand Kant's division of duties of right and ethical duties as a system of duties designed to resolve conflicts of grounds of obligation. The Universal Principle of Right which flows from the fundamental, "innate" right to maximal freedom for each equal to the same freedom for all and in turn governs all acquired right, is that "Any action is **right** if it can coexist with everyone's freedom in accordance with a universal law, or if on its maxim the freedom of choice of each can coexist with everyone's freedom in accordance with a universal law" (*MM*, DR, Introduction, C, VI:230). Satisfaction of this principle is a necessary condition for treating others as ends in themselves, namely as enjoying the freedom to set their own ends (see *MM*, DV, Introduction, section VIII, VI:392) to the greatest extent possible consistent with everyone enjoying that freedom. For reasons that Kant does not fully spell out, coercive enforcement is permissible only to prevent people from treating *other* people as mere means rather than ends; the enforcement of the perfect duty to treat *oneself* as an end and not merely as a means, for example, the prohibition of suicide, has to be left to the virtue of each rather than to the collective agency of a judicial system. So the duties of right plus the perfect duties to self (and perhaps the duties of respect to others, such as not to treat them with contempt)[20] can be considered our duties to establish a systematic union of persons as ends in themselves, and then the remaining, imperfect duties to self and others can be considered the duties to establish a systematic union of the particular ends of persons, with the internal hierarchical relationship between the duty of self-perfection and duties of love toward others already mentioned.

We can also specify the outcome of the satisfaction of our system of duties in systematic terms, namely the systematic union of persons as ends in states and of states in a worldwide league of nations, on the one hand, and the systematic realization of the particular ends of all persons in the highest good possible in the world, on the other. In Kant's vision of a state, there will be hierarchical relations within units such as families and households and horizontal relations among such units, but also hierarchical relations between a sovereign electorate and a legislature that is in some way representative of and beholden to it, and then between that legislature and the executive and judiciary that are its agents, although the latter relations are complicated by the need for a monopoly of coercion in the executive and independence from political pressure in the judiciary.[21] Still, a state can be understood as a complex system of persons, or roles for persons. Likewise, Kant argues that the final end of justice can be achieved only by a worldwide league of nations, which can also be understood as a kind of system. Kant's vision of the highest good possible in the world as "universal happiness combined with and in conformity with the purest morality throughout the world" (*TP*, VIII:279) – I here leave out his alternative vision of the highest good as the combination of individual morality with individual happiness in "a world that is future for us" (*CPuR*, A811/B839), as I believe Kant largely did too after 1790[22] – is also a vision of a system, one in which the individual pursuit of happiness is subordinated to the moral law but, in accordance with the moral law itself, the individual ends of persons are supported and promoted insofar as they are compatible with one another. Kant explicitly calls this a "system of self-rewarding morality" in the *Critique of Pure Reason* (A810/B838), although there he also says that it "is only an idea, the realization of which rests on the condition that **everyone** do what he should." But in this regard the highest good in the world as a systematic realization of ends is on a par with the complete realization of a system of scientific concepts in our own minds and of a complete system of scientific laws in God's mind, which is however also only an idea in our own minds.[23]

All of this has only been a sketch of Kant's vision of a system of our duties on the subjective side as grounding a system of justice and happiness on the objective side. I will now conclude with a comment about Kant's vision of the unification of the systems of nature and freedom into a single system of nature and freedom.

14.4 Conclusion: The System of Nature and Freedom

As I suggested at the outset, Kant's ultimate philosophical ambition was the recognition that human beings and "all other natural things constitute a system of ends in accordance with fundamental principles of reason" (*CPJ*, §83, V:429). Kant argues for this unification of the two systems of nature and freedom in the "Critique of the Teleological Power of Judgment." His conception of this system of systems involves two assumptions. On the one hand, he argues that any proposed hierarchical purposive relations among beings in nature will be entirely arbitrary – grass may exist just to nourish cattle who exist just to nourish human beings, but human beings might just as well exist merely to care for cattle who in turn exist only to maintain the grass – unless some member of such a system has an intrinsic and unconditional value that can be served by the other members; and the only candidate for such a being is humankind itself, in virtue of their humanity, although not as merely natural beings but as possessing the supersensible capacity of freedom and therefore the capacity for morality:

> [T]he final end cannot be an end that nature would be sufficient to produce in accordance with its idea, because it is unconditioned ..... we have in the world only a single sort of beings whose causality is teleological, i.e., aimed at ends and yet at the time so constituted that the law in accordance with which they have to determine ends is represented by themselves as unconditioned and independent of natural conditions but yet as necessary in itself. The being of this sort is the human being, though considered as noumenon:[24] the only natural being in which we can nevertheless cognize, on the basis of its own constitution, a supersensible faculty (**freedom**) and even the law of the causality together with the object that it can set for itself as the highest end (the highest good in the world).
>
> (*CPJ*, §84, V:435)

On the other hand, Kant assumes that nature and human beings insofar as they are part of nature must always operate in accordance with laws, not miracles, and those laws, as we have seen, he assumes must comprise a system. That nature works only in accordance with general laws and that whatever is to be accomplished in nature must be accomplished through its laws, whether what is to be accomplished is the object of divine or of human intention, was part of Kant's philosophical vision from very early on. It is a central part of the argument of *The Only Possible Basis for a Demonstration of the Existence of God* in 1763 that God realizes his intentions through the laws that define the possibilities for nature, not through individual interventions into the course of nature (see, e.g.,

II:118–119). It is the argument for the postulate of the existence of God in the *Critique of Practical Reason* that God authors the laws of nature so that they are consistent with the human achievement of morality. And it is also the argument of the second *Critique* that we can represent human choice, whether it is the choice of good or evil, as efficacious only *through* the laws of nature – although now we must imagine the "whole chain of appearances, with respect to all that the moral law is concerned with, [as] depend[ing] upon the spontaneity of the [human] subject as a thing in itself" (*CPracR*, V:99) rather than upon God's original act of grounding all the possibilities for nature. The tension between the possibility of a single act of divine spontaneity and multiple acts of genuine human spontaneity is perhaps the deepest of all the tensions in the *Critique of Practical Reason* and in Kant's philosophy in general.[25]

Kant's distinction between the ultimate end that can be due to nature itself, namely the development of the culture of discipline, and the final end of the development of human freedom makes clear that there is a fundamental difference between the subsystems of nature and of morality – the latter cannot be presupposed to exist apart from our efforts, but is possible only through our efforts. It is actually in the "Critique of the Aesthetic Power of Judgment" that Kant is clearest about the relation between the system of nature and the system of freedom. This is in his account of our "intellectual interest" in the existence of natural beauty, where he argues that we take such an interest – pleasure at the *existence* of natural beauty and not in the mere *representation* of natural beauty – because in such beauty nature "show[s] some trace or give[s] a sign that it contains in itself some sort of ground for assuming a lawful correspondence of its products with our satisfaction that is independent of all interest ... this interest is moral, and he who takes such an interest in the beautiful in nature can do so only insofar as he has already firmly established his interest in the morally good" (*CPJ*, §42, V:300). That is, nature does not do our moral work for us, but it contains a ground of the possibility of our success in realizing our moral goals. We have to strive to realize those goals – the establishment of the systematic union of persons as ends in themselves and of their lawful ends – but we can only rationally strive to do what is possible, and nature must make the realization of this system possible, but only make it possible. It gives us a hint that it does this in its creation of natural beauty, which, unlike the beauty of art, we cannot and do not make ourselves.[26]

Of course, there is effort involved in our realization of the goal of scientific research also – we have to believe that nature contains a system of laws and of objects in order for it to be rational for us to strive for a systematic representation of those laws and objects, but we still have to construct our representation of those systems by our own efforts. But in this case it is the *representation* of nature that we must construct, while in the moral case it is the *empire of ends itself* that we must construct – that is, while in the case of inquiry our task is to know the natural world, in the case of morality our task is to transform the natural world into a moral world (*CPuR*, A809/B837). That remains a vital difference between the two systems, and one that must be reflected in their unification as well.

Notes

1 In quoting Kant, I follow the convention introduced at the beginning of this volume. Additionally, I use the following abbreviations: *CPJ* = *Critique of the Power of Judgment*; *CPracR* = *Critique of Practical Reason*; *CPuR* = *Critique of Pure Reason*; *FI* = *First Introduction to the Critique of the Power of Judgment*; *G* = *Groundwork for the Metaphysics of Morals*; *MM* = *Metaphysics of Morals*; DR = Doctrine of Right; DV = Doctrine of Virtue; *TP* = *On the Common Saying: That May Be Correct in Theory, but It is of No Use in Practice*.
2 See Guyer (2000, 2001).
3 In several papers, Eric Watkins discusses the formal *similarity* between (Kant's conceptions of) natural laws and moral laws, but does not discuss Kant's ambition of unifying the two systems of natural and moral laws into a single system of systems. See Watkins (2014, 2017).
4 One of the few writers who notices Kant's tri-level distinction among concepts, laws, and objects, although only in the case of nature, not morality, is Peter McLaughlin, who distinguishes between the "classificatory system of species and genera," "a system of physical laws," and "a physical system unified by laws" (McLaughlin 2014, 569–570); the last refers to what I refer to as nature itself.
5 Morrison (2008) distinguishes between the "logical" and "transcendental" conceptions of laws in nature in Kant, as between the representation of laws and the ontological fact of laws. Of course, as Morrison recognizes, this distinction is complicated by the fact that according to Kant we ourselves are supposed to make nature, i.e., constitute the ontological level of laws. I would propose that the ontological level of the laws of nature is reflected in the *receptivity* of sensibility, which provides the material for our constitution of nature, even though the form of nature is supposed to be provided by the *spontaneity* of the understanding and its categories,
6 As Lucy Allais has conclusively documented, Kant uses the term "appearances" (*Erscheinungen*) in two senses, one connoting the mental states or representations by which we represent things, the other things as we represent them by means of those mental states; see Allais (2015, Chapters 1–2). I am using "appearances" in the second sense here. In general, I will not be worrying about the transcendental ideality of nature in this paper.
7 See note 4.
8 I have given my detailed account of the System of Principles in Guyer (1987, Part III, Chapters 6–11). For a recent account, see Westphal (2021, 132–133, 153–179).
9 That it connotes the level of objects (as appearances, not things in themselves) rather than of our representations of them is not how Kant initially defines "transcendental"; he defines it rather as connoting the conditions of the possibility of synthetic *a priori* knowledge (e.g., B40, A85/B117). But by this Kant means synthetic *a priori* knowledge of the objects of experience, so the

transcendental concerns the possibility of knowledge *of objects*, though of objects as appearances rather than as things in themselves. (Putative) knowledge of the latter would be transcendent, not transcendental, though Kant sometimes uses the latter term when he should use the former.

10 A recent, immensely detailed study of the Transcendental Dialectic of the first *Critique*, emphasizing its inheritance from Pyrrhonian skepticism, is Proops (2021). Proops discusses Kant's view that his critique of theoretical metaphysics leaves room for "rational belief" in the objects of metaphysics on moral grounds only briefly in his conclusion (pp. 457–459).
11 Kant thus refers to our *concepts* of space and time in the Transcendental Aesthetic of the first *Critique* (A22–26/B37–42, A30–33/B46–49) even though space and time themselves are supposed to be pure intuitions (and the forms of empirical intuitions): we can have concepts of intuitions, obviously.
12 Following Anscombe (1957).
13 I give my reasons for preferring this translation of *Reich* in Guyer (2022).
14 I have discussed this in more detail elsewhere; see Guyer (2002, 2014).
15 Meier (1764, §8, p. 21).
16 Meier (1764, §19, pp. 42–43).
17 For example, Meier (1762–1774).
18 With the exception of his rejection of the concepts of duties to God, which was always part of his critique of the tradition of moral philosophy from Pufendorf and Wolff to Baumgarten and Meier.
19 See Herman (2021).
20 See Kant, *MM*, DV, §§36–44, VI:458–468.
21 See Kant, *MM*, DR, §49, VI:316–318.
22 See Guyer (2020, Chapter 4).
23 See Guyer (2000).
24 If only Kant had said here "though not considered merely as phenomenon," that is, not merely as part of nature but as free and rational also, without that implying existence at a distinct ontological plane!
25 See Insole (2013).
26 Although, as has often been pointed out, much of what we take as natural beauty, such as a pastoral countryside, is often a joint product of non-human forces and human activity – though then again, insofar as we intervene in nature, such as through farming, building scenic outlooks, etc., we are also part of nature ourselves.

Bibliography

Allais, Lucy. 2015. *Manifest Reality*. Oxford: Oxford University Press.
Anscombe, G.E.M. 1957. *Intention*. Oxford: Basil Blackwell.
Guyer, Paul. 1987. *Kant and the Claims of Knowledge*. Cambridge: Cambridge University Press.
Guyer, Paul. 2000. "The Unity of Nature and Freedom: Kant's Conception of the System of Philosophy." In *The Reception of Kant's Critical Philosophy*, edited by Sally S. Sedgwick, 19–53. Cambridge: Cambridge University Press. Reprinted in *Kant's System of Nature and Freedom: Selected Essays*, 277–313.
Guyer, Paul. 2001. "From Nature to Morality: Kant's New Argument in the 'Critique of Teleological Judgment'." In *Architektonik und System in der Philosophie Kants*, edited by Hans Friedrich Fulda and Jürgen Stolzenberg, 375–404.

Hamburg: Felix Meiner Verlag. Reprinted in *Kant's System of Nature and Freedom: Selected Essays*, 314–42.
Guyer, Paul. 2002. "Kant's Deduction's of the Principles of Right." In *Kant's Metaphysics of Morals: Interpretative Essays*, edited by Mark Timmons, 24–62. Oxford: Oxford University Press. Reprinted in *Kant's System of Nature and Freedom: Selected Essays*, 198–242.
Guyer, Paul. 2005. *Kant's System of Nature and Freedom: Selected Essays*. Oxford: Clarendon Press.
Guyer, Paul. 2014. *Kant*. 2nd ed. London: Routledge.
Guyer, Paul. 2020. *Reason and Experience in Mendelssohn and Kant*. Oxford: Oxford University Press.
Guyer, Paul. 2022. "The Empire of Ends." 2022 de Gruyter Kant Lecture. *Proceedings and Addresses of the American Philosophical Association* 96, 204–37.
Herman, Barbara. 2021. *The Moral Habitat*. Oxford: Oxford University Press.
Insole, Christopher J., 2013. *Kant and the Creation of Freedom: A Theological Problem*. Oxford: Oxford University Press.
McLaughlin, Peter. 2014. "Transcendental Presuppositions and Ideas of Reason." *Kant-Studien* 5: 554–72.
Meier, Georg Friedrich. 1764. *Allgemeine practische Weltweisheit*. Halle: Carl Hermann Hemmerde.
Meier, Georg Friedrich Meier. 1762–1774. *Philosophische Sittenlehre*, 5 vols., 2nd ed. Halle: Carl Hermann Hemmerde.
Morrison, Margaret. 2008. "Reduction, Unity, and the Nature of Science: Kant's Legacy?" In *Kant and Philosophy of Science Today: Royal Institute of Philosophy Supplement 63*, edited by Michela Massimi, 37–62. Cambridge: Cambridge University Press.
Proops, Ian. 2021. *The Fiery Test of Critique: A Reading of Kant's Dialectic*. Oxford: Oxford University Press.
Watkins, Eric. 2014. "What Is, for Kant, a Law of Nature?" *Kant-Studien* 105: 471–90.
Watkins, Eric. 2017. "Kant on the Unity and Diversity of Laws." In *Kant and the Laws of Nature*, edited by Michela Massimi and Angela Breitenbach, 11–29. Cambridge: Cambridge University Press.
Westphal, Kenneth R. 2021. *Kant's Critical Epistemology: Why Epistemology Must Consider Judgment First*. Abingdon and New York: Routledge.

Index

a priori 7–9, 12n9, 13, 14n2, 37n14, 40, 44, 46–48, 50, 55–58, 62n33, 63n52, 64n52, 82, 91n26, 98, 103, 105, 114–117, 119, 125, 144n15, 144n16, 150–151, 153, 155, 157–160, 163–165, 168n17, 169n23, 174, 176–178, 183–188, 190, 207–208, 218, 221, 223, 225, 231n7, 237, 242–244, 248n18, 255–256, 260, 262–264, 272n13, 279–281, 284–285, 291, 294n18, 302, 304, 320–322, 333n9; grounding 9, 189, 194; see also analogies of experience; anticipations of perception
abstraction 43–44, 104, 254, 261–265
Aether 9, 193–208, 209n2–9, 210n14–16, 211n23–27, 211n30, 211n34–36
affinity, Principle of 124
aggregate 1, 10, 12, 87, 97–98, 201, 226–227, 229, 234, 236, 241, 247n14, 250n32–33, 262, 268, 271n11, 279, 302, 312n23, 313n26
Alexander the Great 26
analogies of experience 8, 153, 156, 166, 168n15, 317, 320
anticipations of perception 159, 317
antinomy 85, 89n4, 91n30
anthropology 3–4, 6, 10, 14n4, 150, 161, 174, 234, 237, 240, 248n16, 254, 264–265
Apollonius 115, 126–133, 139, 144n17, 144n19, 144n21
appearance 43, 91n30, 114, 118–119, 125, 141, 142n4, 144n16, 162, 186, 225, 232n18, 266, 315, 317, 332, 333n6, 333n9
architectonic 2, 48, 57–58, 61n26, 241–242, 245; of pure reason 21–24, 242, 244, 296; see also system, of the sciences; systematicity; unity, architectonic
astronomy 39, 86–87, 91n26, 115, 120–122
astronomer 75, 77, 85, 90n15
attention 254, 261–264, 266
axioms of intuition 159, 317

Baldwin, Thomas 90n24
Baumgarten, Alexander 25, 36n6, 56, 232n19, 277, 282, 293n1, 334n18
beauty: of art 332; intellectual interest in 332; of nature 332, 334n26
Blumenbach, Johann Friedrich 187–189
Boerhaave, Hermann 195, 197–199, 201, 206, 209n6, 209n11, 209n12, 210n20
Bolzano, Bernard 111n4
Borelli, Giovanni Alfonso 132
Boyle, Robert 192, 196–198, 209n7, 209n9
Brahe, Tycho 76
Bruno, Giordano 77, 90n14

Cassini, Giovanni Domenico 85
causal (explanation, factors, grounds, models, powers, processes, relations) 9, 78, 101, 125, 154–156, 168n13, 182, 192, 195, 197–198, 203–208, 208n1, 212n34, 259, 285, 287, 290–291; law 155–156

causality 9, 125, 153–157, 168n14, 168n17, 245, 322, 331
chemistry 8–9, 149, 163, 167n2, 169n21, 173–174, 180, 182, 185–190, 194, 208, 232n27, 232n29, 280
choice 267, 303, 316, 329, 332
civil society 238–239, 243
Civilization 265, 268–269
cognition: as an essential end of reason 6, 24–25, 29–32; of oneself 29–30
cohesion 187, 193–194, 196, 198–201, 203–204, 210n16, 211n30
Comet 71–72, 75, 77, 89n7, 124, 319
concept 11, 41, 99, 101, 103, 105–108, 110, 111n10, 114, 117, 119–120, 125, 132, 151–152, 158, 164, 182, 184, 205, 218, 229–230, 231n10, 259, 261–264, 282–286, 289–291, 293n1, 293n6–7, 317–321, 323–324, 333n4, 334n11; a priori 151, 165, 184, 215, 218, 231n7; empirical 118, 164, 316, 318, 320, 326; fundamental/simple 6, 39–54, 56–58, 60n21, 61n32, 62n36, 62n40, 63–64n52, 263, 292; mathematical 102, 111n11, 114, 117, 119, 125, 132, 143n7, 218; metaphysical 84; moral 22, 264; philosophical 218–219; pure 110, 112n21, 117, 119, 125, 144n15, 158, 292; rational 152, 197, 279, 296–297, 311n5, 326; of reason 86, 111n3, 140, 142n4, 148, 215, 242, 264; of science 3, 8, 13, 163, 166n1, 175–176, 215, 219, 226, 249n31–32; scientific 71, 279, 327, 330; of systematicity/of a system 3, 6–7, 13, 14n5, 70, 97, 110n2, 249n31, 324; transcendental 105, 117; of the understanding 105, 119, 125, 152, 155, 292; see also philosophy, school concept of; philosophy, worldly/world concept of
construction 10, 46, 61n26, 102, 111n12, 114–115, 120–121, 125–127, 132–142, 143n7, 144n17, 145n29, 146n35, 165

Copernicus, Nicolaus 70, 74–75, 77–79, 82, 86
cosmogony 7, 69–70, 72–73, 87, 88n2
cosmology 3–4, 6–7, 69–70, 74, 77, 79, 81–84, 87–88, 88n2, 89n4, 89n13, 90n22, 150, 175, 224, 229, 232n18, 232n19
course of events 236
Crusius, Christian August 5, 22, 25–35, 36n7–9, 36n11–12, 37n13, 37n16, 58, 206, 210n20, 249n31
cultivation 240
culture 24, 243, 249n29, 250n43, 265, 268–269, 332
cyclop 10, 254–257, 259–260, 263–264

d'Alembert, Jean le Rond 88
definition 71–72, 218–219, 227–230, 230n4, 231n5, 231n8; of concepts 43, 46, 61n28, 101, 111n11, 123, 127, 132, 134, 138, 145n24, 231n7, 282–283; of empire of ends 328; of interests/ends of reason 24–25, 27, 36n6, 36n9; in philosophy/limits of 9, 215–216, 218–219, 232n26; of physics 8–9, 163–166, 216, 221–228, 230; of principles 176–177; of science (in general) 1, 8, 12, 70, 88, 216; of special sciences 7–8, 12–13, 23, 51, 59n8, 148, 150, 161–163, 166, 215–220, 223, 228–230, 232n28, 311; of a *systema mundi* 71–73, 89n7; of the understanding 7
derivation account 183–186
Desargues, Girard 130–131
Descartes, René 8, 56, 60n16, 126, 128–129, 134–136, 138–139, 141, 144n23, 150, 192, 210n19, 281
duties: of right 326, 329; of virtue 299
dynamic 54, 56, 202, 205

Eberhard, Johann August 102, 104, 107, 110, 126, 128, 132–133, 142
elasticity 193–194, 197–202, 204, 208, 210n21
empire of ends 11, 324, 328, 333
end: absolutely highest 27, 36n9; chief 22, 24, 27–31, 34; final 21,

23–29, 32–35, 36n6, 36n8, 237, 239, 241–245, 246n2, 247n12, 250n37, 271n12, 315, 325, 330–332; formal final 26; highest 22–23, 27, 36n9, 331; middle final 26–27; more distant final 26; objective final 26; subjective final 26; ultimate 26–27, 242, 244, 316, 332
ends: auxiliary 26, 36n8; chief 26; coordinate final 27; essential 12n7, 21–25, 29, 34–35, 152, 232n28, 257; of humanity 10, 116, 152, 254, 256–258, 260–261, 264–265, 270–271; persons as ends in themselves 328–330, 332; of science 7, 12n7, 232n27; subalternate 22–23, 35; subordinate final 27
Erxleben, Johann Christian 199–200, 204, 210n19, 211n26
ethics 11, 142n4, 161, 268, 297–298, 301–310, 311, 326
Euler, Leonard 196, 198–201, 203, 210n13–14, 214n19, 211n24
event 154, 159, 192, 235–236, 286–289, 316
explanation 11, 14n3, 31, 69, 72, 75, 79, 82, 86, 89n12, 91n27, 96, 149–150, 178–180, 186, 189–190, 192–194, 200, 205, 208, 247n12, 260, 278, 284, 287–288, 291, 293, 306, 323; aethereal 9, 194, 199–200, 202, 209n6, 210n16, 211n26; mechanical 173–174, 178–182, 198, 209n6, 260; teleological 173; *see also* scientific reasoning
eye of philosophy 4, 10, 254–255

fire 194, 197–203, 206, 209n11–12
focus imaginarius 131, 145n26, 216
Fontenelle, Bernard le Bovier de 90n14
force 26, 48, 50–54, 56, 72, 79, 87, 91n30, 116, 120, 122, 124–125, 144n14, 150, 176, 193–194, 196–198, 200–202, 207, 210n19, 210n21, 222–225, 227; fundamental 84, 186–187, 194, 199; derivative 188, 199, 203–205, 208, 209n8, 211n29, 211n30

Galilei, Galileo 71, 121, 124, 143n8, 192, 209n10, 227
Gatterer, Johann Christoph 235, 241, 245, 246n6, 247n9, 247–248n14, 248n15
Gehler, Johann Samuel Traugott 181, 187
Gensichen, Johann Friedrich 84, 91n25
germ 4, 238, 259–260, 262
God, cognition of 29–30, 32, 35
gravity 72–73, 109, 164, 199, 201

Hales, Stephen 195, 197–198, 201, 209n6
Halley, Edmond 90n15
happiness 22, 24, 28–29, 34, 240, 243–244, 301, 325–326, 330
heuristics/heuristic device 7, 82, 88, 323
historical knowledge 59n7, 254, 255–259, 271n4; *see also* philosophy, of history; system, of history
historiography 237, 241, 245, 246n1, 246n5, 247n8, 247n12
highest good 22–25, 28, 31–35, 242, 244, 250n37, 251n46, 325, 330–331
human nature 238–239, 242–243, 247n12, 264–266
human race/species 235, 237–239, 244, 248n23, 249n29, 250n33, 251n45, 254, 258–259, 261–262, 265, 267–269
humanity 10, 22, 25, 28, 86, 116, 152, 240, 254, 256–258, 260–262, 264–267, 269–271, 271n10, 271n12, 282, 284, 289–290, 331
Husserl, Edmund 111n4
Huygens, Christiaan 90n14, 198
hypothesis 69, 76, 79, 84, 86, 90n13, 150, 164, 198, 207; *see also* scientific reasoning

idea: of reason 1, 5, 119, 140–141, 217, 225, 242, 265; of a science 149, 215–216, 220, 225, 228, 230; *see also* concept, of reason; schema, schematization
imagination 75, 78, 83–84, 86–87, 119, 135

Index   339

impenetrability 158, 186
inertia 72, 158–159, 168n19, 238–239, 243

Karsten, Wenceslaus Johann Gustav 180–181
Kepler, Johannes 84–85, 122, 130, 143n9, 145n24, 164
knowledge, philosophical 254

La Hire, Philippe de 90n15, 131
Lambert, Johann Heinrich 3, 6, 21–22, 36n2, 39–58, 58n–59n1, 59n2–3, 59n6–9, 59n11, 60n12–17, 60n19–21, 60n24, 61n25–32, 62n33–34, 62n36–41, 62–63n42, 63n43–51, 63–64n52, 64n53, 167n8, 177–178, 249n31
law: of freedom 323; of mechanics 8, 159–160, 166, 167n10; of nature 11, 157, 183–184, 245, 259–260, 287, 316–324, 332, 333n5; transcendental 318, 333n5
Leibniz, Gottfried Wilhelm 56, 60n16, 63n46, 89n6, 131, 150, 154, 167n10, 210n20, 277, 281–282, 293n1
light 78, 80–81, 180, 193–194, 196–203, 206, 211n24, 211n30
logic: formal 42, 104, 151, 292, 293n6; subject-matter of 7, 95–96, 99–100, 103–104
logica: artificialis 96, 104–105, 108; *naturalis* 95, 104–105, 108

Mairan, Jean Jacoues d'Ortous 85
mathematics 3–4, 6–7, 14n2–3, 39, 51, 57, 100–103, 114–117, 120–125, 128, 131, 136, 142, 142n2, 145n28, 149–152, 161–162, 174, 218, 229, 230n4, 250n35, 255, 257, 270
matter 56, 63n49, 149–150, 158–159, 163–166, 168n19, 175–176, 181, 186–187, 192, 194, 196–204, 206–208, 209n8, 210n14, 210n16, 210n19–21, 211n30, 220, 222–225, 227–228, 231n14, 292, 324
maxim 11, 14n2, 86, 99, 118, 143n6, 156, 178, 211n31, 251n45, 260, 302–303, 307–310, 315–316, 323–324, 328–329

mechanics 8, 50, 54, 56, 69–70, 74, 87, 140, 154, 156, 158–160, 166, 167n10, 168n14, 16n19; *see also* explanation, mechanical
mechanism 178, 180–181, 184, 208n1, 250n40, 260; *see also* explanation, mechanical
Meier, Georg Friedrich 21, 105, 112, 277, 281–282, 293n1, 327, 328, 334n15, 334n16, 334n17, 334n18
metaphysics 3–4, 8, 21, 24, 34, 40, 54–55, 57–58, 61n28, 69, 81–83, 89n6, 90n16, 117, 150–152, 161–163, 165–166, 167n6, 173–177, 189, 218, 231n10, 260, 287, 291, 299, 311n2, 311n5, 334n10; of freedom 152, 245; of morals 4, 264, 296, 311n4, 325; of nature 4, 149, 152, 158, 167n7, 245, 292; *see also* philosophy
method: demonstrative/mathematical 41, 117, 175, 219–20, 231n9; of science 1, 4, 8, 13, 14n3, 88, 110, 115, 129, 134, 138, 146n35, 151, 161, 165, 175, 179, 180, 217, 219–220, 223, 228, 232n28, 234, 249n32, 285, 301, 308–309
Milky Way 69, 73, 77–78, 80–81, 83–84, 90n21
Mill, John Stuart 95
motion 44–45, 48, 50, 53–54, 60n24, 69, 71–73, 75–79, 82, 84–86, 89n7, 89n12, 90n15, 91n26, 91n29, 115, 120, 124–125, 150, 154, 157–159, 164, 167n10, 168n15, 175, 177, 190, 197–199, 204, 209n8, 209n11, 210n14, 210n19–20, 290
moralization 265, 168

natural predispositions 238, 259, 262, 266, 269
natural science 4–5, 8, 10–11, 14n2, 34, 39, 69, 84, 116–117, 120–122, 131, 139, 142n4, 148–149, 151–154, 157–166, 166n1, 168n16, 173–174, 178–179, 183–186, 188, 190, 194, 205–206, 208, 212n34, 221, 247n9, 257, 260, 277–278, 280–281, 285–286, 288–289, 292–293
nebular hypothesis 69, 150, 164

nebulous stars 80–81, 90n20–21
Newton, Isaac 8, 56, 69–72, 74, 84–85, 87, 88n2, 115, 121–124, 126, 131–132, 134–142, 145n30, 146n34, 146n35, 149–154, 156–157, 159–160, 167n10, 168n15, 169n23, 177–179, 183, 187, 195–196, 198–199, 201, 206, 209n2, 209n6–9, 210n14, 210n20, 249n26, 260
nutrition 181–182, 188–189

Opus Postumum 9, 187, 194–195, 201, 203, 208, 211n30, 215–216, 221, 223–225, 227–230, 230n1, 231n10, 232n17, 232n27
organon 39, 41–42, 45, 48, 51, 55–58, 58n1, 59n2, 59n6, 62n36, 62n38, 177

Pascal, Blaise 131, 210n21
Peirce, Charles Sanders 78
perspective 78, 84, 90n16, 130–131, 236–237; *see also* standpoint/standpoint changes
phenomenology 43, 51, 56
philosophy: of history 193, 234, 237, 240, 245, 246n2–4, 247n8, 247n12, 248n17–19, 248n23, 250n34, 251n44, 267; school concept of 23, 36n4, 257; of science 1, 13, 14n1–2, 58, 90n13, 195, 208; transcendental 3, 8, 149–150, 152–153, 157–158, 160–161, 163, 166, 167n6, 168n15–16, 224, 228, 232n14, 244, 248n18; worldly/world concept of 23–24, 151–152, 254, 257–258, 261, 264, 271; *see also* knowledge, philosophical
phoronomy 47, 54, 56, 159
plan 4, 61n26, 235, 249, 256, 271n12
planetary orbits 72–73
positive unreason 83
postulates of pure practical reason 316
principle 8, 11, 50–51, 56, 97–98, 100, 102–3, 106–109, 111n3, 111n7, 111n10, 112n17, 123–124, 135, 143n6, 176, 178–179, 257, 263–264, 278, 280, 284, 298–300, 304–305, 321, 326–327, 329; of causality 154–156, 158, 164–165, 168n14, 168n17; constitutive 118; of ethics 302–304; logical 92n32, 211n31, 285, 318–320; metaphysical 159, 165; of morality 32, 298, 300, 306, 310, 312n14, 325; of reason 8, 118–120, 125, 139–42, 146n38, 306; regulative 118, 141, 146n38, 288; of right 329; transcendental 8, 150, 152–154, 156–161, 164–165, 168n17, 285, 319–320
projective geometry 115, 126, 129–131, 138
providence 236, 247n13, 249n28
prudence 31, 251n45
Ptolemy 74–75, 89n11
Pufendorf, Samuel von 11, 299–301, 304–305, 310, 334n18

race(s) 10, 254, 258–261, 267–270, 271n5, 271n10, 272n15
racism 267, 272n16
real connection 236
reason 37n14, 111, 114, 151, 242, 257, 265, 282, 294n10; artists of 33, 242; cosmopolitan concept of 242, 244; end/interest of 6, 11, 21, 23–25, 28–32, 34, 36n6, 241–242, 245, 249n29, 256–257, 302, 315; government of 2; idea (and schema) of 1, 5, 8, 12, 114–115, 119–120, 125, 139–142, 146n37, 215, 217, 225, 230, 242, 249n28, 264–266, 318, 331; maxim of 257; object of 34, 103–104, 288; sciences of 3, 151, 161; self-knowledge of 103, 151, 166, 257, 306; systematicity/unity of 4, 7, 166, 220, 278–281, 309, 325–326, 331; *see also* positive unreason; scientific reasoning; use of reason
Reid, Thomas 305, 308, 313n28, 313n29
repulsion 57, 73, 79, 82, 186–187, 194, 197, 238
Richards, Robert J. 179
Rousseau, Jean-Jacques 3, 21–22, 37n13, 257–258, 265–266

schema, schematization 1, 51, 114, 118–120, 125, 140, 142, 144n15, 146n38, 148–149, 158, 193, 215, 217, 219, 220, 242; see also idea; reason, idea of
science: aim of a 232n28; hierarchy of 175–177, 189 (see also system, of the sciences); object of a 223; see also concept, of a science; definition, of a science; idea, of a science; method, of science; natural science; systematicity
scientific reasoning 74, 78–79, 81, 83, 87–88; see also concept, scientific; explanation; method, of science
Senebier, Jean 181, 188
sense: inner 226; outer 178, 224–225
skill, doctrine of 31, 35
Solar system 69, 72–73, 75, 77–80, 89n7, 90n17, 111n6, 164
Spalding, Johann Joachim 36n5, 246n3, 249n23
spontaneity 105, 332, 333n5
Stahl, Georg 207, 227
standpoint/standpoint changes 74–75, 77–88; see also perspective
Stoics 70
syllogism 10–11, 42, 60n15, 110, 111n3, 140, 277–278, 281–286, 289–292, 293n9, 294n18
system: 70, 87–88, 97–98, 110n2, 111n3, 111n10, 146n39, 152–153, 279–280; vs. aggregate (see aggregate); doctrinal 6–7, 11, 70, 95–96, 160, 222–225 (see also systematicity, logical vs. real); of categories 149, 152–154, 158; of ends 300–305, 215, 323–324; of freedom 316, 332; of history 10, 224, 236, 249–250n32; Lambert's metaphysical 39–40, 52; of laws (incl. best system) 183–185, 259, 316, 320–323; of maxims 316, 324; of morals 296–300, 310–311, 323–326; of nature 11, 97, 222–223, 288, 315–317, 330–332; ontic 7, 11, 70–72, 74, 96 (see also systematicity, logical vs. real); of principles 11, 84, 149, 333; of pure reason 166; of the sciences 2, 6, 12, 40, 48–49, 57–58, 162–163,

270 (see also architectonic; systematicity, outer (=external)); Solar 69, 72–75, 77–78, 89n7; teleological/purposive 242, 270 (see also explanation, teleological); see also concept, of systematicity/of a system; standpoint
systematicity: inner (=internal) s. of a science 2, 4, 160–163, 216–218, 229–230; logical vs. real 10–11, 99–102, 205, 319 (see also system, doctrinal; system, ontic; system, of nature); outer (=external) s. of a science 2, 160–163, 174, 216–218, 230; see also concept, of systematicity/of a system

teleology 10, 176, 179–180, 134, 137–238, 247n9, 248n15
temporal connection 236
the unconditioned 243, 282–284, 286–288
Torricelli, Evangelista 227
transcendental 155, 159–160, 163–164, 168–169n19, 178, 222–223, 231n14, 316, 318, 333–334n9; see also concept, transcendental; law, transcendental; philosophy, transcendental; principle, transcendental
transition (from metaphysical foundations of natural science to physics) 124, 208, 220–222, 225–228, 230, 232n29, 245, 311n2, 322
truth 1, 29, 36n11, 39–43, 51, 55–56, 60n13, 79, 86, 99, 101, 104, 106, 153, 155, 163, 167n11, 168n13, 282, 283–284, 291, 293n1

understanding 39, 42, 51, 55, 84, 99, 110, 116, 118–119, 142–143n5, 153, 184, 205, 259, 278, 283; common 263; concepts/categories of the 125, 152, 155, 231, 317–318, 320–323, 333n5; empirical use of the 116, 184; faculty of the 7, 12, 95–97, 101, 103–109, 114; logical use of the 99; real use of the 99; speculative 263

unity, architectonic 21, 23, 36n3–4, 241–242, 244; *see also* architectonic; system; systematicity
universal gravitation, law of 121
universal history 235–236, 246n1, 247n8, 248n18, 262; *see also* philosophy, of history; world history
use of reason 14n3; common 32; discursive vs. intuitive 114, 116–117, 151; dogmatic 257; hypothetical 74, 76, 78–79, 84; logical vs. real 10–12, 99–100, 111n9, 111n10, 151, 278, 281–282, 286, 288–289, 293n6; practical 242, 244–245, 296–301, 305–309, 311, 316, 318; regulative 14n2, 91n26, 118, 156, 288, 318; transcendental 117; speculative/theoretical 260, 316; *see also* reason

virtue, cognition of 29–30
vocation (of the human being) 21, 24, 36n5, 237, 240–242, 244–245, 248n17, 249n23–24, 265

wisdom, doctrine of 24, 312n20
Wolff, Christian 3, 8, 21, 36n1, 37n13, 42–44, 46, 56, 58, 60n18, 61n28, 91n30, 112n17, 126, 167n10, 173–178, 189, 210n16, 210n20, 219–220, 231n5, 231n8–9, 277, 281–282, 293n1, 334n18
world 10–11, 28–29, 71, 73–74, 80–81, 85, 90n16, 109, 122, 140–141, 146n37, 146n39, 155, 225, 232n14, 232n18, 232n19, 235–236, 238, 256–257, 261, 264, 266, 271n12, 286–293, 318, 325, 330–331, 333; moral world 11, 35, 333; new world 269; *see also* philosophy, worldly/world concept of
world history 235–236, 247n11; *see also* philosophy, of history; universal history
Wren, Christopher 90n21
Wright (of Durham), Thomas 71, 89n8

For Product Safety Concerns and Information please contact our EU
representative GPSR@taylorandfrancis.com
Taylor & Francis Verlag GmbH, Kaufingerstraße 24, 80331 München, Germany

www.ingramcontent.com/pod-product-compliance
Ingram Content Group UK Ltd.
Pitfield, Milton Keynes, MK11 3LW, UK
UKHW021118310725
461390UK00004B/23